The Structure of Values and Norms

Formal representations of values and norms are employed in several academic disciplines and specialties, such as economics, jurisprudence, decision theory, and social choice theory. Professionals in these disciplines will find much of interest in the philosophical issues treated in *The Structure of Values and Norms*.

Sven Ove Hansson closely examines such foundational issues as how the values of wholes relate to the values of their parts, the connections between values and norms, how values can be decision-guiding, and the structure of normative codes. Models of changes in both preferences and norms are offered, as well as a new method to base the logic of norms on that of preferences.

Hansson has developed a unified formal representation of values and norms that reflects both their static and their dynamic properties. This treatment, carried out in terms of both informal value theory and precise logical detail, will contribute to the clarification of basic issues in the philosophical theory of values and norms.

Sven Ove Hansson is Professor of Philosophy at the Royal Institute of Technology in Stockholm, Sweden. He is also Editor-in-Chief of *Theoria*.

Cambridge Studies in Probability, Induction, and Decision Theory

General editor: Brian Skyrms

Ellery Eells, *Probabilistic Causality*
Richard Jeffrey, *Probability and the Art of Judgment*
Robert C. Koons, *Paradoxes of Belief and Strategic Rationality*
Cristina Bicchieri and Maria Luisa Dalla Chiara (eds.), *Knowledge, Belief, and Strategic Interactions*
Patrick Maher, *Betting on Theories*
Cristina Bicchieri, *Rationality and Coordination*
J. Howard Sobel, *Taking Chances*
Jan von Plato, *Creating Modern Probability: Its Mathematics, Physics, and Philosophy in Historical Perspective*
Ellery Eells and Brian Skyrms (eds.), *Probability and Conditionals*
Cristina Bicchieri, Richard Jeffrey, and Brian Skyrms (eds.), *The Dynamics of Norms*
Patrick Suppes and Mario Zanotti, *Foundations of Probability with Applications*
Paul Weirich, *Equilibrium and Rationality*
Daniel Hausman, *Causal Asymmetries*
William A. Dembski, *The Design Inference*
James M. Joyce, *The Foundations of Causal Decision Theory*
Yair Guttmann, *The Concept of Probability in Statistical Physics*
Joseph B. Kadane, Mark B. Schervish, and Teddy Seidenfeld (eds.), *Rethinking the Foundations of Statistics*
Phil Dowe, *Physical Causation*

The Structure of Values and Norms

SVEN OVE HANSSON
Royal Institute of Technology
Stockholm, Sweden

CAMBRIDGE UNIVERSITY PRESS
Cambridge, New York, Melbourne, Madrid, Cape Town, Singapore, São Paulo

Cambridge University Press
The Edinburgh Building, Cambridge CB2 8RU, UK

Published in the United States of America by Cambridge University Press, New York

www.cambridge.org
Information on this title: www.cambridge.org/9780521792042

First published 2001
This digitally printed version 2007

A catalogue record for this publication is available from the British Library

Library of Congress Cataloguing in Publication data

Hansson, Sven Ove, 1951–
The structure of values and norms / Sven Ove Hansson.
p. cm. – (Cambridge studies in probability, induction, and decision theory)
Includes bibliographical references (p.) and indexes.
ISBN 0-521-79204-5
1. Values. 2. Norm (Philosophy) I. Title. II. Series.
BD232 .H297 2001
121′.8–dc21 00-063047

ISBN 978-0-521-79204-2 hardback
ISBN 978-0-521-03723-5 paperback

Contents

Preface

This book has grown out of my attempts to develop a unified formal representation of values and norms that reflects both their static and their dynamic properties. The subject is important for at least two reasons. First, formalized treatment may contribute to the clarification of certain issues in the basic philosophical theory of values and norms, such as the relation between the values of wholes and the values of their parts, the connections between values and norms, and the structure of normative codes.

Second, formal representations of values and norms are employed in several academic disciplines and specialties, notably economics, jurisprudence, decision theory, and social choice theory. Some of the foundational issues of these subjects are closely connected to philosophical issues that will be treated in this book.

The philosophical theory of values and norms has much to learn from the problems it encounters in applications. I hope it can in return provide safer – or at least better understood – foundations for these applications.

Chapter 1 offers some preliminary remarks on the usefulness and the limitations of formalization in this branch of philosophy. Chapter 2 introduces exclusionary preferences, that is, preferences that refer to mutually exclusive alternatives. In Chapter 3, a representation is introduced for preference states, and in Chapter 4 various operations of change on preference states are investigated.

The term *combinative preferences* will be used to refer to preferences with relata that are not necessarily mutually exclusive. The strategy adopted here is to derive combinative preferences from exclusionary preferences. In Chapter 5, the basic principles for such derivations are discussed, and in Chapters 6 and 7 two types of derivations are explored. The first of these is intended to capture preferences that represent pairwise comparisons, including ceteris paribus preferences. The second aims at representing decision-guiding preferences.

Chapter 8 is devoted to monadic (one-place) value predicates, such as 'good' and 'bad.'

The next five chapters are devoted to norms. The normative language and some basic idealizations are introduced in Chapter 9. Chapter 10 deals with that fraction of normative discourse that involves no change in the situation or perspective, so that a fixed set of alternatives can be employed. In Chapter 11, this framework is extended to cover two cases that require changes in the set of alternatives, namely normative counterfactuals and the resolution of normative conflicts. In Chapter 12, formal representations are given of normative rules. Two types of dynamic processes for a system of such rules are investigated: changes of the system and its application to factual situations. In Chapter 13, a special class of normative rules (potestative rules) are used to classify legal relations.

In Chapter 14, some final comments are offered on how the results reported in the earlier chapters can be used, both in philosophy and in the social sciences.

The dependencies between the chapters are summarized in Figure 1. All formal proofs have been deferred to a separate proofs section.

Chapter 1 contains material from Hansson 1994 and 2000a. Sections 2.4 and 2.5 are based on Hansson 1997a, and Section 2.6 is based in part on Hansson 1993a. Chapters 3 and 4 are an improved version of Hansson 1995. Chapter 6 is an improved version of Hansson 1996a; some of its ideas were already presented in Hansson 1989a. Chapter 8 is an improved version of Hansson 1990a. Chapters 9 and 10 contain material from Hansson 1997d, which builds on ideas from Hansson 1990c and 1991b, and to some extent Hansson 1988b and 1988c. Material from Hansson 1999b has been included in Sections 11.1 and 11.2. Chapter 12 contains much material from Hansson and Makinson 1997. Chapter 13 is based on Hansson 1990b and 1996c. A first stage of the endeavours leading up to that chapter was reported in Hansson 1986b.

My work with this book has been supported by a grant from the Bank of Sweden Tercentenary Foundation, which is gratefully acknowledged.

I would like to thank David Makinson for generously letting me include material from a joint paper in Chapter 12, at the same time warning the reader that I have changed it to a point where David can

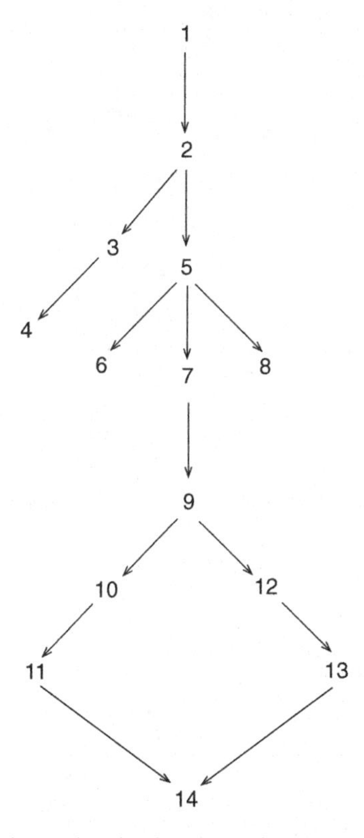

Figure 1. The dependencies between the chapters of this book.

no longer be held responsible. This work has benefitted from discussions over the years with more people than I can remember. Discussions at the seminar of Practical Philosophy at Lund University have been particularly useful. Special thanks go to Sven Danielsson, Wlodek Rabinowicz, Hans Rott, and Wolfgang Spohn for penetrating comments on earlier versions – in Wlodek's case, two earlier versions – of the manuscript.

Sven Ove Hansson
Stockholm, June 2000

Prolegomena

1

Formalization

Formalized philosophy is by no means uncontroversial. Obviously, this book would not have been written had not its author believed that useful insights can be gained from formal treatments of philosophical issues. The purpose of this chapter is to point out some of the advantages, but also some of the limitations, of formalization.

1.1 FORMALIZATION AND IDEALIZATION

A representation in formal language is always the outcome of a simplification for the sake of clarity, or in other words an *idealization*. To idealize in this sense means to perform a "deliberate simplifying of something complicated (a situation, a concept, etc.) with a view to achieving at least a partial understanding of that thing. It may involve a distortion of the original or it can simply mean a leaving aside of some components in a complex in order to focus the better on the remaining ones."[1]

Idealization – not necessarily in formal language – is omnipresent in science, and it seems to be so in philosophy as well. Many, probably most, of the crucial concepts in philosophical discourse originate through idealizations from nonphilosophical language.[2] As one example, it is common in moral philosophy to regard "John ought to

[1] McMullin 1985, p. 248. It is important to distinguish this sense of idealization from that of depicting something as better than it actually is. The two senses relate to different meanings of 'ideal,' namely on the one hand "[s]omething existing only as a mental conception" and on the other something that is "perfect or supremely excellent in its kind". (OED) On the two senses of 'ideal,' see Hansson 1999d. On the role of idealization in philosophy, see also Hansson 1994.

[2] I prefer 'nonphilosophical' to the more common 'prephilosophical,' since there is no reason to believe that philosophy has, in its more than two and a half millennia of existence, been devoid of impact on thinking and language outside of philosophy.

. . . ," "It is a duty for John to . . . ," and "John has an obligation to . . ." as synonymous, in spite of the fact that there are occasions when common usage would accept one or two of these phrases but not the other(s).[3] In this and similar cases, philosophers (tacitly) assume that there is, or can be constructed, a more fundamental and more straight-lined concept behind the embellished meanings of words and phrases in nonregimented natural language.

True, idealization is only one part of the transformation of elements from nonphilosophical language through which philosophical terminology is shaped. The construction of philosophical language also involves the creation of new distinctions and of terms that have no obvious counterparts in nonphilosophical language. Hence, philosophical terminology differs from nonspecialized language in two ways. First, it uses some words in different, idealized ways (e.g., 'good,' 'value,' 'permission'). Second, it employs some linguistic innovations of its own (e.g., 'consequentialism,' 'deontic').

Some philosophers have wished to philosophize in "prephilosophical" language. In my view, this is an illusory undertaking, since nontrivial philosophical insights, with few exceptions, require more precision than what is immediately available in nonregimented language.[4]

In other academic disciplines, the relationship between specialized terminology and the nonspecialized concepts from which they originated may be fairly unproblematic. Physicists who theorize about heat or gravitation do not have to refer back to the nonphysicist's concepts of warm and cold, or light and heavy, in order to justify their theoretical constructions. These scientific concepts have their own justifications, derived from experiments and other exact observations. Philosophers operating with concepts such as goodness or permission are not in this fortunate situation. These philosophical concepts have no justification apart from their capability of clarifying the corresponding nonphilosophical concepts. Hence, on one hand we have to deviate from the general-language meanings of our key terms in order to obtain the precision necessary for philosophical analysis; but on the other hand, if we deviate so far as to lose contact with general-language meanings, then the rationale for the whole undertaking will be lost.

[3] See Section 9.2 and the references given there.

[4] By 'nontrivial,' I mean nontrivial against the background of what has already been said by previous philosophers.

Since different idealizations can clarify different aspects of one and the same concept, it is futile to search for definite or uniquely correct philosophical analyses or explications. Different types and styles of idealizations of one and the same concept should be seen as complements rather than as competitors.

Therefore, a defence of formalized philosophy need not – and in my view could not reasonably – proceed by showing that formalization provides us with the one and only correct approach to philosophical problems. It is sufficient to show that some philosophical problems have aspects that can be clarified with formal methods.

Formalization in philosophy typically results from an idealization in two steps: first from common language to a regimented philosophical language and then from that regimented language into symbolic language.[5] More often than not, most of the idealization takes place in the first of these steps. Therefore, what makes treatments in symbolic language special is not their distance to ordinary discourse (which can be surpassed by treatments in regimented natural language). Rather, what makes them special is the mathematical skill they require and the characteristic types of questions they give rise to.[6]

1.2 THE VIRTUES AND DANGERS OF FORMALIZATION

In the natural sciences, formalized theories have the preponderant advantage of being correlated with empirical measurements and thus being testable in a more exact way than informal theories. No such mensural correlation is available for formal theories in philosophy, such as theories of values and norms.[7] Therefore, the claims that can

[5] On some occasions, there may be more than two distinguishable steps of idealization. In a letter to the author (April 19, 2000), David Makinson pointed out that this is true of the formal representations of legal relations that are discussed in Chapter 13. In this case, there are three steps of idealization, passing through (1) ordinary language, (2) legal discourse, (3) Hohfeld's typology, and (4) formal representations of Hohfeld's distinctions.

[6] It is interesting to note that the additive model used in utilitarianism, although controversial in other respects, seems to have escaped the negative reaction against mathematical representation that other formal models have encountered in some philosophical quarters. One reason for this may be that the mathematical skill it requires is so elementary. However, there is no reason why ease of mastering should be decisive for the accuracy of a formal model.

[7] In this respect, formalized value theory can be compared to pre-Galilean physics. Many medieval physicists employed formal models of physical phenomena, but they did not use these models to predict the outcomes of measurements. See Livesey 1986 and the references given there.

be made for formalization in philosophy are weaker than those that can be made for formalization in the empirical sciences. In philosophy, the major virtue of formalization is the same as that of idealization in informal languages: Isolating important aspects helps to bring them to light. As compared to informal idealization, formalized treatments tend to be helpful in at least four more specific ways.

First, formalization incites definitional and deductive economy. It brings forth questions about the interdefinability of concepts and about minimizing the set of primitive principles of inference. In this book, formalized methods will be used to investigate the interdefinability of evaluative and normative concepts and also, within the first group, that between dyadic and monadic value concepts.

Second, formalization serves to make implicit assumptions visible. For example, in informal discourse on preferences it is often tacitly taken for granted that a well-defined set of alternatives (an alternative set) exists, consisting of the objects to be compared. In formal models of preferences, this assumption has to be made in a precise and explicit manner.

Third, formal theories can support delicate structures that would be much more difficult to uphold and handle in the less unambiguous setting of an informal language. Symbolic treatment has made it possible to penetrate some philosophical issues more deeply than would otherwise have been possible. One of the best examples of this is the relation between truth and language. It is difficult to see how Tarski's semantical analysis of the notion of truth could have been developed in a nonformalized setting. In this book, the discussion of different ways to derive preferences over parts from preferences over wholes draws heavily on distinctions that are readily available in formal languages but next to unattainable in nonformal language.

Fourth, formalization stimulates strivings for completeness. The rigorousness of a formal language is, for instance, necessary to make it meaningful to search for a complete list of valid principles of inference. Often enough, this search may uncover previously unnoticed philosophical problems. One example of this is the study of nonmonotonic inference. The introduction of rigorous formal notation (in particular, a nonmonotonic consequence relation) has led to much more thorough and extensive studies of patterns of nonmonotonic inference and of the relationships between these patterns.[8] In this book, some of the results

[8] Gabbay 1985; Makinson 1993.

on the properties of goodness and badness are examples of such strivings for completeness.

Obviously, formalization also has its dangers.[9] In order to construct a workable formal model, the number of primitive notions has to be reduced to a very minimum. It is often cumbersome to include an additional factor into an already existing formal model. Partly for this reason, the philosophical logician runs the risk of becoming mentally locked in the world of one or a few formal models and therefore neglecting aspects of the real world not covered in these models. Another danger is spending too much time on problems that are mere artifacts of the formal model rather than on more general philosophical problems that the model can be helpful in elucidating. The so-called deontic paradoxes, which arise in certain models of deontic logic but not in informal normative discourse, are examples of this.

1.3 WHY LOGIC?

Formalization in philosophy is in practice virtually synonymous with formalization in logical language.[10] Some of the pioneers of formal logic allotted to logic a unique status in philosophy. Bertrand Russell, for one, maintained that "every philosophical problem, when it is subjected to the necessary analysis and purification, is found either to be not really philosophical at all, or else to be, in the sense in which we are using the word, logical."[11]

Although this book conforms with philosophical tradition in employing logical formalism, I do not subscribe to such special claims for logic. There are no a priori grounds why logical languages should be better suited than other symbolic languages for modelling discourse on norms and values, nor, for that matter, for modelling real-world phenomena to which norms and values refer. The relative usefulness of logic is an open question, and treatments of the same subject matter in other types of formal languages should be welcomed.

[9] Hansson 2000a.

[10] There is no clear demarcation between logic and mathematics. Arguably, much if not most of mathematics can be reconstructed into some form of logic. Here, a logical language means a symbolic language of one of the types that are taught in logic courses and in textbooks on logic.

[11] Russell 1914, p. 14. Russell counted value theory among the "not really philosophical" topics, but his general view is compatible with the inclusion of value theory in the realms of logic. Cf. Davis 1966.

Against this background, it is not as problematic as some have thought to use truth-valued logic for modelling (subject matter expressed by) sentences that are not true or false.[12] A normative sentence such as

(1) Jane ought to help her brother.

cannot be true or false (or at least, for the sake of argument, let us assume that it cannot). Therefore, a formal sentence in a truth-valued language (such as $O\alpha$, where O stands for "ought" and α for "Jane helps her brother") cannot, strictly speaking, represent sentence (1). It can, however, represent the following sentence:

(2) There is a valid norm to the effect that Jane ought to help her brother.

where validity is relative to some moral code or standard. We can assume that there is a one-to-one correspondence between sentences of the type represented by (1) and those of the type represented by (2). Truth and falsity are, of course, fully applicable to sentences of the latter type. Therefore, to the extent that a system of deontic logic adequately mirrors the properties of sentences such as (2), it also mirrors – somewhat more indirectly – those of sentences such as (1).[13]

1.4 A TRADE-OFF BETWEEN SIMPLICITY AND FAITHFULNESS

To formalize philosophical subject matter means to reduce it to a simplified formal model in order to get a clear view of some major aspects at the expense of others.[14] Different formalizations may capture different features. As an example, probabilistic and nonprobabilistic theories of belief seem to capture different properties of human belief systems.

Philosophical or scientific model-making is always a trade-off between simplicity and faithfulness to the original. In philosophy, the subject matter is typically so complex that a reasonably simple model will have to leave out some philosophically relevant features. This makes it possible to devise a counterargument – typically in the

[12] See Makinson 1999 and the references given there.
[13] Cf. Bengt Hansson 1969.
[14] Cf. Merrill 1978, esp. pp. 305–311.

form of a counterexample – that seemingly invalidates the model. However, even if such a counterargument convincingly discloses an imperfection in the model, this is not necessarily a sufficient reason to give up the model. If the counterargument cannot be neutralized without substantial losses of simplicity, then an appropriate response may be to continue using the model, bearing in mind its weaknesses.[15]

As an example of this, it is assumed throughout this book that logically equivalent expressions can be substituted for each other. This assumption makes way for certain counterintuitive inferences, such as the *revenger's paradox*.[16] Let p_1, p_2, and q be mutually exclusive expressions, such that p_1 *or* p_2 *or* q is logically true. Then

$$\text{If } Obligatory(not\text{-}p_1) \text{ then } Obligatory(p_2\text{-}or\text{-}q)$$

is logically true. Now let p_1 signify that John kills his wife's murderer, p_2 that he kills only persons other than his wife's murderer, and q that he kills nobody at all. It follows that if John ought not to kill his wife's murderer, then he ought to kill either only persons other than his wife's murderer, or no one at all.

Intersubstitutivity has similar effects on the logic of values. Let p denote that you receive €100 tomorrow, q that you receive €50 tomorrow, and r that you are robbed of all the money that you have the day after tomorrow. Presumably, you prefer p to q. By intersubstitutivity, you then also prefer $(p\&r) \vee (p\&\neg r)$ to q. However, the direct translation of $(p\&r) \vee (p\&\neg r)$ into natural language does not seem to be preferable to the direct translation of q into natural language. The disjunctive formulation of the comparison redirects attention, and the impression is created that each of the disjuncts is claimed to be preferred to q.

These and other counterintuitive inferences can only be avoided by giving up intersubstitutivity, thereby losing much of the simplicity and

[15] This view is by no means uncontroversial among philosophers. Nicholas Rescher (1968, p. 238) maintains that a formal analysis of an informal concept "must, if acceptably executed, end with results that are fully compatible with ordinary conceptions." According to Hector-Neri Castañeda, the simpler of two theories should only be chosen if they account for exactly the same data; he thus gives faithfulness absolute priority over simplicity (Castañeda 1984, see in particular pp. 241–244).

[16] The paradox was introduced in Hansson 1991a, to which the reader is referred for a somewhat more thorough discussion.

logical strength of the formal structure. It is, on balance, better for most purposes to endure the somewhat strange consequences of intersubstitutivity than to pay the high price for getting rid of them.[17]

1.5 FORMALIZING CHANGE

There are several ways in which formal logic can be used to express changes. In what may be called *time-indexed models*, a (discrete or continuous) variable is employed to represent time. The object of change (such as a state of the world, state of affairs, state of mind, state of belief, value state, etc.) can then be represented as a function of this variable, so that a state of the world (etc.) is assigned to each point in time. A further development of this framework is to make it nondeterministic by allowing for a bundle of functions, typically structured as a branching tree.

A quite different mode of representation is that of *input-assimilating models*.[18] In such models, the object of change (such as a state of belief) is exposed to an input (such as a new piece of information) and is changed as a result of this. No explicit representation of time is included in models of this type. Instead, the characteristic mathematical constituent is a function that to each pair of a state and an input assigns a new state. (Nondeterminism can be achieved by replacing the function by a relation.)[19] From the early 1980s and onward, input-assimilating models of belief states and databases have been the subject of a rapidly growing number of studies.[20]

Input-assimilating models have the advantage of focussing on the causes and mechanisms of change. They are tailored to exhibit the effects of external causes on systems that are changed only in response to external causes ("inputs") and are otherwise stable. This makes them extremely well suited to represent changes in most types of computerized systems, such as databases. It also makes them tolerably well suited to represent important aspects of changes in human states of mind. It is, at least for some purposes, a reasonable idealization to dis-

[17] An interesting exception is the formal representation of free-choice permission. See Section 9.1.

[18] Cf. Hansson 1999c, Section 1.3. In game theory, extensive form games may be seen as time-indexed and normal form games as input-assimilating.

[19] Lindström and Rabinowicz 1989, 1991; Segerberg 1995.

[20] A seminal paper was Alchourrón, Gärdenfors, and Makinson 1985. For an overview, see Hansson 1999c.

regard those changes in a person's beliefs or values that have no immediate external cause in order to focus better on the mechanisms of externally caused changes.[21] Input-assimilating models are also suitable to represent legal codes. At least ideally, a legal code remains the same unless a decision is taken to change it in a specified way.

In this book, input-assimilating models of changes in values and norms will be explored. However, there is no reason to believe that these are the only models that can shed light on the dynamics of values and norms. Other constructions, such as time-indexed models, may bring out important aspects that input-assimilating models hide from sight. The relative merits and demerits of different formal approaches can only be judged after each of them has been fairly thoroughly investigated.

[21] Among the most important internally generated processes are those that aim at making a state of mind coherent or making it stable against further rational deliberations (reflective equilibrium). Recently, some attempts have been made to model such processes in otherwise input-assimilating models. See Hansson 1997b and Olsson 1997.

PART I

Values

2

Exclusionary Preferences

From a logical point of view, the major value concepts of ordinary language can be divided into two categories. The *monadic* (classificatory) value concepts, such as 'good,' 'very bad,' and 'worst,' evaluate a single referent. The *dyadic* (comparative) value concepts, such as 'better,' 'worse,' and 'equal in value to,' indicate a relation between two referents.[1] Chapters 2–7 are devoted to the dyadic concepts and Chapter 8 to the monadic ones.

In Chapters 2–4, the objects of preferences are assumed to be mutually exclusive, that is, none of them is compatible with, or included in, any of the others. No further assumptions are made about their internal structure. They may be physical objects, types or properties of such objects, states of affairs, possible worlds – just about anything. Preferences over a set of mutually exclusive relata will be referred to as *exclusionary*. The condition of mutual exclusivity is removed in Chapters 5–7.

2.1 SOME BASIC CONDITIONS

The values held and endorsed by an individual are complex and intricately connected with that individual's beliefs and emotions.[2] The very process of isolating her values from the rest of her mind involves a

[1] This classification is not exhaustive. For some purposes, it is useful to have three- and four-termed value predicates, such as "if x, then y is better than z" (conditional preferences) and "x is preferred to y more than z is preferred to w." Neither of these will be treated in this book. For a valuable discussion of four-termed preference relations, see Packard 1987.

[2] In moral philosophy, many categorizations of preferences have been made that will not be further referred to here, since it is not clear how they affect the *formal* properties of preferences: intrinsic vs. extrinsic preferences, self-regarding vs. other-regarding preferences, synchronic vs. asynchronic preferences, etc. For a useful list of such distinctions, see Fehige and Wessels 1998, pp. xxv–xxvi.

considerable idealization, and the further process of expressing these isolated values in terms of a binary relation takes idealization one step further. It can hardly be a reasonable criterion of rationality that these two steps can be taken in only one way, which involves no distortion of the original. Therefore, we should not expect the preferences of a rational being (and still less those of an actual being) to have a single, correct formalization (such as, perhaps, a unique preference relation over possible worlds). Instead, we should expect different formalizations to be capable of capturing various parts of what may be called the value component of a person's state of mind. Different types of preference relations may be needed for different purposes.

Before introducing the logical apparatus, we need to make two idealizing assumptions, both of which serve to abstract from certain instabilities of actual preferences. To the extent that value statements are precise, they are relative to some more or less explicit *criterion* (standard of evaluation). The best car on sale is not necessarily the best car for me to buy. A good philosopher may be a bad mother, etc. Most criteria for value statements can be subsumed under categories such as instrumental, aesthetic, ethical value,[3] but an exhaustive description of the criterion that is applied in a particular case will typically have to go far beyond such summary categories. As a limiting case, value statements may be intended to include all aspects, in other words, to represent an evaluation that takes everything into account ("synoptic" values).[4]

If vacillation between criteria is allowed, then "counterexamples" can easily be contrived against virtually any proposed logical condition for preferences. ("Ms. Jones is a better professor than Ms. Smith since she conducts more interesting research. But Ms. Smith is a better professor than Ms. Jones since she teaches better. Therefore, both are better than the other, and we have shown that betterness is not asymmetric.") In order to get the logical investigation going, we must make an assumption of *criterial constancy*: All value statements in the formal language represent values according to one and the same criterion.

The values that we entertain and express depend not only on how we conceive the alternatives and on the criteria that we apply, but also

[3] von Wright 1963b.

[4] Rescher 1968, p. 293. Rawling (1990, p. 495) uses the term 'categorical preferences' for essentially the same concept as Rescher's 'synoptic preferences.'

on the method of elicitation – the procedure by which we are induced to develop and express values. For instance, a person may arrive at a different pattern of preferences if she chooses between pairs of alternatives than if she matches alternatives by adjusting a feature of one alternative to make it equal in value to another, given alternative.[5]

From the point of view of rational reconstruction, this is a disturbing feature of human preferences. Suppose that with one elicitation procedure I have come to the conclusion that I prefer X to Y, whereas with another such procedure I have found myself to prefer Y to X. I cannot then really consider myself to have settled the value comparison between X and Y. Should I have the opportunity and the ability to resolve the issue by deliberation, I can reasonably be expected to do so, by giving up either one or both of the two opposite preferences.

Although actual agents may make opposite (inconsistent) value statements, it will therefore be assumed, as a minimal rationality requirement, that such inconsistencies have been eliminated. In other words, the value statements under study have been subjected to sufficient (reflection-guided) adjustment to have attained consistency. This may be seen as a requirement of a *reflective equilibrium with respect to the logical structure* of preferences: Further deliberations on the logical structure alone will lead to no changes.[6]

2.2 THE COMPARATIVE VALUE CONCEPTS

The two fundamental comparative value concepts are 'better' (strict preference) and 'equal in value to' (indifference).[7] The relations of preference and indifference between alternatives will be denoted by the symbols $>$ and $\equiv$, respectively.

In accordance with a long-standing philosophical tradition, $A>B$ will be taken to represent "B is worse than A" as well as "A is better than B."[8] This is not in exact accordance with ordinary English. We tend to use 'better' when focusing on the goodness of the higher ranked of the

[5] In the psychological literature, this is referred to as "preference reversal" or as a "lack of procedure invariance." See: Tversky et al. 1988, p. 371; Schkade and Johnson 1989, pp. 203–204; Goldstein and Einhorn 1987; Johnson et al. 1988.

[6] On reflective equilibria, see Rawls 1971, pp. 48–51 and Tersman 1993.

[7] Halldén 1957, p. 10.

[8] "Worse is the converse of better, and any verbal idiosyncrasies must be disregarded" (Brogan 1919, p. 97).

two alternatives and 'worse' when emphasizing the badness of the lower ranked one.[9] There may also be other psychological or linguistic asymmetries between betterness and worseness.[10] However, the distinction between betterness and converse worseness can only be made at the price of a much more complex formal structure. The distinction does not seem to have enough philosophical significance to be worth this complexity, at least not in a general-purpose treatment of the subject.

When describing the preferences of others, we tend to use the word 'preferred.' The word 'better' is used when we express our own preferences and also when we refer to purportedly impersonal evaluations. Although these are important distinctions, it is gainful to abstract from them in a logical investigation. The preferences referred to in this book are preferences of rational individuals.[11]

The following four properties of the two exclusionary comparative relations will be taken to be part of the meaning of the concepts of (strict) preference and of indifference:

(1) If A is better than B, then B is not better than A.
(2) If A is equal in value to B, then B is equal in value to A.
(3) A is equal in value to A.
(4) If A is better than B, then A is not equal in value to B.

It follows from (1) that preference is irreflexive, that is, that A is not better than A. These four will be taken to be the basic properties of a logic of value comparisons.

Definition 2.1. *A* (triplex) comparison structure *is a triple* $\langle \mathcal{A}, >, \equiv \rangle$, *in which* $\mathcal{A}$ *is a set of alternatives and* $>$ *and* $\equiv$ *are relations in* $\mathcal{A}$ *such that for all* $A, B \in \mathcal{A}$:

(1) $A>B \rightarrow \neg(B>A)$ (asymmetry of preference)
(2) $A \equiv B \rightarrow B \equiv A$ (symmetry of indifference)
(3) $A \equiv A$ (reflexivity of indifference)
(4) $A>B \rightarrow \neg(A \equiv B)$ (incompatibility of preference and indifference)

[9] Halldén 1957, p. 13; von Wright 1963a, p. 10; Chisholm and Sosa 1966a, p. 244.
[10] Tyson 1986, p. 1060. Cf. also Houston et al. 1989.
[11] To the extent that impersonal betterness exists in a form accessible to such individuals, its properties may be close to those of pairwise preferences, as introduced in Section 2.3. At any rate, the principle of comparison-cost avoidance (Section 2.4) does not seem to be plausible for impersonal betterness.

Furthermore:

$$A \geq B \leftrightarrow A > B \vee A \equiv B \text{ (weak preference)}$$

The intended reading of ≥ is 'at least as good as' (or more precisely: 'better than or equal in value to'). As is well known, ≥ can replace (strict) preference and indifference as primitive relations in comparison structures:[12]

Observation 2.2. *Let* $\langle \mathcal{A}, >, \equiv \rangle$ *be a triplex comparison structure, and let* ≥ *be the union of* > *and* ≡. *Then:*

(1) $A > B \leftrightarrow A \geq B \ \& \ \neg(B \geq A)$
(2) $A \equiv B \leftrightarrow A \geq B \ \& \ B \geq A$

The choice of primitives (either ≥ alone or both > and ≡) is a fairly inconsequential choice between formal simplicity (≥) and conceptual clarity (> and ≡).[13] The following is an alternative to Definition 2.1:

Definition 2.3. *A* (duplex) comparison structure *is a pair* $\langle \mathcal{A}, \geq \rangle$, *in which* $\mathcal{A}$ *is a set of alternatives and* ≥ *is a reflexive relation on* $\mathcal{A}$. *The derived relations* > *and* ≡ *are defined as follows:*

$$A > B \text{ if and only if } A \geq B \text{ and } \neg(B \geq A)$$
$$A \equiv B \text{ if and only if } A \geq B \text{ and } B \geq A$$

It will be seen that the defined relation ≥ of Definition 2.1 is reflexive and that the defined relations > and ≡ of Definition 2.3 satisfy conditions (1)–(4) of Definition 2.1. It follows that the two definitions are interchangeable. Given our definitions, the four conditions of Definition 2.1 are in combination equivalent to the reflexivity of weak preference.

In order to simplify notation, chains will be contracted. Hence, $A \geq B \geq C$ abbreviates $A \geq B \ \& \ B \geq C$, and $A > B > C \equiv D$ abbreviates $A > B$ &

[12] On the choice of primitives in preference logic, see Burros 1976.

[13] As was pointed out to me by Wlodek Rabinowicz, there are interpretations of strict and weak preference according to which the former is not definable in terms of the latter. For an example, let v be a function that to each alternative A assigns an interval of real numbers. We may then have $A > B$ if and only if $min(v(A)) > max(v(B))$ and $A \geq B$ if and only if both $max(v(A)) \geq max(v(B))$ and $min(v(A)) \geq min(v(B))$. Given these definitions, > is not definable in terms of ≥. However, as will be argued in Section 2.6, to base preferences on underlying value intervals is not a plausible construction.

$B>C$ & $C \equiv D$. Furthermore, $>^*$ stands for $>$ repeated any finite nonzero number of times (and similarly for the other relations). Thus, $A>^*C$ denotes that either $A>C$ or that there are $B_1, \ldots B_n$ such that $A>B_1$ & $B_1>B_2$ & $\ldots$ $B_{n-1}>B_n$ & $B_n>C$.

2.3 A CENTRAL DILEMMA IN PREFERENCE LOGIC

In the search for further logical properties of preferences, we have to confront a certain contrariety in the intuitive notion of preference. There are two properties that we have strong reasons to ascribe to preferences and that yet turn out to be difficult to reconcile.

One of these is *pairwiseness*: The preference relation has only two relata. A sentence such as $A \geq B$, $A>B$, or $A \equiv B$ refers only to the two alternatives A and B. Therefore, its validity should depend exclusively on the properties of A and B and should not be influenced by those of other elements of the set of alternatives. It should make no difference if we compare A and B as part of our deliberations on the elements of the set $\{A,B,C\}$ or as part of our deliberations on the elements of $\{A,B,D,E\}$. Hence, pairwiseness can also be expressed as invariance under (1) additions to the alternative set, and (2) subtractions from the alternative set of elements other than the two relata.

The other property is *choice-guidance*: Preferences over an alternative set should have such a structure that they can be used to guide our choice among the elements of that set. More precisely, they should allow us to identify the best alternative(s).

Choice-guidance refers to choosing at once among all the elements of the alternative set, not to choosing among pairs of two elements. Hence, the two principles in combination require that comparisons of only two alternatives at a time be sufficient to determine a choice among all the alternatives. It is to be expected that the contact between the two principles will be far from frictionless. For an example, consider my comparisons of the overall quality of three TV channels. When I compare Channel 1 to Channel 2, the dominant difference is that the first has a more reliable news programme. Therefore, I consider Channel 1 to be better than Channel 2. The major difference that I am aware of between Channels 2 and 3 is that the former has more extensive coverage of musical events. Therefore, I consider Channel 2 to be better than Channel 3. Finally, the striking difference between Channels 3 and 1 is that the former shows more recently produced cinema films, for which I prefer it to the other. Hence, in pairwise terms, I prefer

Channel 1 to Channel 2, Channel 2 to Channel 3, and Channel 3 to Channel 1.[14]

This preference pattern does not violate criterial constancy, since one and the same criterion, such as overall quality, may be used for all three comparisons.[15] What differs is the selection of facts used to judge this criterion. (The reason for this selection of facts may be lack of knowledge. I may, for instance, never have watched the news on Channel 3, so I therefore cannot include that issue in any comparison other than that between Channels 1 and 2.)

This set of (pairwise) preferences is not inconsistent or otherwise problematic, unless we want to use it to guide a choice among the alternatives. In fact, this is quite unlikely to happen. We seldom have to choose the best TV channel; instead, we choose among individual programmes. However, we can imagine situations in which one has to make a choice among the three channels. A whimsical landlord can decide that I am only allowed access to one TV channel, or – more felicitously – I can be appointed juror in a competition between TV channels. Then the uselessness of this set of preferences for purposes of choosing will be obvious enough: No alternative can reasonably be chosen, since each of them is worse than some of the others.

There are two possible major approaches to this central dilemma in preference logic. One is to require that pairwise preferences be adjusted so that they become structurally fit for choice-guidance. The other is to distinguish between two types of preferences: pairwise and choice-guiding.[16]

The first approach is the more common one. Particularly in economic theory, the conventional approach is to take preference and

[14] This insight is not new. Larry Temkin, among others, has noted that "if *different* factors were relevant for comparing different outcomes, it could be true that even if A were better than B, and B better than C, in terms of the relevant factors for comparing *those* alternatives, A might *not* be better than C, in terms of the relevant factors for making *that* comparison" (Temkin 1996, p. 193). The above example is similar in structure to Schumm's (1987) Christmas ornament example.

[15] This, like most other criteria used in value comparisons, may be conceived as the aggregation of a number of subcriteria. Similar examples can be constructed that refer to overall ethical or aesthetic evaluations.

[16] The distinction between preferences that satisfy pairwiseness and preferences that satisfy choice-guidance is related to, but not identical with, Howard Sobel's (1997) distinction between 'pairwise preferences' and 'preferences *tout court*.' Sobel's 'pairwise preferences' are defined in terms of hypothetical choice, and his 'preferences *tout court*' represent welcomeness. Cf. also Temkin's concept of "essentially pairwise comparative" ideals (1997, p. 304).

choice to be interdefinable: Choice is treated as revealed preference, and preference as an expression of hypothetical choice.[17] In this framework, all preferences are required to be choice-guiding.

In favour of this approach, it can be maintained that our intuitions about the structure of preferences are determined by paradigmatic cases in which preferences are used to guide choices. By analogy or habit of mind, we expect the nonparadigmatic cases to satisfy requirements whose rationale is, strictly speaking, only present in the paradigmatic cases. For this reason, logical principles that derive from the requirements of choice-guidance can also be applied to preferences that are not choice-guiding.[18]

Against this and in favour of the second approach, it can be said that there are cases when preferences are decidedly non–choice-guiding. For example, suppose that I have a ticket in a lottery. One of the prizes is a gift voucher that is worth €3,000 in the local grocery store. Another prize is a beautiful antique set of chess pieces. I prefer winning the set of chess pieces to winning the voucher. This preference cannot be reconstructed as hypothetical choice, since there is no way to choose what you win in a lottery. (And indeed, if I were to choose between the chess pieces and the voucher, I would choose the voucher, out of responsibility for my family.)

Hence, even if there is a tendency in nonphilosophical usage to conflate the two types of preferences, the importation of this conflation into philosophical analysis is not necessarily commendable. It should be the purpose of philosophical analysis to increase clarity. This purpose is in general better served by separating than by amalgamating half-intermixed but conflicting concepts. Therefore, in this book pairwise and choice-guiding preferences will be developed separately.

Both exclusionary and combinative preferences can be either pairwise or choice-guiding. In this chapter, we are concerned with exclusionary preferences. I am not aware of any reasonable logical requirements on pairwise exclusionary preferences in addition to the reflexivity of weak preference, the defining property introduced in Section 2.2.[19] To the contrary, choice-guidance implies further demands

[17] For discussions of the relation between choice and preference, see Sen 1973, 1993; Broome 1978; Reynolds and Paris 1979; Hansson 1988a, 1992a.

[18] I went somewhat too far in maintaining this in Hansson 1997a.

[19] This refers to principles for preferences over a fixed alternative set.

on exclusionary preferences, which will be explored in the next two sections. The properties of combinative preferences, pairwise and choice-guiding, will be discussed in Chapters 5–7.

2.4 MINIMAL PRINCIPLES FOR CHOICE-GUIDING PREFERENCES

Since we now turn to preferences intended to guide decisions or choices, the following principle is a self-evident starting point for our deliberations:

Choice-guidance. *The logical properties of preferences should be compatible with the use of preferences as guides for choosing among the elements of the alternative set.*

In addition to this, we need to invoke another principle, which derives from the fact that preferences do not come to us out of thin air. The acquisition of preferences often costs time and effort.[20] A rational subject should avoid comparison costs that are not matched by the expected advantages of having more determinate or more well-founded preferences.[21] This is what the following principle demands:

Comparison-cost avoidance.[22] *Excessive comparison costs are avoided.*

[20] I refer here primarily to the costs involved in obtaining knowledge about the empirical facts on which preferences are based. The formation of preferences may also involve "computational" costs, but these are difficult to treat in a model with full intersubstitutivity of logically equivalent expressions. Cf. Spohn 1997, pp. 295–299; 2000.

[21] Cf. Halldin 1986. Halldin does not, however, distinguish between decision costs and comparison costs.

[22] It might be maintained that decision-theoretic arguments such as this are inappropriate in preference theory since they put the cart before the horse. Traditionally, preference theory is seen as one of the ground-pillars on which decision theory is built. To use decision-theoretic arguments at the foundations of preference theory would therefore, or so a critic might say, disturb the hierarchical structure of disciplines. Howard Sobel seems to defend such a hierarchical structure when criticizing "Hansson's proposal for adjusting *preferences* to facilitate stable choices," which he sees as "a strange reversal of the natural order, which order would have actions constrained by credences and preferences, rather than having credences and preferences fiddled with to facilitate actions informed by them" (Sobel 1997, pp. 69–70). In my view, this hierarchical structure is a chimerical one.

It is an immediate consequence of comparison-cost avoidance that choice-guiding preferences cannot in general be taken to satisfy completeness ($A \geq B \vee B \geq A$).[23] This conclusion is at variance with tradition in formal studies of preference, but it conforms with actual patterns of preference.[24] As an example, in the choice between three brands of canned soup, *A*, *B*, and *C*, I clearly prefer *A* to both *B* and *C*. As long as *A* is available, I do not need to make up my mind whether I prefer *B* to *C*, or *C* to *B*, or consider them to be of equal value. Similarly, a voter in a multiparty election does not have to rank the parties or candidates that she does not vote for.[25]

For a preference relation to be choice-guiding, it must supply at least one alternative that is eligible, that is, one that can reasonably be chosen. The minimal formal criterion for eligibility is that the chosen alternative be no worse than any other alternative:[26]

Weak eligibility.[27] *There is at least one alternative A such that for all B,* $\neg(B>A)$.

Preference patterns that do not satisfy weak eligibility are not effective choice-guides. Suppose, for instance, that a subject has the three alternatives *A*, *B*, and *C* to choose between. If she strictly prefers *A* to

The interdependence of fundamental concepts is a pervasive feature of conceptual analysis. Neither values nor decisions are conceptually or genetically prior to the other. We need each of these groups of concepts in order to clarify the other.

[23] Completeness may also be violated for reasons unrelated to comparison costs. You may, for instance, consider the preservation of Acropolis and that of the giant panda to be incomparable, so that no amount of further deliberation can make the comparison possible.

[24] In most studies of preference logic, it has been assumed that weak preference is complete. (A notable early exception is Armstrong 1950.) This is also a standard assumption in applications of preference logic to economics and to social decision theory. In economics, it may reflect a presumption that everything can be "measured with the measuring rod of money" (Broome 1978, p. 332).

[25] These examples also show that comparability need not be transitive, i.e., *B* may be comparable to both *A* and *C* without *A* being comparable to *C* (Cf. Davidson et al. 1955, p. 146).

[26] Weaker criteria, such as to take an alternative that is not worse than all other alternatives, may be useful as criteria of acceptability, but it does not seem appropriate to call them criteria of eligibility or choice. On criteria weaker than weak eligibility, see Chapter 8.

[27] In this and the following definitions in this section, explicit mention of the underlying alternative set has been omitted for the sake of brevity.

B, B to C, and C to A, then the choice supported by her preferences will depend on the stage at which she halts her deliberations. As has already been indicated, preferences with this structure are not useful to guide choices.

An alternative A such that for all B, $\neg(B>A)$, will be called a *weakly eligible alternative*. (Even if the subject's choice is guided by a preference relation, a weakly eligible alternative need not be a *chosen* alternative. She may pick one out of several weakly eligible alternatives.)[28]

Let A be a weakly eligible alternative, and let B be an alternative that is *not* (weakly) eligible. Furthermore, suppose that A and B are comparable – that either $A>B$, $A\equiv B$, or $B>A$. Then, by the definition of weak eligibility, $B>A$ does not hold. It would also be strange for A and B to be equal in value, that is, for $A\equiv B$ to hold. If preferences are choice-guiding, then two alternatives should not be considered to be of equal value if one of them is eligible and the other is not. We may therefore conclude, as a consequence of the principle of choice-guidance, that if A but not B is weakly eligible, then A and B are not equal in value. In an equivalent formulation:

Top-transitivity of weak eligibility. *If* $A\equiv B$, *and* $\neg(C>A)$ *for all* C, *then* $\neg(C>B)$ *for all* C.

I consider top-transitive weak eligibility to be a minimal criterion for rational choice-guiding preferences.

An alternative may be weakly eligible even though there are alternatives to which it has not been compared. This corresponds to actual patterns of choice-guiding preferences. There are about ten brands of tomato ketchup in the supermarket. I do not know that any of these is better than the one that I buy, but there are quite a few that I have never tasted. In this case, comparison costs are so high (compared to the expected gains from making a better choice) that I am satisfied with only weak eligibility.

However, there are other situations in which the subject wants to have compared the chosen alternative to all other alternatives, in spite of the comparison costs involved. In particular, this holds when (1) all incomparabilities are defeasible, and (2) the expected comparison costs

[28] Cf. Ullmann-Margalit and Morgenbesser 1977.

are low in relation to the expected gains from a more well-informed decision. In such cases, there should be at least one weakly eligible alternative that is comparable to all other alternatives, or equivalently:

Strong eligibility. *There is at least one alternative A such that for all B, $A \geq B$.*

An alternative A such that for all B, $A \geq B$ will be called a *strongly eligible alternative.*[29]

Suppose that A is strongly eligible and that B is not. Choice-guiding preferences should not then put A and B on an equal footing, in other words, it should not be the case that $A \equiv B$. In an equivalent formulation:

Top-transitivity of strong eligibility. *If $A \equiv B$, and $A \geq C$ for all C, then $B \geq C$ for all C.*

Both weak and strong eligibility allow for the existence of more than one (weakly, respectively strongly) eligible alternative. In general, this is just as it should be. Choice-guiding preferences may leave us with several alternatives, each of which may be chosen. However, on some occasions we may wish the preference relation to provide us with *exactly* one eligible alternative. In particular, this may occur when it is important to be capable of a quick decision, so that hesitation between several optimal alternatives must be avoided. Then at least the first of the following two conditions should be satisfied:

Exclusive weak eligibility. *There is exactly one alternative A such that for all B, $\neg(B > A)$.*

Exclusive strong eligibility. *There is exactly one alternative A such that for all B, $A \geq B$.*

2.5 RESTRICTABILITY

The sets of alternatives that our preferences refer to are not immutable. New alternatives can become available, and old ones can be lost. Such changes may force us to renewed deliberations.

[29] H. G. Herzberger (1973, p. 197) used the terms "liberal maximalization" and "stringent maximalization" for essentially the same concepts that I call weak and strong eligibility.

Reopening an issue can be uneconomical in terms of comparison costs. In other words, comparison costs may be so structured that the cost of reconsidering an issue is high, whereas the marginal costs of making one's deliberations sufficient for possible future changes of the alternative set are low. In such cases, the subject may wish to perform her original deliberations thoroughly enough to minimize the risk of having to reopen the issue. There are two principal means to achieve this, namely (1) to include potential new alternatives in one's original set of alternatives, and (2) to make sufficient comparisons to ensure that (weak or strong) eligibility holds not only for the original set of alternatives but also if one or several of the original alternatives are lost. If no alternative is considered to be exempt from possibly being lost in the future, then it may be a cost-minimizing strategy to pursue one's deliberations until (weak or strong) eligibility holds for all nonempty subsets of the original alternative set.

It must be emphasized that the preference relation best suited for guiding choices among a certain set of alternatives need not be a suitable guide for choosing among a particular subset of that set – not even if comparison costs are negligible.[30] This can be expressed more precisely with the concept of restrictability. A property (such as a rationality postulate) will be said to hold *restrictably* for a set of alternatives if and only if it holds for all its nonempty subsets. It can then be established that the property of being a suitable guide for decisions need not be a restrictable property of a preference relation.

Two types of examples showing this are well known from the literature on choice functions. First, the alternative set may carry information, as in Amartya Sen's example: "[G]iven the choice between having tea at a distant acquaintance's home (x), and not going there (y), a person who chooses to have tea (x) may nevertheless choose to go away (y), if offered – by that acquaintance – a choice over having tea (x), going away (y), and having some cocaine (z)."[31] Second, choice may be positional. In a choice between a big apple, a small apple, and an orange, you may choose the big apple, but in a choice between only the two apples you may nevertheless opt for the smaller one.[32]

[30] Tullock's (1964) famous argument against preference intransitivity was based on the assumption that the same preference relation guides choices among the elements of a set and among those of any of its subsets. For a criticism, see Anand 1993.

[31] Sen 1993, p. 502. See also Kirchsteiger and Puppe 1996.

[32] Anand 1993, p. 344. On positional choice, see Gärdenfors 1973.

Hence, restrictability is an important special case, not a universal requirement.

As will be seen from the following theorem, if the eligibility properties are required to hold restrictably, then we obtain rationality criteria of the more well-known types, such as completeness, acyclicity, and variants of transitivity.

Theorem 2.4. *Let $\geq$ be a relation over some finite set $\mathcal{A}$ with at least two elements. Then:*

(1) $\geq$ satisfies restrictable weak eligibility if and only if it satisfies acyclicity.[33]
(2) $\geq$ satisfies restrictable strong eligibility if and only if it satisfies completeness and acyclicity.
(3) $\geq$ satisfies restrictable top-transitive weak eligibility if and only if it satisfies acyclicity and PI-transitivity.
(4) $\geq$ satisfies restrictable top-transitive strong eligibility if and only if it satisfies completeness and transitivity.
(5) $\geq$ satisfies restrictable exclusive weak eligibility if and only if it satisfies completeness, acyclicity, and antisymmetry.[34]
(6) $\geq$ satisfies restrictable exclusive strong eligibility if and only if it satisfies completeness, acyclicity, and antisymmetry.

Acyclicity is the property $A >^* B \rightarrow \neg(B > A)$. PI-transitivity is $A > B$ & $B \equiv C \rightarrow A > C$. Antisymmetry is $A \geq B$ & $B \geq A \rightarrow A = B$.

As the theorem shows, a dynamic property such as restrictability can be used to derive static properties such as acyclicity and transitivity. Therefore, static preference logic should not be seen as prior to dynamic principles. To the contrary, (informal) intuitions about both the static and the dynamic properties of preferences are useful for the development of a formal account of their static properties.

As was argued in Section 2.4, top-transitive weak eligibility is a minimal requirement on a choice-guiding preference relation. Depending on the structure of the comparison costs, additional logical requirements may be warranted. Two such additional requirements are

[33] Another formulation of this part of the theorem is given as theorem 1 in Jamison and Lau 1973.

[34] This part of the theorem essentially coincides with theorem 4 of Sen 1971. The interpretation is different, since Sen takes a choice function as primitive, and his preference relation is the ranking that can be derived from that function.

Table 1. *How the rationality requirements on choice-guiding exclusionary preferences depend on the decision-theoretical preconditions of choice.*

	*Net **loss** from covering all possible future losses of alternatives*	*Net **gain** from covering all possible future losses of alternatives*
*Net **loss** from considering all currently available alternatives*	top-transitive weak eligibility	restrictable top-transitive weak eligibility (= acyclicity + PI-transitivity)
*Net **gain** from considering all currently available alternatives*	top-transitive strong eligibility	restrictable top-transitive strong eligibility (= completeness + transitivity)

particularly plausible, namely (1) top-transitive strong eligibility, and (2) restrictability. They give rise to the four major possibilities that are summarized in Table 1.

In this book, no other property than reflexivity will be assumed to hold in general for the exclusionary preference relation $\geq$. Instead, the properties needed for each particular purpose will be specified.

2.6 OTHER APPROACHES TO RATIONAL PREFERENCES

Before leaving the logic of exclusionary preference relations, a few words should be said about three common types of arguments for logical principles of rational preferences that I have refrained from appealing to: composite alternatives, indiscernibility, and direct appeals to intuition.

Composite alternatives appear most notably in the money-pump argument for transitivity of strict preference, which originates with F. P. Ramsey.[35] For an example, suppose that a stamp collector has cyclic preferences with respect to three stamps, denoted *a*, *b*, and *c*. She prefers *a* to *b*, *b* to *c*, and *c* to *a*. Following Ramsey, we may assume that there is an amount of money, say 10 cents, that the collector is prepared to pay for exchanging *b* for *a*, *c* for *b*, or *a* for *c*. She comes into a stamp dealer's shop with stamp *a*. The stamp dealer offers to trade her *a* for *c*, if she pays 10 cents. She accepts the deal.

[35] Ramsey 1931, p. 182.

We can let $\langle x,v \rangle$ denote that the collector owns stamp x and has paid v cents to the dealer.[36] Hence, she has moved from the state $\langle a,0 \rangle$ to the state $\langle c,10 \rangle$. With similar offers from the stamp dealer, she can be induced to move through the following sequence of states, each of which is preferred to its immediate predecessor:

$\langle a,0 \rangle$
$\langle c,10 \rangle$
$\langle b,20 \rangle$
$\langle a,30 \rangle$
$\langle c,40 \rangle$
$\langle b,50 \rangle$
$\langle a,60 \rangle$
. . .

In this way, the collector will be bereft of all her money, to no avail.

A major weakness in this argument against cyclic preferences is that it only allows the stamp collector to make binary choices, following the agenda of the cunning dealer. In real-life situations, as soon as she has discovered her opponent's strategy, she will instead make a choice over the full set of composite alternatives, $\{\langle a,0 \rangle, \langle c,10 \rangle, \langle b,20 \rangle, \langle a,30 \rangle, \langle c,40 \rangle, \langle b,50 \rangle, \langle a,60 \rangle \ldots\}$. (This may require some adjustment of the original preference relation, in order to achieve at least weak eligibility.)[37] Hence, the unlimited loss of money does not follow from the application of a preference relation to a fixed alternative set. Instead, it follows from the successive application of the preference relation to a selected sequence of sets of (two) alternatives, in such a way that the agent is presumed to choose "without consideration of what he did at previous choice points and without consideration of what he can anticipate doing at subsequent choice points."[38] Therefore, this is not a valid argument for acyclicity.[39]

[36] This notation differs from that used in the rest of the book. It has been chosen to highlight the compositional nature of the alternatives.

[37] Note that weak eligibility does not imply acyclicity since it allows for cycles not involving any of the weakly eligible alternatives.

[38] McClennen 1990, p. 97.

[39] If the agent is not resolute, i.e., she may deviate from the chosen strategy, then she may be vulnerable to the money pump. On resoluteness, see McClennen 1990 and Rabinowicz 2000. For a review of arguments of the money-pump type, see Hansson 1993a.

Please note that, although I do not accept the money-pump argument against the cyclic pattern $c>a>b>c$, I consider this an inadmissible pattern for choice-guiding preferences over the set $\{a,b,c\}$ of alternatives. The reason for this is that weak eligibility is violated.[40]

Another common type of argument in preference logic appeals to imperceptible differences between alternatives. For example, assume that there are 1,000 cups of coffee, numbered C_0, C_1, C_2, . . . up to C_{999}. Cup C_0 contains no sugar, cup C_1 one grain of sugar, cup C_2 two grains, etc. Since I cannot taste the difference between C_0 and C_1, they are equally good in my taste, $C_0 \equiv C_1$. For the same reason, we have $C_1 \equiv C_2$, $C_2 \equiv C_3$, etc. all the way up to $C_{998} \equiv C_{999}$.

If indifference is transitive, then it follows from $C_0 \equiv C_1$ and $C_1 \equiv C_2$ that $C_0 \equiv C_2$. Furthermore, it follows from $C_0 \equiv C_2$ and $C_2 \equiv C_3$ that $C_0 \equiv C_3$. Continuing this procedure, we obtain $C_0 \equiv C_{999}$. However, this is absurd since I can clearly taste the difference between C_0 and C_{999} and like the former much better. In cases like this (with insufficient discrimination), it does not seem plausible for the indifference relation to be transitive.

Although examples can certainly be constructed in which the structure of preferences is determined in this way by indiscernability, this can hardly be considered a typical pattern of real-life preferences. It seems to me that a disproportionately strong emphasis on these rare cases has led to the uncritical acceptance of constructions that rely on the (generally speaking implausible) assumption that there is some underlying complete preference ordering, from which the actual preferences can be obtained through the application of a veil of indiscernability.[41]

Finally, logical properties may be argued for by direct appeal to intuition. Savage and others have argued in this way in favour of transitivity. According to Savage, whenever I find a PPP-cycle ($A>B$, $B>C$, and $C>A$) among my own preferences, "I feel uncomfortable in much the same way that I would do when it is brought to my attention that

[40] The same cyclic pattern can be acceptable if embedded in a larger alternative set, such as $\{a,b,c,d\}$, with $d>c>a>b>c$, $d>a$, and $d>b$.

[41] The most important such constructions are semiorders and interval orders. For an overview and key references, see Section 2.7 in Hansson 2000b. Some mathematical properties are treated in Hansson 1993b. For an interesting critical discussion of the intuitive plausibility of semiorders and interval orders, see Danielsson 1998. It should be observed that in addition to their use in preference theory, these classes of orderings also have an important role in measurement theory.

some of my beliefs are logically contradictory. Whenever I examine such a triple of preferences on my own part, I find that it is not at all difficult to reverse one of them. In fact, I find on contemplating the three alleged preferences side by side that at least one of them is not a preference at all, at any rate not any more."[42] There is also some empirical evidence that when people are faced with their own intransitivities, they tend to modify their preferences to make them transitive.[43]

In my view, direct appeal to unanalyzed intuition should only be used as a last resort. An intuition like the one referred to by Savage should not be taken at face value. Instead, we should search for its possible motivations. In this case, I am aware of two reasonable such motivations. One is that preference can be conceived as (at least ideally) derivable from differences in the quantity of some entity such as goodness or utility. With this assumption, it can be argued that weak preference should be an ordering (a transitive and complete relation). In my opinion, however, this is a much too strong, and indeed question-begging, assumption. Preference logic should be built on as weak and uncontroversial assumptions as possible.

The other basis for the intuition that Savage refers to is one that we have already investigated in some detail, namely, that we expect preferences to have choice-guiding capacity. However, as we saw in Section 2.4, the logical principles that follow in the general case from choice-guidance (namely top-transitive weak eligibility) do not imply acyclicity, which Savage argues for with this example.

[42] Savage 1954, p. 21.

[43] Tversky 1969. Among children, intransitivity of preference is much more common. See Bradbury et al. 1984.

3

Preference States

In Chapter 2, it was taken for granted that a subject's state of mind with respect to preferences can be adequately represented by a binary relation. For some purposes – and in particular for studies of changes in preference – a somewhat more sophisticated mode of representation will turn out to be useful. The purpose of this chapter is to introduce a more general format for the representation of *preference states*. A preference state is an idealized state of mind with respect to value comparisons of a particular set of alternatives.

To see why preference relations are not a fully general representation of preference states, it is sufficient to note that if "A is at least as good as B" is validated by a preference relation, then so is either "A is better than B" or "A and B are of equal value." (If $A \geq B$ holds, then so does either $A>B$ or $A \equiv B$.) Furthermore, a preference relation cannot validate a sentence such as $A>C \vee B>C$ without also validating either $A>C$ or $B>C$, and it cannot validate $A \geq B \vee B \geq A$ ("A and B are comparable") without validating either $A \geq B$ or $B \geq A$. In the models to be introduced in this chapter, these restrictions will be removed.

It is possible to interpret this generalization as representing the subject's lack of knowledge about her own preferences. However, this is not the intended interpretation. The framework to be presented here was developed to represent lack of determinateness in the preferences themselves. My preferences may be such that I hold A be to at least as good as B, but it is not settled (and not merely so that I do not know) whether or not I also hold A to be better than B. I can, for instance, opine that Alfvén was at least as good a composer as Berwald, without being committed to having an opinion on whether he was better or just equally good.

An obvious and reasonably general way to increase the expressive power is to represent each preference state by the set of sentences about preferences that it validates. This construction is similar to that of a belief set (corpus) that consists of all the sentences that the subject believes in or is committed to believe in. In analogy to belief sets, a set consisting of all the sentences (in a given language) that represent preferences held by the subject will be called a *preference set.* In order to define this notion more precisely, we need to extend the formal apparatus.

The object language will consist of atomic preference sentences of the form $A \geq B$ and their truth-functional combinations. It is convenient in this context to assume that there is a universal set $\mathcal{U}$ of potential alternatives.[1] A preference relation should have a subset of that set as its domain.

Definition 3.1. *An* alternative *is an element of the set* $\mathcal{U} = \{A, B, C \ldots\}$ *of alternatives. Nonempty subsets of* $\mathcal{U}$ *are denoted* $\mathcal{A}, \mathcal{B}, \mathcal{C} \ldots$ *and are called* sets of alternatives.

The object language $\mathcal{L}_{\mathcal{U}}$ *consists of the sentences that are formed according to the following rules:*

(1) If $X, Y \in \mathcal{U}$*, then* $X \geq Y \in \mathcal{L}_{\mathcal{U}}$.
(2) If $\alpha \in \mathcal{L}_{\mathcal{U}}$*, then* $\neg\alpha \in \mathcal{L}_{\mathcal{U}}$.
(3) If $\alpha, \beta \in \mathcal{L}_{\mathcal{U}}$*, then* $\alpha \vee \beta \in \mathcal{L}_{\mathcal{U}}$, $\alpha \& \beta \in \mathcal{L}_{\mathcal{U}}$, $\alpha \rightarrow \beta \in \mathcal{L}_{\mathcal{U}}$*, and* $\alpha \leftrightarrow \beta \in \mathcal{L}_{\mathcal{U}}$.

$X>Y$ *is an abbreviation of* $(X \geq Y) \& \neg(Y \geq X)$*, and* $X \equiv Y$ *of* $(X \geq Y) \& (Y \geq X)$.

The following definition introduces some useful notation:

Definition 3.2. *Let* $\mathcal{A} \subseteq \mathcal{U}$ *and* $S \subseteq \mathcal{L}_{\mathcal{U}}$*. Then:*

$S \uparrow \mathcal{A}$ *("S restricted to* $\mathcal{A}$*") consists of all elements of S that are logically equivalent with some element of* $\mathcal{L}_{\mathcal{U}}$ *in which both arguments of each instance of* $\geq$ *are elements of* $\mathcal{A}$*. Furthermore,* $S \downarrow \mathcal{A}$ *("S excepting* $\mathcal{A}$*") consists of all elements of S that are logically equivalent with some*

[1] As an additional measure of convenience, linguistic expressions denoting alternatives will be called 'alternatives.'

element of $\mathcal{L}_{\mathcal{U}}$ in which no instance of ≥ has an argument that is an element of $\mathcal{A}$.

Thus, $S{\uparrow}\mathcal{A}$ consists of those sentences in S that only refer to alternatives in $\mathcal{A}$, and $S{\downarrow}\mathcal{A}$ of those sentences in S that do not refer to any alternative in $\mathcal{A}$.

Rationality postulates, such as transitivity and weak eligibility, are usually assumed to be common to all preference states in a particular framework. Therefore, they should be treated in the same way as logical axioms, namely as background conditions that are not changed when the preference state is changed. Formally, they will be represented by *preference postulates*, such as $X>Y \,\&\, Y>Z \rightarrow X>Z$.

In order to make Definition 3.2 sufficiently precise, we need to specify the logic referred to in that definition. This can be done by designating the consequence operator of the logic. The minimal consequence operator for our purposes is the classical truth-functional consequence operator Cn_0, such that for each set S in $\mathcal{L}_{\mathcal{U}}$, $\mathrm{Cn}_0(S)$ is the set of truth-functional consequences of S. Rationality postulates will be represented by stronger consequence operators. Let T be a set of preference postulates. Then Cn_T is the operator such that $\mathrm{Cn}_T(S)$ consists of the logical consequences that can be obtained from S, using both truth-functional logic and the postulates in T. The following definition and theorem provide a precise construction of Cn_T and a verification that the construction works.

Definition 3.3. *Let T be a set of sentences in $\mathcal{L}_{\mathcal{U}}$. $s(T)$ is the set of substitution-instances of elements of T. Furthermore, Cn_T is the operation on subsets of $\mathcal{L}_{\mathcal{U}}$ such that for all $S \subseteq \mathcal{L}_{\mathcal{U}}$:*

$$\mathrm{Cn}_T(S) = \mathrm{Cn}_0(s(T) \cup S)$$

where Cn_0 is the operator of truth-functional consequence over $\mathcal{L}_{\mathcal{U}}$.

As an example, if $(X{\geq}Y \vee Y{\geq}X) \in T$ and $\neg(A{\geq}B) \in S$, then $B{\geq}A \in \mathrm{Cn}_T(S)$. (Clearly, $A{\geq}B \vee B{\geq}A \in s(T)$, and the rest follows truth-functionally.)

Observation 3.4. *For any set $T \subseteq \mathcal{L}_{\mathcal{U}}$, Cn_T is a consequence operator, that is, it satisfies:*

(1) $A \subseteq \mathrm{Cn}_T(A)$ (inclusion)
(2) *If* $A \subseteq B$, *then* $\mathrm{Cn}_T(A) \subseteq \mathrm{Cn}_T(B)$ (monotony)
(3) $\mathrm{Cn}_T(A) = \mathrm{Cn}_T(\mathrm{Cn}_T(A))$ (iteration)

Furthermore, it satisfies:

(4) $\mathrm{Cn}_0(A) \subseteq \mathrm{Cn}_T(A)$ (supraclassicality)
(5) $\psi \in \mathrm{Cn}_T(A \cup \{\chi\})$ *if and only if* $(\chi \rightarrow \psi) \in \mathrm{Cn}_T(A)$. (deduction)
(6) *If* $\chi \in \mathrm{Cn}_T(A)$, *then* $\chi \in \mathrm{Cn}_T(A')$ *for some finite subset* A' *of* A. (compactness)

In what follows, various rationality criteria will be treated as optional background conditions that can be added to the general framework. However, one exception will be made: The (weak) preference relations that are used to construct preference states will be required to be reflexive. There are two reasons for this exception. First, reflexivity is a weak and uncontroversial condition (cf. Section 2.2). Second, as will soon be seen, reflexivity makes it possible to simplify the representation of preference states.

From a formal point of view, the adoption of reflexivity as a general principle means that consequence relations over $\mathcal{L}_{\mathcal{U}}$ will include not only Cn_0, but also the minimal reflexive consequence operation, $\mathrm{Cn}_{\{X \geq X\}}$.

An adequate representation of a preference state should convey information about (1) the set of alternatives, and (2) the preference sentences (in $\mathcal{L}_{\mathcal{U}}$) about these alternatives that are endorsed by the subject. A preference state can therefore be represented by an ordered pair $\langle \mathcal{A}, S \rangle$, with $\mathcal{A}$ a set of alternatives and S a preference set that is a (logically closed) subset of $\mathcal{L}_{\mathcal{U}} \uparrow \mathcal{A}$. However, since weak preference is assumed to be reflexive, we have $\langle \mathcal{A}, S \rangle = \langle \{X \mid X \geq X \in S\}, S \rangle$. We can therefore, without any loss of information, omit explicit reference to the set of alternatives. Preference states can then be represented by sets of sentences as follows:

Definition 3.5. *For any subset* S *of* $\mathcal{L}_{\mathcal{U}}$, $|S| = \{X \mid X \geq X \in S\}$.

Definition 3.6. *Let* $\{X \geq X\} \subseteq T \subseteq \mathcal{L}_{\mathcal{U}}$. *A set* $S \subseteq \mathcal{L}_{\mathcal{U}}$ *is a* T-consistent preference set *if and only if:*

(1) $S = (\mathrm{Cn}_T(S)) \uparrow |S|$, *and*
(2) *For all* $\alpha \in S$, $\neg\alpha \notin S$.

Representation of preference states by preference sets will be called *sentential representation.*

Just like belief sets, preference sets as defined here are closed under logical consequence. Hence, if A, B, and C are elements of the alternative set, a person who subscribes to the preference sentence $A \geq B$ is assumed to also subscribe to the sentence $A \geq B \vee A \geq C$. Furthermore, if transitivity is one of the background conditions and she subscribes to both $A \geq B$ and $B \geq C$, then we assume that she also subscribes to $A \geq C$. The logical closure of the preference set may be seen as the outcome of a reflective equilibrium in the sense introduced in Section 2.1. Somewhat more modestly, we may interpret a preference set as representing, not the set of actually endorsed preference sentences, but the set of preference sentences that the agent is committed to endorse.[2]

3.2 PREFERENCE MODELS

There is also another credible representation of preference states, namely as sets of preference relations.[3] Let $\mathbf{R}$ be a set of preference relations. We may stipulate that a particular formal sentence, such as $A \geq B$, is validated by $\mathbf{R}$ if and only if it is validated by all elements of $\mathbf{R}$. In this framework, as well, $A \geq B$ may be validated without either $A > B$ or $A \equiv B$ being validated. (Let $\mathbf{R}$ consist of two relations, one of which validates $A > B$ and the other $A \equiv B$.) This construction will be called the *relational representation* of preference states, and a set $\mathbf{R}$ of preference relations that represents a preference state will be called a *preference model*. The formal development of this idea is fairly straightforward. We need the following notation:

Definition 3.7. *Let* $R \subseteq \mathcal{U} \times \mathcal{U}$ *and* $\mathcal{B} \subseteq \mathcal{U}$*. Then:*

(1) $|R| = \{X \mid \langle X,X \rangle \in R\}$
(2) $R \uparrow \mathcal{B} = R \cap (\mathcal{B} \times \mathcal{B})$
(3) $R \downarrow \mathcal{B} = R \cap ((|R| \setminus \mathcal{B}) \times (|R| \setminus \mathcal{B}))$

Explicit reference to the set of alternatives can be omitted, given the innocuous assumption that weak preference is reflexive.

[2] This distinction was introduced for belief sets by Isaac Levi (1974, 1977, 1991).
[3] This construction is closely related to Levi's representation of indeterminate preferences as sets of utility functions. See Levi 1974 (p. 409), 1980, 1986.

Definition 3.8. *Let R be a reflexive subset of $\mathcal{U}\times\mathcal{U}$. The set $[R]$ of sentences in $\mathcal{L}_{\mathcal{U}}\uparrow|R|$ is defined as follows, where α and β are elements of $\mathcal{L}_{\mathcal{U}}\uparrow|R|$.*

(1) If $A, B \in |R|$, then $A\geq B \in [R]$ if and only if $\langle A,B\rangle \in R$.
(2) $\neg\alpha \in [R]$ if and only if: $\alpha \notin [R]$.
(3) $\alpha\vee\beta \in [R]$ if and only if: either $\alpha \in [R]$ or $\beta \in [R]$.
(4) $\alpha\&\beta \in [R]$ if and only if: both $\alpha \in [R]$ and $\beta \in [R]$.
(5) $\alpha\rightarrow\beta \in [R]$ if and only if: if $\alpha \in [R]$ then $\beta \in [R]$.
(6) $\alpha\leftrightarrow\beta \in [R]$ if and only if: $\alpha \in [R]$ if and only if $\beta \in [R]$.

A sentence α is validated *by R if and only if $\alpha \in [R]$.*
For any set $T \subseteq \mathcal{L}_{\mathcal{U}}$, R is T-obeying *if and only if $s(T)\uparrow|R| \subseteq [R]$.*

Definition 3.9. *A* preference model *is a nonempty set* $\mathbf{R}$ *of reflexive relations that have the same nonempty subset of $\mathcal{U}$ as their common domain.*

$|\mathbf{R}|$ is the common domain of the elements of $\mathbf{R}$. The set $[\mathbf{R}]$ of sentences in $\mathcal{L}_{\mathcal{U}}\uparrow|\mathbf{R}|$ is defined as follows:

$$[\mathbf{R}] = \cap\{[R] \mid R \in \mathbf{R}\}$$

A sentence α is validated *by $\mathbf{R}$ if and only if $\alpha \in [\mathbf{R}]$. For any set $T \subseteq \mathcal{L}_{\mathcal{U}}$, $\mathbf{R}$ is* T-obeying *if and only if $s(T)\uparrow|\mathbf{R}| \subseteq [\mathbf{R}]$.*

Observation 3.10. *Let $\mathbf{R}$ be a preference model and let $T \subseteq \mathcal{L}_{\mathcal{U}}$. Then $\mathbf{R}$ is T-obeying if and only if each element of $\mathbf{R}$ is T-obeying.*

Hence, every preference model $\mathbf{R}$ gives rise to a set $[\mathbf{R}]$ of preference sentences. This feature makes it possible to relate preference models to preference sets in an exact way. The two representations of preference states are both intuitively plausible. Fortunately, we do not have to determine which of them is the more plausible, since they can be shown to be interchangeable.

Theorem 3.11. *Let $\{X\geq X\} \subseteq T \subseteq \mathcal{L}_{\mathcal{U}}$. A set $S \subseteq \mathcal{L}_{\mathcal{U}}$ is a T-consistent preference set if and only if there is a T-obeying preference model $\mathbf{R}$ such that $S = [\mathbf{R}]$.*

The choice between sentential and relational representation of preference states is therefore a matter of expository convenience and lucidity. In what follows, relational representation will be used since it allows for simple and natural definitions of operators of change.

3.3 INTERSECTIBLE RATIONALITY POSTULATES

Rationality criteria for choice-guiding preferences can be applied to preference states, just as they were applied to (single) preference relations in Chapter 2. With respect to preference models, such properties can be required to be intersectible in the following sense:

Definition 3.12. *A property of preference models is* intersectible *if and only if: If it holds for all elements of a preference model* **R**, *then it holds for* **R**.

Furthermore, it is strictly intersectible *if and only if: If it holds for at least one element of a preference model* **R**, *then it holds for* **R**.

To see why intersectibility is a plausible property for postulates of rationality, we can interpret the elements of the preference model **R** as those preference relations that are compatible with the agent's preferences. Suppose that **R** satisfies a certain rationality postulate P (such as transitivity), and furthermore suppose that the agent modifies her preference state to allow for some additional preference relations R_1, ... R_n. (More will be said about such changes in Chapter 4.) If each of $R_1, \dots R_n$ satisfies P, then we should expect the resulting preference model $\mathbf{R}\cup\{R_1, \dots R_n\}$ to do the same.

Strict intersectibility does not have at all the same intuitive plausibility as intersectibility *simpliciter*. As we will soon see, however, some rationality postulates turn out to satisfy this stronger variant.

When rationality postulates are applied to preference models, the distinction between object language and metalanguage turns out to have a surprising impact. To see this, let us consider the property of completeness. It can be expressed in either of the following two ways:

(1) For all $X, Y \in |R|$: $(X \geq Y) \vee (Y \geq X) \in [R]$ (completeness)
(2) For all $X, Y \in |R|$: Either $(X \geq Y) \in [R]$ or $(Y \geq X) \in [R]$ (*COMPLETENESS*)

The difference is of course that in (1), the disjunction inherent in the definition of completeness is included in the object language, whereas in (2) it is found in the metalanguage. On the level of individual preference relations, the distinction is inconsequential: (1) and (2) are equivalent. This equivalence vanishes, however, when we transfer the two definitions to the framework of preference models:

(1′) For all $X, Y \in |\mathbf{R}|$: $(X\geq Y)\vee(Y\geq X) \in [\mathbf{R}]$ (completeness)
(2′) For all $X, Y \in |\mathbf{R}|$: Either $(X\geq Y) \in [\mathbf{R}]$ or $(Y\geq X) \in [\mathbf{R}]$ (*COMPLETENESS*)

For a simple demonstration that (1′) and (2′) are not equivalent, consider the following preference model:

$$\mathbf{R} = \{R_1, R_2\}, \text{ with}$$
$$R_1 = \{A\geq A, B\geq B, A\geq B\}, \text{ and}$$
$$R_2 = \{A\geq A, B\geq B, B\geq A\}.$$

It follows from $A\geq B \in [R_1]$ that $(A\geq B)\vee(B\geq A) \in [R_1]$, and similarly from $B\geq A \in [R_2]$ that $(A\geq B)\vee(B\geq A) \in [R_2]$. Hence, $(A\geq B)\vee(B\geq A) \in [R_1]\cap[R_2] = [\mathbf{R}]$. Since we also have $A\geq A \in [\mathbf{R}]$ and $B\geq B \in [\mathbf{R}]$, it can be concluded that (1′) holds. However, we also have $B\geq A \notin [R_1]$, and hence $B\geq A \notin [R_1]\cap[R_2] = [\mathbf{R}]$, and similarly $A\geq B \notin [R_2]$, so that $A\geq B \notin [R_1]\cap[R_2] = [\mathbf{R}]$. From this we may conclude that (2′) does not hold.

The following three observations categorize some of the major rationality postulates according to whether or not they satisfy intersectibility. Note that several postulates appear in two variants, depending on the division of sentential connectives between object language and metalanguage.

Observation 3.13. *The following properties are strictly intersectible:*

(1) If $X\geq Y, Y\geq X \in [\mathbf{R}]$, *then* $X = Y$. (antisymmetry)
(2) There is no series $X_1, \dots X_n$ *such that* $X_1>X_2, X_2>X_3, \dots X_{n-1}>X_n, X_n>X_1 \in [\mathbf{R}]$. (acyclicity)
(3) There is at least one X *such that for all* Y, $Y>X \notin [\mathbf{R}]$. (*WEAK ELIGIBILITY*)

Observation 3.14. *The following properties are intersectible:*

(1) If $X \in |\mathbf{R}|$, *then* $X\geq X \in [\mathbf{R}]$. (reflexivity)
(2) If $X, Y \in |\mathbf{R}|$, *then* $(X\geq Y)\vee(Y\geq X) \in [\mathbf{R}]$. (completeness)
(3) $(X\geq Y)\&(Y\geq Z)\rightarrow(X\geq Z) \in [\mathbf{R}]$ (transitivity)
(4) If $X\geq Y \in [\mathbf{R}]$ *and* $Y\geq Z \in [\mathbf{R}]$, *then* $X\geq Z \in [\mathbf{R}]$. (*TRANSITIVITY*)
(5) $(X\equiv Y)\&(Y\equiv Z)\rightarrow(X\equiv Z) \in [\mathbf{R}]$ (transitivity of indifference)

(6) *If* $X \equiv Y \in [\mathbf{R}]$ *and* $Y \equiv Z \in [\mathbf{R}]$*, then* $X \equiv Z \in [\mathbf{R}]$. (*TRANSITIVITY OF INDIFFERENCE*)

(7) $(X > Y) \& (Y > Z) \rightarrow (X > Z) \in [\mathbf{R}]$ (quasi-transitivity)

(8) *If* $X > Y \in [\mathbf{R}]$ *and* $Y > Z \in [\mathbf{R}]$*, then* $X > Z \in [\mathbf{R}]$. (*QUASI-TRANSITIVITY*)

(9) $(X > Y) \& (Y \equiv Z) \rightarrow (X > Z) \in [\mathbf{R}]$ (PI-transitivity)

(10) *If* $X > Y \in [\mathbf{R}]$ *and* $Y \equiv Z \in [\mathbf{R}]$*, then* $X > Z \in [\mathbf{R}]$. (PI-*TRANSITIVITY*)

(11) *If* $X \equiv Y \in [\mathbf{R}]$ *and* $\neg(Z > X) \in [\mathbf{R}]$ *for all Z, then* $\neg(Z > Y) \in [\mathbf{R}]$ *for all Z.* (top-transitivity of weak eligibility)

(12) *If* $X \equiv Y \in [\mathbf{R}]$ *and* $X \geq Z \in [\mathbf{R}]$ *for all Z, then* $Y \geq Z \in [\mathbf{R}]$ *for all Z.* (top-transitivity of strong eligibility)

Observation 3.15. *The following properties are nonintersectible:*

(1) *If* $X, Y \in |\mathbf{R}|$*, then either* $X \geq Y \in [\mathbf{R}]$ *or* $Y \geq X \in [\mathbf{R}]$. (*COMPLETENESS*)

(2) *There is at least one X such that for all Y,* $X \geq Y \in [\mathbf{R}]$. (strong eligibility)

(3) *There is at least one X such that for all Y,* $\neg(Y > X) \in [\mathbf{R}]$. (weak eligibility)

(4) *If* $X \equiv Y \in [\mathbf{R}]$ *and* $Z > X \notin [\mathbf{R}]$ *for all Z, then* $Z > Y \notin [\mathbf{R}]$ *for all Z.* (*TOP-TRANSITIVITY OF WEAK ELIGIBILITY*)

Hence, most of the rationality postulates discussed in Chapter 2 satisfy intersectibility when applied to preference models. The only major exception is strong eligibility. (Note that the negative results in parts 1, 3, and 4 of Observation 3.15 are paralleled by positive results for other variants of the same properties in Observation 3.13, part 3 and Observation 3.14, parts 2 and 11.) These results add to the credibility of preference models as representations of preference states.

4

Changes in Exclusionary Preferences

It is a salient feature of preferences that they change, both in response to external stimuli and as a result of mental processes. Many social phenomena cannot be understood unless changes in preferences are included in the analysis.

Economists have noted that the preferences of a consumer depend on her past consumption, either through habituation or through need for variety.[1] Consumers' preferences may also be influenced by various social factors, such as advertising, propaganda, and conformity with the habits and opinions of others.[2] Preference change also has an essential role in the emergence of morality and of social cooperation; cooperation becomes individually rational if individuals develop preferences that reflect internalized commitments to social values.[3] In this way, preference change is related to central issues in moral and social philosophy. As an example of this, John Stuart Mill's theory of the higher pleasures can be seen as "an attempt to analyze the question of how one would want one's character or system of preferences to develop over time."[4] Furthermore, preference change is an important factor to take into account in the analysis of certain decision paradoxes. An unusually clear example is the Deterrence Dilemma: "[U]nless you, a nuclear super-power, intend to retaliate if attacked by another . . . and unless this would guide you in an attack, you will be attacked."[5] Under these conditions, a subject with pacifist preferences would best serve these preferences by revising them and acquiring a preference for retaliation.[6]

[1] von Weizäcker 1971; Becker 1992.
[2] For useful reviews, see Day 1986 and Bowles 1998.
[3] Raub and Voss 1990, p. 82.
[4] McPherson 1982, p. 260.
[5] MacIntosh 1992, p. 505.
[6] On preference change in decision paradoxes, see also Cave 1998.

In spite of the prevalence of preference change, previous formalized studies of preferences have been restricted to their static properties. This chapter provides a formal framework for the investigation of different types of changes in exclusionary preferences. Rationality criteria for operations of change will be proposed, that can be imposed in addition to "static" rationality criteria of the conventional type. The representation of preference states as preference models, introduced in Chapter 3, will be used throughout this chapter. In Section 4.1, the basic framework for preference change is introduced in an informal manner, and four elementary types of change are identified. The subsequent four sections are devoted to each of these types of change: revision (4.2), contraction (4.3), subtraction (4.4), and addition (4.5).

4.1 THE BASIC FRAMEWORK

It is a natural starting point to investigate which concepts and methods can be adopted from the much more thoroughly explored field of belief change. The dominant models of belief change are *input-assimilating* (cf. Section 1.5). They describe how the subject's belief state is transformed upon assimilation of an input, which typically represents some factual information. These models are also *reductionist* in the sense that complex changes are assumed to be decomposable into sequences of changes of a few basic types.[7] Furthermore, most of these models are *sentential*. The belief state is represented by a set of sentences that are identified with the sentences believed to be true. Inputs take the form of sentences (or possibly sets of sentences)[8] to be removed from, or added to, the set of believed sentences.

Sentential representation of beliefs is far from unproblematic. Actual epistemic agents are moved to change their beliefs largely by nonlinguistic inputs, such as sensory impressions. Sentential models of belief change (tacitly) assume that such primary inputs can, in terms of their effects on belief states, be adequately represented by sentences. Thus, when I see a hen on the roof (a sensory input), I adjust my belief state *as if* I modified it to include the sentence "there is a hen on the roof" (a linguistic input).

[7] Levi 1991, p. 65; Hansson 1999c, Sections 1.3 and 1.6.
[8] Hansson 1989b, 1993d. Fuhrmann and Hansson 1994.

The approach to preference change to be presented here shares these basic assumptions. The framework is *input-assimilating*, that is, it represents processes in which a preference state is transformed by an input into a new preference state. Admittedly, not all preference changes in real life take place in response to a well-determined input. Neither do all belief changes. In both cases, the input-assimilating model is based on the simplifying assumption that the cause(s) of a change can be represented in the form of an input. The fact that actual preference change often tends to be sudden rather than gradual contributes somewhat to the realism of this assumption.[9]

Furthermore, just as in belief change theory, *sentential representations* will be used. Actual subjects change their preferences by miscellaneous mechanisms. In the formal model, however, inputs take the form of sentences that represent preferences. If I grow tired of my favourite brand of mustard, *A*, and start to like brand *B* better, then this will be represented by a change with the sentence "*B* is better than *A*" as an input.[10] Hence, the model to be introduced here focuses on how a preference state is adjusted in order to accommodate linguistically specified new patterns of preference. The relation between actual incentives to change and idealized sentential inputs will not be covered here. (Arguably, the development of larger models of preference dynamics that include the motive forces of change is even more urgent than the corresponding task in belief dynamics.)

A primary task in the development of a logic of preference change is to identify the major types of (elementary) changes to which preference states can be subjected. Again, we can use concepts that have been developed for belief change.[11] In the dominating models of belief change, all changes are assumed to be ultimately decomposable into two major types of changes:

(1) Removal of a belief from a belief set (*contraction*).
(2) Incorporation of a belief into a belief set.

It is assumed that all operations of change, however complex, can be reconstructed as a series of contractions and incorporations. This does

[9] Mistri 1996, p. 336.
[10] The sentential input approach is particularly adequate when preference change is the result of an attempt, either by the subject or by some external agent, to achieve a specified preference pattern. This is exemplified by preference changes induced in decision paradoxes, as referred to above.
[11] Alchourrón, Gärdenfors, and Makinson 1985.

not mean that actual changes always take place in such stepwise fashion, only that their outcomes are such that they could have been obtained in that way.

There are two major types of belief incorporations: *revisions* and *expansions*. Revisions are consistency-preserving at the price of giving up previous beliefs if this is necessary to maintain consistency. Expansions are belief-preserving at the price of giving rise to an inconsistent belief set if the old and the new information are not logically compatible.

Just like beliefs, preferences can be given up or acquired. Following the terminology of the belief change literature, these two types of operations will be called *contraction* and *revision*, respectively. Revision is consistency-preserving. No operation that can give rise to an inconsistent preference state will be considered here.

Preference states can also change through the addition of new alternatives or the loss of old ones. The former type of change typically takes place when a new alternative is found to be available, and the latter takes place when some old alternative is found to be no longer available and is therefore excluded from consideration. These operations will be called *addition* and *subtraction*, respectively. They complete our list of elementary types of preference change:

(1) *Revision*: **R** is changed to include some preference sentence in $\mathcal{L}_{\mathcal{U}}\uparrow|\mathbf{R}|$.
(2) *Contraction*: **R** is changed to exclude some preference sentence.
(3) *Addition*: An alternative is added to $|\mathbf{R}|$.
(4) *Subtraction*: An alternative is removed from $|\mathbf{R}|$.

Complex changes are assumed to be decomposable into changes of these four types. As an example, suppose that a new alternative B is discovered and is found to be better than all the original alternatives. The resulting change is assumed to be decomposable into a sequence of two changes: (1) addition of B, and (2) revision by the sentence $B>A_1$ & $B>A_2$ & . . . & $B>A_n$, where $A_1 \dots A_n$ are the original alternatives.[12] Just as for belief change, the formal decomposition does not require that the change *actually* takes place in these two consecutive steps. It only requires that its outcome be the same *as if* it had taken place in

[12] If the original set of alternatives is infinite, then (multiple) revision by the set $\{B>A_1, B>A_2 \dots\}$ should be performed instead.

that way. Nevertheless, this reduction to four elementary types of change is far from unproblematic. Like the corresponding reduction for belief change, it provides us with a manageable formal structure, but not without losses in realism. In particular, sudden large changes in preference (or belief) may be difficult to represent adequately as sequences of changes of the elementary types.

With these limitations in mind, we can now take up the task of developing formal accounts of the four elementary types of preference change.

4.2 REVISION

An operator of revision is an operator that, given a preference state and an input sentence α, assigns a new preference state in which α holds. Just as some belief revisions are belief-contravening, some preference revisions force us (on pain of inconsistency) to give up previous preferences. In both cases, we wish the operation of change to be minimal in the sense of not involving any changes not occasioned by the incorporation of the input sentence.[13]

Most formal accounts of belief revision operate by assigning priorities to previous beliefs. The elements (or subsets) of the belief set are assumed to be ordered according to their retractability or degree of epistemic importance. It is commonly assumed that this ordering is prior to, and independent of, the epistemic input. In other words, the epistemic inputs that give rise to revision do not provide any new information about the priorities to be made when deciding what previous beliefs to remove in order to make room for the new belief specified by the input.[14] This may be a plausible assumption in epistemic contexts. In studies of preference change, however, the corresponding assumption is not at all plausible.

There are, roughly speaking, two ways to change one's preferences in order to accommodate a new preference represented by a sentence

[13] In a discussion of the intentional acquisition of preferences, in situations like the Deterrence Dilemma, MacIntosh (1992, p. 524) observed that the new preferences should be "different only in the case where the P[aradoxical] C[hoice] S[ituation] obliges one to intend an action not maximizing on the old [preferences]. The revision must be the minimum mutilation required."

[14] Glaister (2000) has sketched operations of belief change in which epistemic inputs may have to be supplemented with priority-related information. For a comment, see Hansson 1999c.

such as $A \geq B$: Either you change the position of A or that of B. In most cases, the primary input that we wish to mirror in the formal model provides us with information about which of these to choose: You get tired of brand A and start to like it less than brand B, which was your previous second choice. You learn that the political party X has changed its policies on unemployment insurance and start to like it more than party Y, and so on. This is an essential feature of preference revision that should be reflected in the formal framework. A distinction should therefore be made between:

(1) revising **R** by $A \geq B$ through changing the position of A,
(2) revising **R** by $A \geq B$ through changing the position of B, and
(3) revising **R** by $A \geq B$.

In order to capture the difference, the revision operator will be indexed by the alternative(s) whose position should preferably be changed. Thus, (1) will be represented by $\mathbf{R} *_A (A \geq B)$. More generally, the format for revision will be:

$$\mathbf{R} *_{\mathcal{B}} \alpha$$

where α is the new sentence to be accepted and $\mathcal{B}$ (the *priority index*) is a set of alternatives that are "loosened" in the revision, that is, their positions should be changed rather than those of other alternatives. $\mathbf{R} *_{\mathcal{B}} \alpha$ is taken to be undefined if α refers to some alternative not in $|\mathbf{R}|$. (Otherwise, we would not have pure revision, but a combination of addition and revision.)

The following two abbreviations will be used: If $\mathcal{B}$ is empty, then we write $\mathbf{R} * \alpha$ instead of $\mathbf{R} *_{\emptyset} \alpha$. If $\mathcal{B}$ is a singleton, $\mathcal{B} = \{A\}$, then we write $\mathbf{R} *_A \alpha$ instead of $\mathbf{R} *_{\{A\}} \alpha$.

Two slightly more general input representations are worth analyzing. First, sets of sentences, rather than single sentences, can be allowed as inputs. Revision can then be denoted $\mathbf{R} *_{\mathcal{B}} \Pi$, where $\Pi \subseteq \mathcal{L}_{\mathcal{U}} \uparrow |\mathbf{R}|$. However, if Π is finite, then it can be replaced by the conjunction of all its elements. In order to see that this replacement is reasonable, it is sufficient to compare the success conditions for the two types of operation. A multiple revision $\mathbf{R} *_{\mathcal{B}} \{\alpha_1, \ldots \alpha_n\}$ is successful if and only if $\{\alpha_1, \ldots \alpha_n\} \subseteq [\mathbf{R} *_{\mathcal{B}} \{\alpha_1, \ldots \alpha_n\}]$, or equivalently $(\alpha_1 \& \ldots \& \alpha_n) \in [\mathbf{R} *_{\mathcal{B}} \{\alpha_1, \ldots \alpha_n\}]$. A single-sentence revision $\mathbf{R} *_{\mathcal{B}} (\alpha_1 \& \ldots \& \alpha_n)$ is successful if and only if $(\alpha_1 \& \ldots \& \alpha_n) \in [\mathbf{R} *_{\mathcal{B}} (\alpha_1 \& \ldots \& \alpha_n)]$. Hence, the success conditions for $\mathbf{R} *_{\mathcal{B}} \{\alpha_1, \ldots \alpha_n\}$ and $\mathbf{R} *_{\mathcal{B}} (\alpha_1 \& \ldots \& \alpha_n)$ coincide. Since no other reasonable conditions seem to separate the two operations, multiple revision can be reduced to single-sentence revision. Admittedly, this does not apply to revision by an infinite set of sentences, but that case seems to be of minuscule practical importance.

The second possible generalization is to replace the input sentence α in $\mathbf{R} *_{\mathcal{B}} \alpha$ by a preference model $\mathbf{R}'$. This would give rise to the format $\mathbf{R} *_{\mathcal{B}} \mathbf{R}'$, meaning "the outcome of revising $\mathbf{R}$ to make it validate the preference sentences validated by $\mathbf{R}'$." This, however, is equivalent with the multiple revision $\mathbf{R} *_{\mathcal{B}} [\mathbf{R}']$. If the alternative set is finite, then $[\mathbf{R}']$ can – for the reasons just discussed – be replaced by a conjunctive sentence $\&[\mathbf{R}']$, so that $\mathbf{R} *_{\mathcal{B}} \mathbf{R}' = \mathbf{R} *_{\mathcal{B}} \&[\mathbf{R}']$. Hence, single-sentence revision can cover this case as well, again excepting the unordinary infinite case. Admittedly, the notation $\mathbf{R} *_{\mathcal{B}} \mathbf{R}'$ has advantages over $\mathbf{R} *_{\mathcal{B}} \alpha$ in terms of mathematical symmetry. On the other hand, it seems to be less directly related to intuitive notions of preference change. Therefore, single-sentence inputs, and the notation $\mathbf{R} *_{\mathcal{B}} \alpha$, will be used here.

As has already been indicated, the outcome of revising $\mathbf{R}$ by α, $\mathbf{R} *_{\mathcal{B}} \alpha$, should be a preference model that validates α and that is, given this, as similar to $\mathbf{R}$ as possible. For the formal development, we need measures of similarity, one for each priority index.

To begin with, let us consider how to measure the similarity between two (single) preference relations. The similarity between R_1 and R_2 is inversely related to the difference between them.[15] Therefore, the similarity between R_1 and R_2 can be defined in terms of a measure that is applied to the symmetric difference between the two. (For any two sets Φ and Ψ, $\Phi \Delta \Psi$, the symmetric difference between Φ and Ψ, is equal to $(\Phi \backslash \Psi) \cup (\Psi \backslash \Phi)$.) Similarity in terms of preferences not involving alternatives in $\mathcal{B}$ should be given absolute priority over similarity in terms of preferences referring to elements of $\mathcal{B}$. Thus, if $(R_1 \Delta R_2) \downarrow \mathcal{B} \subset (R_3 \Delta R_4) \downarrow \mathcal{B}$, then R_1 is more similar to R_2 than is R_3 to R_4. A variety of measures can be constructed that satisfy this criterion. One natural construction employs a metric function μ in the following way:[16]

Definition 4.1. *A* maximizing measure *on $\mathcal{P}(\mathcal{U} \times \mathcal{U})$ is a function μ that assigns a real number to all elements of $\mathcal{P}(\mathcal{U} \times \mathcal{U})$, and such that, for all $\Phi, \Psi \subseteq \mathcal{U} \times \mathcal{U}$:*

$$\textit{If } \Phi \subset \Psi, \textit{ then } \mu(\Phi) < \mu(\Psi)$$

15 Hansson 1992b.

16 As a somewhat simplistic example, $\mu(\Phi)$ may be equal to the number of elements of Φ. This construction was used in Hansson 1995.

*For any set $\mathcal{B} \subseteq \mathcal{U}$, the $\mathcal{B}$-*ordering *based on* μ, $\sqsubseteq_{\mathcal{B},\mu}$ *(with the indices omitted whenever convenient) is the ordering of pairs of elements of $\mathcal{P}(\mathcal{U}\times\mathcal{U})$ such that:* $\langle R_1,R_2\rangle \sqsubseteq \langle R_3,R_4\rangle$ *holds if and only if either*

$$\mu((R_1 \Delta R_2)\downarrow\mathcal{B}) < \mu((R_3 \Delta R_4)\downarrow\mathcal{B})$$

or

$$\mu((R_1 \Delta R_2)\downarrow\mathcal{B}) = \mu((R_3 \Delta R_4)\downarrow\mathcal{B}) \text{ and } \mu((R_1 \Delta R_2) \leq \mu((R_3 \Delta R_4).$$

$\sqsubset$ *is the strict part of* $\sqsubseteq$.

Definition 4.1 can be straightforwardly extended to allow us to compare a preference relation to a preference model (i.e., to a set of preference relations). The "distance" between R and $\mathbf{R}$ should be equal to the "distance" between R and the element of $\mathbf{R}$ that it is closest to.[17] (The distance between Bonn and Austria is equal to the distance between Bonn and the part of Austria that is closest to Bonn.) In formal language:

Definition 4.2. $\langle R_1,\mathbf{R}\rangle \sqsubseteq \langle R_2,\mathbf{R}\rangle$ *holds if and only if there is some* $R' \in \mathbf{R}$ *such that, for all* $R'' \in \mathbf{R}$:

$$\langle R_1,R'\rangle \sqsubseteq \langle R_2,R''\rangle.$$

We now have the tools needed to define the operator of revision.[18] $\mathbf{R} *_{\mathcal{B}}\alpha$ should validate α, thus $\alpha \in [R]$ for all $R \in \mathbf{R} *_{\mathcal{B}}\alpha$. Furthermore, every element R of $\mathbf{R} *_{\mathcal{B}}\alpha$ should be maximally similar to $\mathbf{R}$, given that it validates α. In other words, $\mathbf{R} *_{\mathcal{B}}\alpha$ should consist of those α-validating preference relations that are maximally similar to $\mathbf{R}$, according to the priority index $\mathcal{B}$.

Definition 4.3. *Let* $\mathbf{R}$ *be a preference model, let* $T \subseteq \mathcal{L}_{\mathcal{U}}$ *and* $\mathcal{B} \subseteq \mathcal{U}$. *Then the operator* $*_{\mathcal{B}}$ *on* $\mathbf{R}$ *is a* $\mathcal{B}$-prioritized preference revision on $\mathbf{R}$ *in the logic determined by* Cn_T *if and only if there is some $\mathcal{B}$-ordering*

[17] Similarly, the distance between $\mathbf{R}$ and $\mathbf{R}'$ should be equal to the shortest distance that there is between some $R \in \mathbf{R}$ and some $R' \in \mathbf{R}'$. For our present purposes, however, we have no need for distances between two preference models.

[18] The construction presented below was introduced in Hansson 1995. Quite similar constructions, but applied to beliefs rather than preferences, have been proposed independently by Rabinowicz (1995) and Schlechta (1997).

$\sqsubseteq \; = \; \sqsubseteq_{\mathcal{B},\mu}$ *such that for all* Cn_T*-consistent sentences* α *in* $\mathcal{L}_{\mathcal{U}}\uparrow|\mathbf{R}|$, $R \in \mathbf{R} *_{\mathcal{B}}\alpha$ *if and only if:*

(1) $\alpha \in [R]$
(2) *R is T-obeying, and*
(3) *there is no T-obeying* R' *with* $\alpha \in [R']$ *such that* $\langle R',\mathbf{R}\rangle \sqsubset \langle R,\mathbf{R}\rangle$.

For the reasons given in Section 4.1, revision by a sentence that refers to alternatives not in $|\mathbf{R}|$ is left undefined. Furthermore, revision by an inconsistent sentence is left undefined. (If needed, a clause can be added that takes care of this case, e.g., by letting $\mathbf{R} *_{\mathcal{B}}\alpha = \mathbf{R}$ if α is Cn_T-inconsistent.)

The following observation establishes some formal properties of the revision operator. Most of them are named in analogy to the corresponding postulates in belief change theory.

Observation 4.4. *Let* **R** *be a preference model. Let* α *and* β *be* Cn_T*-consistent sentences in* $\mathcal{L}_{\mathcal{U}}\uparrow|\mathbf{R}|$ *and let* $*_{\mathcal{B}}$ *be a* $\mathcal{B}$*-prioritized preference revision on* **R** *for some* $\mathcal{B} \subseteq \mathcal{U}$*. Then:*

(1) $\mathbf{R} *_{\mathcal{B}}\alpha$ *is a T-obeying preference model.* (closure)
(2) $\alpha \in [\mathbf{R} *_{\mathcal{B}}\alpha]$. (success)
(3) *If* $\alpha \in [\mathbf{R}]$, *then* $\mathbf{R} = \mathbf{R} *_{\mathcal{B}}\alpha$. (vacuity)
(4) *If* α *and* β *are logically equivalent, then* $\mathbf{R} *_{\mathcal{B}}\alpha = \mathbf{R} *_{\mathcal{B}}\beta$. (intersubstitutivity)[19]
(5) *If* $\neg\alpha \notin [\mathbf{R}]$, *then* $\mathbf{R} *_{\mathcal{B}}\alpha = \{R \in \mathbf{R} \mid \alpha \in [R]\}$.
(6) *If* $\neg\alpha \notin [\mathbf{R}]$, *then* $[\mathbf{R} *_{\mathcal{B}}\alpha] = Cn_T([\mathbf{R}] \cup \{\alpha\})\uparrow|\mathbf{R}|$.
(7) *If* $\neg\beta \notin [\mathbf{R} *_{\mathcal{B}}\alpha]$, *then* $(\mathbf{R} *_{\mathcal{B}}\alpha) *_{\mathcal{B}}\beta = \mathbf{R} *_{\mathcal{B}}(\alpha\&\beta)$.
(8) $\mathbf{R} *_{\mathcal{B}}(\alpha\vee\beta)$ *is equal to either* $\mathbf{R} *_{\mathcal{B}}\alpha$, $\mathbf{R} *_{\mathcal{B}}\beta$, *or* $\mathbf{R} *_{\mathcal{B}}\alpha \cup \mathbf{R} *_{\mathcal{B}}\beta$. (disjunctive factoring)
(9) $(\mathbf{R}_1\cup\mathbf{R}_2) *_{\mathcal{B}}\alpha$ *is equal to either* $\mathbf{R}_1 *_{\mathcal{B}}\alpha$, $\mathbf{R}_2 *_{\mathcal{B}}\alpha$, *or* $\mathbf{R}_1 *_{\mathcal{B}}\alpha \cup \mathbf{R}_2 *_{\mathcal{B}}\alpha$.

4.3 CONTRACTION

In belief change theory, there is a curious asymmetry between the formal and the informal importance of the two major operators of

[19] The term 'extensionality' is commonly used for analogous properties in belief revision theory. The postulate introduces an extensionality principle in the sense explained by Marcus (1960). However, this terminology does not tally with other usages of 'extensionality,' such as that of Carnap (1956).

change. In most of the more formally well-developed accounts, operators of contraction have the central role and operators of revision are treated as secondary to, and derived from, operators of contraction. With regard to informal examples, the relation between the two operators is reversed. Good examples of belief revision, that is, of the incorporation of new beliefs, are readily available. It is much more difficult to find credible examples of (pure) belief contraction – of how a belief is lost without any new belief being added to the set of beliefs. In most cases, the retraction of a belief is provoked by the acquisition of some other belief that forces the old one out. The best examples of pure belief contraction seem to be what may be called "contraction for the sake of argument" ("mind-opening contraction").[20] In order to give a belief p a hearing although it contradicts one's present state of belief, one contracts the belief set by $\neg p$.

In the dynamics of preference, analogous cases of contraction for the sake of argument may occur. Hypothetical arguments about preferences may contain statements such as "Even if we do not assume that A is better than B . . . ." In addition, another, and probably more important, case can be made for pure contraction: lost grounds for preferences. If you lose the only reason that you had for holding that $A \geq B$, then you should retract $A \geq B$ from the set of preferences that you endorse, without adding anything else to it.[21]

To contract your preference state by α means to open it up for the possibility that $\neg\alpha$. No possibilities are lost in this process, but new ones are added. Therefore, the contracted preference model should be a superset of the original model. The new elements should validate $\neg\alpha$, and given this they should be as similar as possible to the original model. We can therefore define contraction in terms of revision:

Definition 4.5. *Let* $*_{\mathcal{B}}$ *be an operator of* $\mathcal{B}$*-prioritized preference revision for some* $\mathcal{B} \subseteq \mathcal{U}$ *in the logic determined by* Cn_T*. The corresponding operator* $\div_{\mathcal{B}}$ *of* $\mathcal{B}$-prioritized preference contraction *is the operator such that, for all sentences* $\alpha \in \mathcal{L}_{\mathcal{U}}\!\uparrow\!|\mathbf{R}|$ *such that* $\alpha \notin \mathrm{Cn}_T(\emptyset)$*:* $\mathbf{R} \div_{\mathcal{B}} \alpha = \mathbf{R} \cup \mathbf{R} *_{\mathcal{B}} (\neg\alpha)$.

[20] Fuhrmann 1991a; Hansson 1999c, Section 2.1.

[21] An analogous case for belief contraction cannot be made. The reason for retracting a belief is typically another belief that is then accepted, so that pure contraction does not occur.

Contraction by logical truths is left undefined. (A convention, such as letting $\mathbf{R}\div_{\mathcal{B}}\alpha = \mathbf{R}$ if $\alpha \in \mathrm{Cn}_T(\varnothing)$, can be adopted whenever it is useful.) Furthermore, contraction by sentences referring to alternatives outside of $|\mathbf{R}|$ is left undefined (but a natural definition would be to let $\mathbf{R}\div_{\mathcal{B}}\alpha = \mathbf{R}$ in this case).

A possible generalization is to allow for multiple contraction, in other words, contraction by a set of sentences instead of a single sentence. Contrary to what we found for multiple revision, multiple contraction cannot easily be subsumed under a single-sentence operation. To see that, consider contraction of $\mathbf{R}$ by the set $\{\alpha_1,\alpha_2\}$. The obvious success condition is that both α_1 and α_2 should be removed, or in more precise terms: $\{\alpha_1,\alpha_2\} \cap [\mathbf{R}\div_{\mathcal{B}}\{\alpha_1,\alpha_2\}] = \varnothing$.[22] There is no truth-functional combination f of two sentences such that it holds in general that $\{\alpha_1,\alpha_2\} \cap [\mathbf{R}\div_{\mathcal{B}}\{\alpha_1,\alpha_2\}] = \varnothing$ if and only if $f(\alpha_1,\alpha_2) \notin [\mathbf{R}\div_{\mathcal{B}}\{\alpha_1,\alpha_2\}]$. (In particular, disjunction, the obvious candidate, does not satisfy this property. It may very well be the case that $\{\alpha_1,\alpha_2\} \cap [\mathbf{R}\div_{\mathcal{B}}\{\alpha_1,\alpha_2\}] = \varnothing$ and $\alpha_1 \vee \alpha_2 \in [\mathbf{R}\div_{\mathcal{B}}\{\alpha_1,\alpha_2\}]$.)[23] Therefore, multiple contraction cannot in general, not even in the finite case, be reduced to single-sentence contraction. One possible construction of multiple contraction is the following:

$$\mathbf{R} \div_{\mathcal{B}} \Phi = \mathbf{R} \cup \bigcup_{\alpha \in \Phi} (\mathbf{R} *_{\mathcal{B}} \neg\alpha)$$

Following the precedent of belief change theory, this first development of preference change will focus on the single-sentence format, and the development of multiple operations is deferred to a later occasion.

The next two observations provide some formal properties of (single-sentence) preference contraction and some connections between the operators of preference contraction and preference revision.

Observation 4.6. *Let* $\mathbf{R}$ *be a T-obeying preference model and* $\div_{\mathcal{B}}$ *a* $\mathcal{B}$-*prioritized preference contraction in the logic determined by* Cn_T. *Let* α *and* β *be* Cn_T-*consistent elements of* $\mathcal{L}_{\mathcal{U}}\!\uparrow\!|\mathbf{R}|$. *Then:*

(1) $\mathbf{R}\div_{\mathcal{B}}\alpha$ *is a T-obeying preference model.* (closure)
(2) $\mathbf{R} \subseteq \mathbf{R}\div_{\mathcal{B}}\alpha$ (inclusion)

[22] An alternative success condition is that $\{\alpha_1,\alpha_2\} \not\subseteq [\mathbf{R}\div_{\mathcal{B}}\{\alpha_1,\alpha_2\}]$ (cf. Fuhrmann and Hansson 1994).

[23] For a simple example, let $\mathbf{R}\div_{\mathcal{B}}\{A{\geq}B,\ B{\geq}A\} = \{\{A{\geq}A, B{\geq}B, A{\geq}B\},\{A{\geq}A, B{\geq}B, B{>}A\}\}$.

(3) *If* $\alpha \notin [\mathbf{R}]$, *then* $\mathbf{R}\div_{\mathcal{B}}\alpha = \mathbf{R}$. (vacuity)
(4) $\alpha \notin [\mathbf{R}\div_{\mathcal{B}}\alpha]$ (success)
(5) *If* α *and* β *are logically equivalent, then* $\mathbf{R}\div_{\mathcal{B}}\alpha = \mathbf{R}\div_{\mathcal{B}}\beta$. (intersubstitutivity)
(6) $\mathbf{R}\div_{\mathcal{B}}(\alpha\&\beta)$ *is equal to either* $\mathbf{R}\div_{\mathcal{B}}\alpha$, $\mathbf{R}\div_{\mathcal{B}}\beta$, *or* $\mathbf{R}\div_{\mathcal{B}}\alpha \cup \mathbf{R}\div_{\mathcal{B}}\beta$. (conjunctive factoring)
(7) $(\mathbf{R}_1\cup\mathbf{R}_2)\div_{\mathcal{B}}\alpha$ *is equal to either* $\mathbf{R}_1\div_{\mathcal{B}}\alpha \cup \mathbf{R}_2$, $\mathbf{R}_1 \cup \mathbf{R}_2\div_{\mathcal{B}}\alpha$, *or* $\mathbf{R}_1\div_{\mathcal{B}}\alpha \cup \mathbf{R}_2\div_{\mathcal{B}}\alpha$.

Observation 4.7. *Let* $\mathbf{R}$ *be a T-obeying preference model, let* $*_{\mathcal{B}}$ *be a* $\mathcal{B}$*-prioritized preference revision on* $\mathbf{R}$ *for some* $\mathcal{B} \subseteq \mathcal{U}$ *in the logic determined by* Cn_T, *and let* $\div_{\mathcal{B}}$ *be the corresponding operator of* $\mathcal{B}$*-prioritized preference contraction. Furthermore, let* α *be a* Cn_T*-consistent sentence in* $\mathcal{L}_{\mathcal{U}}\!\uparrow\!|\mathbf{R}|$. *Then:*

(1) *If* $\alpha \in [\mathbf{R}]$, *then* $\mathbf{R} = (\mathbf{R}\div_{\mathcal{B}}\alpha) *_{\mathcal{B}}\alpha$ (recovery)
(2) $\mathbf{R} *_{\mathcal{B}}\alpha = (\mathbf{R}\div_{\mathcal{B}}(\neg\alpha)) *_{\mathcal{B}}\alpha$ (Levi identity 1)
(3) $[\mathbf{R} *_{\mathcal{B}}\alpha] = (\text{Cn}_T([\mathbf{R}\div_{\mathcal{B}}(\neg\alpha)] \cup \{\alpha\}))\!\uparrow\!|\mathbf{R}|$ (Levi identity 2)
(4) $\mathbf{R}\div_{\mathcal{B}}\alpha = \mathbf{R} \cup \mathbf{R} *_{\mathcal{B}}(\neg\alpha)$ (Harper identity 1; *by definition*)
(5) $[\mathbf{R}\div_{\mathcal{B}}\alpha] = [\mathbf{R}] \cap [\mathbf{R} *_{\mathcal{B}}(\neg\alpha)]$ (Harper identity 2)

The recovery postulate in part (1) is similar to the recovery postulate in belief change theory (if $p \in \mathbf{K}$, then $\mathbf{K} = \mathbf{K}\div p + p$). A conspicuous difference is that the recovery postulate for belief contraction is expressed in terms of expansion, whereas recovery for preference contraction is expressed in terms of revision. The reason for this is that no expansion operator has been defined for preference models.

Recovery is probably the most commonly criticized postulate in belief change theory and has often been said to be too strong.[24] It can be taken, though, as a first approximation of the requirement that contraction be performed with minimal losses of previous beliefs. If recovery is satisfied, then the expulsion of α has taken place with such small losses of original beliefs that everything will be reconstructed if α is reinstated.

It is one of the hallmarks of the AGM framework for belief change that contraction and revision are closely knit together by the Levi identity ($\mathbf{K}*p = \text{Cn}((\mathbf{K}\div\neg p)\cup\{p\})$) and the Harper identity ($\mathbf{K}\div p = \mathbf{K} \cap (\mathbf{K}*\neg p)$). The operators of revision and contraction

[24] See the references given in Hansson and Rott 1998 and Hansson 1999c.

defined above for preference models are connected in essentially the same way.

4.4 SUBTRACTION

Subtraction simply means that an alternative is withdrawn from consideration. It can be defined as follows:

Definition 4.8. *The operator* $\ominus$ *of* subtraction *is the operator such that for each preference model* **R** *and each* $A \in \mathcal{U}$:

$$\mathbf{R}\ominus A = \{R\downarrow\{A\} \mid R\in \mathbf{R}\}$$

The following are some major properties of subtraction:

Observation 4.9. *Let* **R** *be a T-obeying preference model and let* $A, B \in \mathcal{U}$. *Then:*

(1) If $A \notin |\mathbf{R}|$, *then* $\mathbf{R}\ominus A = \mathbf{R}$. (vacuity)
(2) $|\mathbf{R}\ominus A| = |\mathbf{R}|\setminus\{A\}$ (success)
(3) $\mathbf{R}\ominus A\ominus B = \mathbf{R}\ominus B\ominus A$ (commutativity)
(4) $\mathbf{R}\ominus A$ *is a T-obeying preference model.* (closure)

Subtraction is a minimal logic-preserving operation that removes an alternative from the alternative set.[25] In actual life, it is possible for the removal of an alternative to have effects other than the minimal effects of subtraction.[26] In accordance with the decomposition principle explained in Section 4.1, such larger changes should be formalized as a series of operations that may contain contraction(s) and revision(s) in addition to subtraction.

It is difficult to find clear real-life examples of subtraction. When restricting your deliberations to a smaller set of alternatives, you typically inactivate rather than give up those earlier preferences that referred to the deleted alternative(s). Suppose, for instance, that your choice among the thirteen main courses on a restaurant menu was reindeer steak. Upon learning that this meal is no longer served, you typically retain your original preferences and can now express them

[25] It is logic-preserving only if there are no existence claims in the logic.
[26] See Section 2.5 on positional choice and the informational value of menus.

hypothetically ("If reindeer steak were served . . ."). Hence, with respect to the availability of intuitive examples, subtraction fares about as badly as belief contraction (whereas preference contraction, as we saw in Section 4.3, is easily exemplified). Like belief contraction, subtraction is nevertheless an operation worth studying as a potential component of larger complexes of change.

It follows from Definition 4.8 that $A \geq A \notin [\mathbf{R} \ominus A]$. However, it must be emphasized that the exclusion of $A \geq A$ when A is subtracted does not involve giving up the conviction that A is equal in value to itself. It is only, in the present formalization, a formal consequence of the exclusion of A from the set of alternatives.

At first sight, it may seem to be a good idea to define $\ominus$ in terms of preference contraction. One might expect $\mathbf{R} \ominus A$ to be constructible as the outcome of contracting $\mathbf{R}$ by all A-referring sentences, either in sequence or in one single step (multiple contraction). Neither of these constructions can be used, however, since the Cn_T-theorems referring to A are retained after any contraction or series of contractions that we perform. In particular, this applies to the Cn_T-theorem $A \geq A$ (which we cannot get rid of without complicating the formal representation of belief states; cf. Section 3.1).

On the other hand, it may be claimed that the framework should be adjusted so that no preference sentences are immune to contraction. However, such a modified framework would probably be less intuitively plausible. There is a difference between (1) giving up the preferences that one has with respect to A while still counting it as an alternative, and (2) excluding A altogether from the set of alternatives. It is an advantage of the formal framework introduced here that it allows us to make this distinction.

4.5 ADDITION

Just like the other three operations, addition (of an alternative) should preserve T-obedience. In general, for $\mathbf{R} \oplus A$, the result of adding A to $\mathbf{R}$, to be T-obeying, it should be the case that $s(T)\uparrow|\mathbf{R} \oplus A| \subseteq [R]$ for every $R \in \mathbf{R} \oplus A$. (Since we have assumed that $X \geq X \in T$, it clearly follows from this that $A \geq A \in [\mathbf{R} \oplus A]$.) Furthermore, $\mathbf{R} \oplus A$ should validate exactly the same sentences about elements of $|\mathbf{R}|$ that are validated by $\mathbf{R}$. In addition to this, it should validate (only) sentences that it must validate in order to be T-obeying. This leads us to the following definition:

Definition 4.10. *The operator* $\oplus$ *of* addition *is the operator such that for each T-obeying preference model* $\mathbf{R}$ *and each* $A \in \mathcal{U}$,

$R \in \mathbf{R}\oplus A$ *if and only if*

(1) $|R| = |\mathbf{R}| \cup \{A\}$,
(2) There is some $R' \in \mathbf{R}$ *such that* $R' = R\uparrow|\mathbf{R}|$, *and*
(3) $\mathrm{s}(T)\uparrow(|\mathbf{R}|\cup\{A\}) \subseteq [R]$.

The following observation confirms the plausibility of Definition 4.10.

Observation 4.11. *Let* $\mathbf{R}$ *be a T-obeying preference model, and let* $A, B \in \mathcal{U}$.

(1) If $A \in |\mathbf{R}|$, *then* $\mathbf{R}\oplus A = \mathbf{R}$. (vacuity)
(2) $|\mathbf{R}\oplus A| = |\mathbf{R}| \cup \{A\}$ (success)
(3) $\mathbf{R}\oplus A\oplus B = \mathbf{R}\oplus B\oplus A$ (commutativity)
(4) $\mathbf{R}\oplus A$ *is a T-obeying preference model.* (closure)
(5) $[\mathbf{R}\oplus A] = (\mathrm{Cn}_T([\mathbf{R}]))\uparrow(|\mathbf{R}| \cup \{A\})$ (accuracy)

A connection similar to the recovery postulate obtains between the operations of addition and subtraction:

Observation 4.12. *Let* $\mathbf{R}$ *be a preference model, and let* $A \in \mathcal{U}$.
If $A \notin |\mathbf{R}|$, *then* $\mathbf{R}\oplus A\ominus A = \mathbf{R}$. (subtractive recovery)

The converse relationship $\mathbf{R}\ominus A\oplus A = \mathbf{R}$ does not hold under any reasonable conditions, since "contingent" preference sentences (those not implied by $\mathrm{Cn}_T(\emptyset)$) are removed by $\ominus$ but are not reintroduced by $\oplus$.

In summary, the results of this chapter show that changes in preference models can be modelled in a way that mirrors the major structures of the dominant AGM model in belief change. To show this is only a beginning in a subject that seems to have as many potential ramifications as belief change.

5

Constructing Combinative Preferences

In Chapters 2–4, we studied exclusionary preferences, that is, preferences that refer to a set of mutually exclusive alternatives that are taken as primitive units with no internal structure. In actual discourse on preferences, we often make statements that transgress these limitations. In a discussion on musical pieces, someone may express preferences for orchestral music over chamber music and also for Baroque over Romantic music. We may then ask how that person rates Baroque chamber music versus orchestral music from the Romantic period. Assuming that these comparisons are all covered by one and the same preference relation, some of the relata of this preference relation are not mutually exclusive.

The logic of such nonexclusionary or *combinative* preferences is the subject of this and the following two chapters. In Section 5.1, the relation between exclusionary and combinative preferences is discussed, and it is proposed that the latter be based on the former. In Section 5.2, a formal representation is introduced that covers both (complete) alternatives and (incomplete) relata.[1] In Section 5.3, two major approaches to combinative preferences are identified. The two approaches are then developed in some detail in Chapters 6 and 7.

5.1 CONNECTING THE TWO LEVELS

It is almost too obvious to be argued that strong connections should be expected to hold between the preferences that refer to a set of (mutually exclusive) alternatives and the preferences that refer to incomplete relata that are associated with these same alternatives. In

[1] The arguments of a combinative preference relation can be called 'alternatives.' In this book, however, they are mostly called 'relata,' in order to avoid confusion with the alternatives of the underlying exclusionary preference relation.

the formal representation, there are two major ways to construct these connections.

One of these is the *holistic* approach, which takes preferences over wholes for basic and uses them to derive combinative preferences. The other approach may be called *aggregative*. It takes smaller units (expressible as incomplete relata) to be the fundamental bearers of value, and the values of complete alternatives are obtained by aggregating the values of these units. A precise aggregative model was developed by Warren Quinn, on the basis of a proposal by Gilbert Harman. In Quinn's model, (intrinsic) values are assigned to certain basic propositions, which come in groups of mutually exclusive propositions. To a conjunction of basic propositions is assigned the sum of the intrinsic values of its conjuncts. Various proposals have been put forward for the calculation of truth-functional combinations of basic propositions, other than conjunction.[2]

As was clarified by Wolfgang Spohn, the essential conditions for the aggregative approach to be workable are that there are isolable, evaluatively independent bearers of value, and that a numerical representation is available in which aggregate value is obtainable through addition of the values of these isolable units.[3] These conditions are satisfied in a utilitarian theory of moral betterness. This was indeed what Quinn had in mind; he considered it "natural to suppose that the most evaluatively prior of all states of affairs are those which locate a specific sentient individual at a specific point along an evaluatively relevant dimension such as happiness, virtue, wisdom, etc. Thus for each pair consisting of an individual and a dimension there will be a distinct basic proposition for each point on that dimension which that individual may occupy."[4]

However, the forms of utilitarianism that lend themselves to this mathematization are not the only reasonable theories of moral value, and there are also nonmoral preference relations for which the aggregative approach does not seem to be at all suitable. Although many different factors may influence our judgment of the overall aesthetic value of a theatre performance, we cannot expect its overall value to be derivable in a mechanical way (such as addition) from these factors. The aesthetic value of the whole cannot be reduced in a

[2] Harman 1967; Quinn 1974; Oldfield 1977; Carlson 1997; Danielsson 1997.

[3] Spohn 1978, pp. 122–129.

[4] Quinn 1974, p. 131. Cf. Harman 1967, p. 799.

summative way into isolable constituents. An analogous argument can be made against applying the aggregative approach to moral value according to intuitionist moral theories. "The value of a whole must not be assumed to be the same as the sum of the values of its parts."[5]

The holistic approach allows us to make use of the results already obtained for exclusionary preferences: An underlying exclusionary preference relation for (mutually exclusive) alternatives can be used to derive preferences over the incomplete relata associated with these alternatives. Mutually exclusive alternatives, and the exclusionary preference relation over them, may be seen as a semantics for the – more syntactic – preferences over (sentences representing) states of affairs. Due to this, and to the implausibility in many cases of the decomposition required in aggregative models, the holistic approach will be followed here.

This is not an unusual choice; the holistic approach has in fact been chosen by most philosophical logicians dealing with combinative preferences. The conventional approach is to take preferences over possible worlds (represented by maximal consistent subsets of the language) as a starting point for deriving preferences over other relata.[6]

Possible world modelling has the advantages of generality and logical beauty, but it also has the disadvantage of cognitive unrealism.[7] In practice, we are not capable of deliberating on anything approaching the size of completely determinate possible worlds.[8] Instead, we restrict our deliberations to objects of manageable size. A more realistic holism should therefore be based on smaller wholes, namely alternatives that cover all the aspects under consideration – but not all the aspects that might have been considered.[9] This approach may be called "myopic holism."

[5] Moore 1903, p. 28.

[6] Rescher 1967; Åqvist 1968; Cresswell 1971; von Wright 1972; von Dalen 1974; von Kutschera 1975; Trapp 1985; Hansson 1989a, 1996a.

[7] In studies of concepts or phenomena that one considers to be independent of cognition, it may be reasonable to abstract from cognitive limitations and use models with completely determinate possible worlds. This applies to some concepts of possibility; R. M. Adams has indeed claimed that "possibility is holistic rather than atomistic, in the sense that what is possible is possible only as part of a possible completely determinate world" (Adams 1974, p. 225). However, this argument for possible world modelling is not applicable to evaluative and normative concepts.

[8] I use the term 'deliberation' in a wide sense that includes processes that are not systematic or even conscious.

[9] It may be seen as an application of Simon's "bounded rationality view" (Simon 1957). Alternatives smaller than possible worlds are referred to in decision theory

Here, all preferences will be assumed to refer to some set of (mutually exclusive) alternatives that are not identified with possible worlds. Exclusionary preferences over these alternatives are taken to be basic, and from them preferences over other relata will be derived.

It must be borne in mind that this construction is a logical reconstruction rather than a faithful representation of actual deliberative or evaluative processes. In everyday life, I prefer chess to boxing *simpliciter*. Only as a result of philosophical reflection do I prefer certain contextually complete alternatives in which I watch or take part in chess to certain other such alternatives in which I watch or take part in pugilism.[10] However, under the condition of a reflective equilibrium introduced in Section 2.1, we may assume that preferences over combinative relata cohere with preferences over mutually exclusive alternatives. This assumption will be made here, as well as the additional assumption that this coherence is so tightly knit that preferences over combinative relata can be reconstructed from the exclusionary preference relation (although, of course, they did not originate that way). The second of these assumptions stands, admittedly, on less safe ground than the first, but, as will be seen in the following chapters, it provides us with the basis for a series of fruitful formal explications of preference.

5.2 REPRESENTING RELATA AND ALTERNATIVES

Our next task is to develop a formal structure in which relata and complete alternatives can be represented. This cannot be done without abstracting from some of the distinctions available in ordinary language. In nonregimented language, all sorts of abstract and concrete entities can serve as the relata of preference relations. Thus, one may prefer butter to margarine, democracy to tyranny, or Bartok's fourth string quartet to his third. In spite of this, logical analyses of combinative preferences have been almost exclusively concerned with relata that represent states of affairs.

This practice is based on the assumption that combinative preferences over other types of entities can be adequately expressed as

as 'small worlds.' See Savage 1954; Simon 1957, pp. 196–200; Toda and Shuford 1965; Toda 1976; Schoemaker 1982, p. 545; Humphreys 1983, p. 24; Mendola 1987, p. 134; Hansson 1993c, 1996d.

[10] Pollock 1983, esp. pp. 413–414; Beck 1941, esp. p. 12.

preferences over states of affairs. R. Lee went as far as saying that "all preferences can be understood in terms of preference among states of affairs or possible circumstances. A preference for bourbon, for example, may be a general preference that one drink bourbon instead of drinking scotch."[11]

It is not quite as simple as that. Some preferences are difficult to reconstruct with states of affairs as relata. As was pointed out to me by Wlodek Rabinowicz, particularly good examples of this can be found in the aesthetic realm. The statement that Bartok's fourth quartet is better than his third cannot be satisfactorily expressed as a preference for the state of affairs that the fourth quartet exists (is played, is listened to, etc.) rather than the third.

However, in most cases preferences like this seem to be reasonably representable as exclusionary. When we compare aesthetic objects, these objects typically form a set of mutually exclusive alternatives. Hence, although Lee's claim is too strong, we may assume that in most cases preferences can either be represented as exclusionary (in which case, alternatives may be taken as primitives) or as combinative over alternatives representable as states of affairs.

For this reason, combinative preferences will be taken to have states of affairs as relata. States of affairs, in turn, will be represented in the usual way by sentences in sentential logic.[12]

Definition 5.1. *$\mathcal{L}$ is a nonempty language that is closed under the truth-functional operations $\neg$ (negation), $\vee$ (disjunction), & (conjunction), $\rightarrow$ (implication), and $\leftrightarrow$ (equivalence).*[13]

In order to express the logical relations between sentences in the formal language, an operator of logical consequence (Cn) will be used. This operator satisfies the three standard conditions (inclusion, monotony, and iteration) and three additional conditions that ensure its compliance with classical sentential logic.

[11] Lee 1984, pp. 129–130. Cf. von Wright 1963a, p. 12, and 1972, pp. 143–144; Trapp 1985, p. 303.

[12] In economic literature, it is more common to let states of affairs be represented by coordinates of vectors. Thus, complete alternatives are represented by n-tuples $\langle x_1, \ldots x_n \rangle$ of features. Typically, each feature (coordinate) corresponds to the quantity of some commodity and takes real numbers as values. The vectorized notation can be translated into sentential language by the use of predicates that represent values assigned to the coordinates.

[13] Cf. Observation 3.4.

Definition 5.2. *The operator* Cn *of logical consequence is a function from and to subsets of $\mathcal{L}$, such that for all such sets S and T and all elements p, q, p′, and q′ of $\mathcal{L}$:*

(1) $S \subseteq \mathrm{Cn}(S)$ (inclusion)
(2) *If* $S \subseteq T$, *then* $\mathrm{Cn}(S) \subseteq \mathrm{Cn}(T)$. (monotony)
(3) $\mathrm{Cn}(S) = \mathrm{Cn}(\mathrm{Cn}(S))$ (iteration)
(4) *If p can be derived from S by classical propositional logic, then* $p \in \mathrm{Cn}(S)$. (supraclassicality)
(5) $(p \rightarrow q) \in \mathrm{Cn}(S)$ *if and only if* $q \in \mathrm{Cn}(S \cup \{p\})$. (deduction)
(6) *If q′ can be obtained from q by substitution of a subformula p of q by p′, and* $(p \leftrightarrow p') \in \mathrm{Cn}(\varnothing)$, *then* $(q \leftrightarrow q') \in \mathrm{Cn}(\varnothing)$. (intersubstitutivity)[14]

A set S is consistent *if and only if there is no sentence p of the language such that both* $p \in \mathrm{Cn}(S)$ *and* $\neg p \in \mathrm{Cn}(S)$. *A sentence q is consistent if and only if {q} is consistent.*

$S \vdash q$ *is used as an alternative notation for* $q \in \mathrm{Cn}(S)$, $p \vdash q$ *for* $q \in \mathrm{Cn}(\{p\})$, *and* $\vdash q$ *for* $q \in \mathrm{Cn}(\varnothing)$.

As was indicated in Chapter 1, intersubstitutivity is an idealization that has been introduced in order to simplify the formal structure.

Sets of sentences in $\mathcal{L}$ can be used to represent states of affairs. Logically equivalent sets represent the same state of affairs, that is, if $\mathrm{Cn}(S) = \mathrm{Cn}(S')$ for some $S, S' \subseteq \mathcal{L}$, then S and S' represent the same state of affairs. Furthermore, single sentences can be used as an alternative representation, with the obvious convention that if $\alpha \in \mathcal{L}$, then α and $\{\alpha\}$ represent the same state of affairs.

Since alternatives are limiting cases of states of affairs, they should also be represented by sets of sentences.[15] Clearly, the set of alternatives should be nonempty.[16] Each alternative should be logically con-

[14] If the language contains no non–truth-functional operators, then intersubstitutivity follows from supraclassicality. However, this does not hold in the general case. As an example, let K be a non–truth-functional operator. Then $(K(p) \leftrightarrow K(p\&(q\vee\neg q))) \in \mathrm{Cn}(\varnothing)$ does not follow from (1)–(5), but it follows from (4) and (6).

[15] It would also be possible to let the alternatives be nonstructured entities and instead introduce a valuation function $\mathcal{V}$ such that for each alternative X, $\mathcal{V}(X)$ is the set of sentences that represent states of affairs that hold in X. The two approaches are equivalent for our purposes, and notational convenience has guided the choice made here.

[16] Arguably, it should have at least two elements. The weaker assumption of nonemptiness is sufficient for the formal developments.

sistent. It is also convenient to express the alternatives as logically closed sets. This is the simplest way to achieve that if $\text{Cn}(A) = \text{Cn}(A')$, then A and A' represent the same alternative. Hence:

Definition 5.3. *A subset* $\mathcal{A}$ *of* $\mathcal{P}(\mathcal{L})$ *is a* sentential alternative set (*a* set of sentential alternatives) *if and only if:*

(1) $\mathcal{A} \neq \emptyset$, *and*
(2) If $A \in \mathcal{A}$, *then* A *is consistent and logically closed* ($A = \text{Cn}(A)$).

This definition is too wide, since it allows for alternative sets such as $\{\text{Cn}(\{p\}), \text{Cn}(\{p,q\})\}$ in which one alternative is a proper subset of another. Such constructions should be excluded, and we also have reasons to exclude alternative sets such as $\{\text{Cn}(\{p\}), \text{Cn}(\{q\})\}$ in which two alternatives are logically compatible. Mutual exclusivity is a characteristic feature of alternatives that distinguishes them from relata in general.[17] These requirements can be summarized as follows:

Definition 5.4. *A subset* $\mathcal{A}$ *of* $\mathcal{P}(\mathcal{L})$ *is a* set of mutually exclusive alternatives *if and only if:*

(1) $\mathcal{A} \neq \emptyset$,
(2) If $A \in \mathcal{A}$, *then* A *is consistent and logically closed* ($A = \text{Cn}(A)$), *and*
(3) If $A, A' \in \mathcal{A}$ *and* $A \neq A'$, *then* $A \cup A'$ *is inconsistent.* (mutual exclusivity)

This definition still allows for an alternative set such as the following:

$$\{\text{Cn}(\{p,q\}), \text{Cn}(\{p,\neg q\}), \text{Cn}(\{\neg p\})\}$$

For a concrete example, let us think of the alternative set containing the following three alternatives, referring to possible ways of spending an evening:

(1) Eating out (p) and going to the theatre (q).
(2) Eating out (p) and not going to the theatre ($\neg q$).
(3) Not eating out ($\neg p$).

[17] Logical incompatibility is actually too strong a condition here. Practical incompatibility would be sufficient, but in order to avoid undue complication of the formal structure, logical incompatibility will be used.

This is a strange set of alternatives, since the third alternative is less specified than the other two. If neither $\mathrm{Cn}(\{\neg p,q\})$ nor $\mathrm{Cn}(\{\neg p, \neg q\})$ has to be excluded from consideration, then the two of them should replace $\mathrm{Cn}(\{\neg p\})$. If only one of them is available, then that one alone should replace $\mathrm{Cn}(\{\neg p\})$. The outcome of amending the set of alternatives in either of these ways is a new alternative set in which all alternatives have been specified in the same respects. This makes it possible to compare them in a more uniform way. Such uniformity is a prerequisite for exhaustiveness in deliberation.[18] The following definition is intended to include the full set of reasonable conditions on an alternative set.

Definition 5.5. *A subset* $\mathcal{A}$ *of* $\mathcal{P}(\mathcal{L})$ *is a* set of contextually complete alternatives *if and only if:*

(1) $A \neq \emptyset$,
(2) If $A \in \mathcal{A}$, *then* A *is consistent and logically closed* ($A = \mathrm{Cn}(A)$), *and*
(3) If $p \in A$, $A \in \mathcal{A}$, *and* $A' \in \mathcal{A}$, *then either* $p \in A'$ *or* $(\neg p) \in A'$. (relative negation-completeness)[19]

Observation 5.6. *Any set of contextually complete alternatives is also a set of mutually exclusive alternatives.*

In what follows, neither mutual exclusivity nor contextual completeness will be taken for granted. Instead, it will be stated for each formal result what types of alternative sets it applies to.

The following notation will be useful:

Definition 5.7. *Let* $\mathcal{A}$ *be a set of sentential alternatives in* $\mathcal{L}$. *The subset* $\mathcal{L}_{\mathcal{A}}$ of $\mathcal{L}$ *is the set consisting exactly of (1)* $\cup\mathcal{A}$ *and (2) the truth-functional combinations of elements of* $\cup\mathcal{A}$.

$\mathcal{L}_{\mathcal{A}}$ *is called the* $\mathcal{A}$*-language. Its elements are the* $\mathcal{A}$*-sentences.*

[18] Such exhaustiveness is not always mandatory, as can be seen from an example proposed by Wlodek Rabinowicz. Let p denote that I go out and q that I wear a tie. Then $\{\mathrm{Cn}(\{p,q\}), \mathrm{Cn}(\{p,\neg q\}), \mathrm{Cn}(\{\neg p\})\}$ is an adequate alternative set, provided that q is value-relevant in the presence of p but not of $\neg p$.

[19] On this property, cf. Hansson 1992b.

Definition 5.8. *Let $\mathcal{A}$ be a set of sentential alternatives, and let p and q be elements of $\cup\mathcal{A}$. Then:*

$\vDash_{\mathcal{A}} q$ denotes that $q \in A$ for all $A \in \mathcal{A}$.
$p \vDash_{\mathcal{A}} q$ denotes that $q \in A$ for all $A \in \mathcal{A}$ such that $p \in A$.
$p \dashv\vDash_{\mathcal{A}} q$ denotes that $p \vDash_{\mathcal{A}} q$ and $q \vDash_{\mathcal{A}} p$.
p and q are $\mathcal{A}$-incompatible if and only if $\vDash_{\mathcal{A}} \neg(p\&q)$.

5.3 PAIRWISE AND DECISION-GUIDING PREFERENCES

The distinction between pairwise and decision-guiding (choice-guiding) preferences was drawn in Section 2.3 in an exclusionary framework. Both types of preferences can be extended to a combinative framework, that is, to one with relata that are not mutually exclusive alternatives. The rationale for this extension is not the same, however, for pairwise and decision-guiding preferences.

The need for an extension of pairwise preferences to a combinative framework can be most clearly seen for ceteris paribus preferences, preferences other things being equal. When saying that you like p better than q ceteris paribus, you indicate that there are other specifications that can be added to both p and q, such that you like, for instance, $p\&r_1\&\dots r_n$ better than $q\&r_1\&\dots r_n$ (but not necessarily better than $q\&\neg r_1\&\dots\neg r_n$). Hence, the very notion of ceteris paribus preferences presupposes the existence of larger alternatives in which the relata are included.

Ceteris paribus preference is decidedly pairwise. When comparing p and q ceteris paribus, we focus on the difference between p and q and determine what effect this difference has when p and q are included in alternatives with other differences minimized. The comparisons between alternatives that are relevant for this purpose are comparisons between p-containing and q-containing alternatives that are otherwise as similar as possible.

The use of combinative preferences for decision-guiding purposes has a quite different motivation. Arguably, an ideal agent with unlimited deliberative capacity would have no need for combinative preferences. She could, speaking figuratively, juggle with any number of balls. There would be no reason for her not to perform all of her deliberations with reference to the full set of contextually complete alternatives. Actual agents, on the other hand, with their limited capacities, will often have use for comprehensive comparisons between groups of alternatives. It is for the purpose of expressing such comparisons that a combinative preference relation is needed. For example, suppose that we are choosing between the three operas and seven spoken dramas that are advertised for this evening in Stockholm. We can then make statements such as "I prefer going to the opera to going to the theatre" or "I prefer going to a comedy rather than a tragic play," and so on. These statements refer to nonexclusive relata rather than to mutually exclusive alternatives.

In addition, as will be seen in Chapter 10, combinative decision-guiding preference relations can be helpful in constructing a logic for norms.

As was already mentioned, when we compare p to q ceteris paribus, we focus on the effects of the differences between p and q. In decision-guiding deliberations, we should instead be interested in finding out the total set of possible effects of choosing p, respectively q. It follows from this that when we compare p to q for decision-guiding purposes, the same p-alternatives should be relevant as when we compare p to some other sentence r for the same purposes. Considerations of similarity have no place here.[20]

Due to these considerations, the derivation of combinative from exclusionary preferences will be quite different for pairwise and decision-guiding preferences. These formal developments are the subject of the following two chapters.

[20] Howard Sobel made a similar comment with respect to a preference relation that mirrors the degree of welcomeness of states of affairs. "[W]hether or not an agent would welcome A more than he would B, depends on how much he would welcome A, on how much he would welcome B, and on nothing else . . . [H]ow much an agent *would* welcome a proposition at a time – as distinct from his *sense* for how much he would welcome it – is not a function of what, if anything, he intends to compare it with" (Sobel 1997, p. 53).

6

Pairwise Combinative Preferences

When discussing with my wife what table to buy for our living room, I said: "A round table is better than a square one." By this I did not mean that irrespective of their other properties any round table is better than any square-shaped table. Rather, I meant that any round table is better (for our living room) than any square table that does not differ significantly in its other characteristics, such as height, sort of wood, finishing, price, etc. This is preference ceteris paribus, or "everything else being equal."

As was indicated in Section 5.3, ceteris paribus preferences are a paradigm type of pairwise preferences. It is the purpose of this chapter to investigate how combinative pairwise preferences can be derived from exclusionary preferences. Examples and intuitive arguments will appeal to the ceteris paribus interpretation, but the formal framework is more general and does not exclude other interpretations.

Section 6.1 is devoted to a fairly detailed investigation of the feature that characterizes combinative in contradistinction to exclusionary preferences, namely that compatible relata can be compared. In Sections 6.2 and 6.3, a general format for deriving combinative preferences is introduced. In Section 6.4, three more specified variants of this format are introduced, and in Sections 6.5 and 6.6 their logical properties are studied.

6.1 HOW TO COMPARE COMPATIBLE ALTERNATIVES

It is obvious from ordinary usage that ceteris paribus preferences are combinative, in other words, they allow for compatible relata.[1] There is nothing strange or unusual with an utterance such as "It is better to have a cat than to have a dog" – although it is both possible and

[1] See also the argument for this given in Section 5.3.

common to have both a cat and a dog.[2] Before we can develop a model of pairwise combinative preferences, we need to analyze and make explicit the conventions that guide our understanding of such utterances.

A child may very well protest against the quoted sentence, saying: "No, it is better to have a dog, if you have a cat too." We perceive this as a sign that the child has misunderstood what it means to make this comparison. Having both a cat and a dog is not under consideration. The sentence expresses a comparison between cat-and-no-dog and dog-and-no-cat. This seems to be a general feature of ceteris paribus comparisons between compatible alternatives. As was noted by Hector Castañeda, "[w]hen St. Paul said 'better to marry than to burn' he meant 'it is better to marry and not to burn than not to marry and to burn.' "[3]

This convention was already observed by Sören Halldén in his pioneering work on preference logic. He concluded: "If we say that it would be better if p than if q, then we mean that it would be better if $p\&\neg q$ than if $q\&\neg p$."[4] The same standpoint was taken by Georg Henrik von Wright in his analysis of ceteris paribus preference.[5] The following has become a standard procedure in preference logic:

Translation procedure 1 (Halldén). *The informal statement "p is better than q" is translated into* $(p\&\neg q)>(q\&\neg p)$*, and "p is equal in value to q" is translated into* $(p\&\neg q)\equiv(q\&\neg p)$.

This is by no means bad as a first approximation. It works in cases such as the one just cited, when the alternatives are compatible and neither of them logically implies the other. It also works when the alternatives are logically incompatible. (Then $p\&\neg q$ is equivalent to p and $q\&\neg p$ to q.)

However, Halldén's translation procedure runs into serious trouble when at least one of p and q logically implies the other. Then it forces us to compare a state of affairs to a contradictory state of

[2] Trapp claimed that "no two relata of a preference relation should be considered to be true in the same possible world" among a certain class of possible worlds that are at most minimally different from the actual world (Trapp 1985, p. 301). For a rebuttal, see Hansson 1989a, pp. 5–6.

[3] Castañeda 1958. Cf. 1 Cor 7:9.

[4] Halldén 1957, p. 28.

[5] von Wright 1963a, pp. 24–25, and 1972, pp. 146–147.

affairs.[6] This problem was observed by Aleksandar Kron and Veselin Milovanović, who decided to accept the translation procedure but left as an open question "what it could mean to prefer a contradiction to something else or to prefer a state of affairs to a contradiction."[7]

The translation procedure runs into serious difficulties when a sentence p is compared to itself; this comparison will be reduced to comparing logical contradiction to itself. Arguably, logical contradiction is equal in value to itself, but this does not seem to be the right reason why a noncontradictory statement p should be equal in value to itself. The right reason must be concerned with comparing p to itself, not contradiction to itself.

A remedy for this breakdown can be found simply by observing how the problematic cases are treated in informal discourse. Let p denote "I work hard and earn a lot of money" and q "I work hard." A case can be made for the viewpoint that p and q are incomparable. However, it should be clear that *if the comparison can be made in a meaningful way*, then p will be preferred to q (provided that you value money).[8] This comparison does not invoke the contradictory state of affairs $p\&\neg q$. Rather, the actual comparison takes place between p and $q\&\neg p$. It would seem correct to say that since $p\&\neg q$ is contradictory, it is not used to replace p.

Similarly, a comparison between p and itself does not involve a comparison between $p\&\neg p$ and itself. Since $p\&\neg p$ is contradictory, it is not used to replace p.

We are thus led to the following definition and translation procedure:

Definition 6.1. *p/q ("p and if possible not q") is equal to p if $p\&\neg q$ is logically contradictory, and otherwise it is equal to $p\&\neg q$.*

Translation procedure 2. *The informal statement "p is better than q" is translated into $(p/q)>(q/p)$, and "p is equal in value to q" is translated into $(p/q)\equiv(q/p)$.*[9]

[6] For expository convenience, I make the breakneck ontological assumption that there is a state of affairs, the "contradictory state of affairs" that corresponds to logically contradictory sentences. On the implausibility of comparisons to contradiction, see also Section 8.3.

[7] Kron and Milovanović 1975, p. 187. Cf. Trapp 1985, pp. 314–318.

[8] Or, at least, it will be weakly preferred. Cf. Section 6.3.

[9] This procedure and the / notation were introduced in Hansson 1989a.

This procedure yields the same result as Halldén's in the two cases when the latter turns out to be satisfactory, namely when p and q are incompatible and when they are compatible and neither of them implies the other. In the remaining cases, namely when one or both of p and q implies the other, the second procedure yields an intuitively more reasonable result than Halldén's procedure.

However, we are not yet finished. The use of logical contradiction in the definition of / leads to undesired results. Let p denote "I go to the moon" and q, "I travel by spaceship." A comparison between p and q will, according to Translation Procedure 2, be conceived as a comparison betwen $p\&\neg q$ and $q\&\neg p$. However, $p\&\neg q$ is not a serious possibility, although it is clearly *logically* possible. The only reasonable way to perform this comparison (outside of certain science fiction contexts) is to compare p to $q\&\neg p$. More generally, p/q should be defined as p not only when $p\&\neg q$ is logically impossible but also when it is for other reasons not to be counted as possible or, more precisely, not included in any element of the alternative set:

Definition 6.2. *$p/_{\mathcal{A}}q$ ("p and if $\mathcal{A}$-possible not q") is equal to $p\&\neg q$ if $p \nvDash_{\mathcal{A}} q$. If $p \vDash_{\mathcal{A}} q$, then $p/_{\mathcal{A}}q$ is equal to p.*

Translation procedure 3. *The informal statement "p is better than q" is translated into $(p/_{\mathcal{A}}q)>(q/_{\mathcal{A}}p)$, and "p is equal in value to q" is translated into $(p/_{\mathcal{A}}q)\equiv(q/_{\mathcal{A}}p)$.*

This is the translation procedure that will be used in the rest of this chapter. It has been tailored to pairwise preferences and should not be used for decision-guiding preferences. (As noted in Section 5.3, when comparing p to q for decision-guiding purposes, we should take the same potential realizations of p into account as when we compare p to some other sentence r. It follows from this that p should not be transformed into $p/_{\mathcal{A}}q$ in one case and into $p/_{\mathcal{A}}r$ in the other.)

6.2 FROM EXCLUSIONARY TO PAIRWISE PREFERENCES

As we have just seen, an (informal) pairwise comparison between the relata p and q should be translated into a formal comparison between the relata $p/_{\mathcal{A}}q$ and $q/_{\mathcal{A}}p$. Therefore, it should be derivable from a comparison between alternatives in which $p/_{\mathcal{A}}q$ is true and alternatives in which $q/_{\mathcal{A}}p$ is true. A pair of alternatives $\langle A_1,A_2\rangle$, such that $p/_{\mathcal{A}}q$ is true in A_1 and $q/_{\mathcal{A}}p$ is true in A_2, will be called a *representation* of the pair $\langle p/_{\mathcal{A}}q,q/_{\mathcal{A}}p\rangle$.

Definition 6.3. *Let $\mathcal{A}$ be a set of sentential alternatives. An element A of $\mathcal{A}$ is a* representation *in $\mathcal{A}$ of a sentence x if and only if $x \in A$.*

An element $\langle A,B\rangle$ of $\mathcal{A} \times \mathcal{A}$ is a representation *in $\mathcal{A}$ of the pair $\langle x,y\rangle$ of sentences if and only if $x \in A$ and $y \in B$.*

A sentence x or a pair $\langle x,y\rangle$ of sentences is representable *in $\mathcal{A}$ if and only if it has a representation in $\mathcal{A}$.*

More concisely, x is representable in $\mathcal{A}$ if and only if $x \in \cup\mathcal{A}$, and $\langle x,y\rangle$ if and only if $x, y \in \cup\mathcal{A}$.

Not all representations of $\langle p/_{\mathcal{A}}q,q/_{\mathcal{A}}p\rangle$ need to be relevant to the comparison between p and q. Those that are relevant will be picked out by a *representation function.*

Definition 6.4. *A* representation function *for a set $\mathcal{A}$ of sentential alternatives is a function f such that for all sentences x and y:*[10]

(1) If $\langle x,y\rangle$ is representable in $\mathcal{A}$, then $f(\langle x,y\rangle)$ is a nonempty set of representations of $\langle x,y\rangle$ in $\mathcal{A}$.
(2) Otherwise, $f(\langle x,y\rangle) = \emptyset$.

Representation functions provide a general format for deriving pairwise combinative preference relations from exclusionary preference relations:

Definition 6.5. *Let $\geq$ be a relation on the set $\mathcal{A}$ of sentential alternatives and f a representation function for $\mathcal{A}$. The weak preference relation $\geq_f$, the f-*extension *of $\geq$, is defined as follows:*[11]

$$p \geq_f q \text{ if and only if } A \geq B \text{ for all } \langle A,B\rangle \in f(\langle p/_{\mathcal{A}}q,q/_{\mathcal{A}}p\rangle).$$

[10] The concept of a representation function was introduced in Hansson 1989a.

[11] Definition 6.5 applies the representation function to a single preference relation. Alternatively, representation functions can be applied to preference models, as defined in Section 3.2. Preference states can then be represented either (1) as pairs $\langle \mathbf{R},f\rangle = \langle\{R_1,\ldots R_n\},f\rangle$ where f is a representation function to be combined with each element of $\mathbf{R}$, or (2) more generally as sets of the form $\{\langle R_1,f_1\rangle,\ldots\langle R_n,f_n\rangle\}$, where each element is a pair consisting of a preference relation and a representation function to be combined with that preference relation.

As can be seen from the definition, $\geq_f$ conforms with Translation Procedure 3. It is proposed as the formal counterpart of the "at least as good as" of the informal language.

For most purposes, it can be assumed that a comparison between p and q and one between q and p are based on comparisons between the same pairs of (mutually exclusive) alternatives. This assumption corresponds to the following symmetry property of representation functions:

Definition 6.6. *A representation function f for a set $\mathcal{A}$ of sentential alternatives is* symmetric *if and only if for all sentences $x, y \in \cup\mathcal{A}$ and all elements A and B of $\mathcal{A}$: $\langle A,B \rangle \in f(\langle x,y \rangle)$ if and only if $\langle B,A \rangle \in f(\langle y,x \rangle)$.*

Another plausible property of a representation function is that reflexive comparisons of states of affairs (comparisons of a state of affairs to itself) should only be represented by reflexive comparisons of alternatives (comparisons of an alternative to itself). This can also be required for comparisons between states of affairs that are coextensive, that is, hold in exactly the same alternatives:

Definition 6.7. *A representation function f for a set $\mathcal{A}$ of sentential alternatives satisfies* weak centring[12] *if and only if for all sentences $x \in \cup\mathcal{A}$ and all elements A_1 and A_2 of $\mathcal{A}$:*

$$\text{If } \langle A_1,A_2 \rangle \in f(\langle x,x \rangle), \text{ then } A_1 = A_2.$$

Furthermore, it satisfies centring *if and only if for all sentences $x, y \in \cup\mathcal{A}$:*

$$\text{If } \langle A_1,A_2 \rangle \in f(\langle x,y \rangle) \text{ and } \vDash_{\mathcal{A}} x \leftrightarrow y, \text{ then } A_1 = A_2.$$

We should expect an extended preference relation to say about the complete alternatives exactly what the underlying exclusionary preference relation says about them. If there is a sentence a that has A as its only representation and a sentence b that has B as its only representation, then $a \geq_f b$ should hold if and only if $A \geq B$ holds. Indeed, this condition holds for all representation functions.

[12] This property was called "centring" in Hansson 1996a.

Observation 6.8. *Let ≥ be a relation on the set $\mathcal{A}$ of sentential alternatives and f a representation function for $\mathcal{A}$. Furthermore, let A and B be elements of $\mathcal{A}$, and a and b sentences such that A is the only representation of a in $\mathcal{A}$, and B the only representation of b in $\mathcal{A}$. Then:* $a \geq_f b$ *if and only if* $A \geq B$.

Corollary. *If $\mathcal{A}$ is a mutually exclusive alternative set, and* $A = \text{Cn}(\{a\})$ *and* $B = \text{Cn}(\{b\})$, *then* $a \geq_f b$ *if and only if* $A \geq B$.

6.3 STRICT PREFERENCE AND INDIFFERENCE

Given an exclusionary preference relation ≥, the representation function provides us with a weak combinative preference relation $\geq_f$. In addition to this, we need combinative relations of indifference and strict preference.

The derivation of a combinative strict preference relation from ≥ and *f* can be performed in at least two different ways. The underlying principles can be seen from the following example: A friend is going to bring a bottle of wine to be served either with a vegetable soup or with a fish soup; you do not know which. You consider vegetable soup to go equally well with red and white wine, but fish soup to go better with white wine. According to one interpretation of "better," it is (other things being equal) better to have the white wine than the red one. (The white wine is always at least as good, and in one case it is better.) However, someone might object to this usage, saying: "No, it is not quite correct to say that it is better to have the white wine. It is undetermined whether it is better or just equally good."

The first of these interpretations corresponds to $>_f$, and the second to $\gg_f$, in the following formal definition:

Definition 6.9. *Let ≥ be a relation on the set $\mathcal{A}$ of sentential alternatives, and f a representation function for $\mathcal{A}$. The two strict preference relations* $>_f$ *and* $\gg_f$ *are defined as follows:*

$$p >_f q \text{ if and only if } p \geq_f q \;\&\; \neg(q \geq_f p)$$
$$p \gg_f q \text{ if and only if } A > A' \text{ for all } \langle A, A' \rangle \in f(\langle p/_{\mathcal{A}}q, q/_{\mathcal{A}}p \rangle)$$

It follows directly from this definition that if *f* is symmetric, then $\gg_f$ is a stronger relation than $>_f$, that is, if $p \gg_f q$ then $p >_f q$, whereas the converse implication does not hold in general. $>_f$ has the substantial advantage of bearing the same relationship to $\geq_f$ as $>$ to $\geq$. However, $\gg_f$ is a sufficiently plausible construction to be worth investigating. In what follows, the two will be studied in parallel.

Two indifference relations, $\equiv_f$ and $\cong_f$, can be defined correspondingly:

Definition 6.10. *Let $\geq$ be a relation on the sentential alternative set $\mathcal{A}$, and f a representation function for $\mathcal{A}$. The two indifference relations $\equiv_f$ and $\cong_f$ are defined as follows:*

$$p \equiv_f q \text{ if and only if } p \geq_f q \;\&\; q \geq_f p$$
$$p \cong_f q \text{ if and only if } A \equiv A' \text{ for all } \langle A, A' \rangle \in f(\langle p/_{\mathcal{A}} q, q/_{\mathcal{A}} p \rangle)$$

However, the difference betwen $\equiv_f$ and $\cong_f$ vanishes if f is symmetric:

Observation 6.11. *Let $\geq$ be a relation on the sentential alternative set $\mathcal{A}$, and f a representation function for $\mathcal{A}$. If f is symmetric, then for all p and q: $p \equiv_f q$ if and only if $p \cong_f q$.*

The relationships between the relations introduced in this and the previous section are summarized in Figure 2.

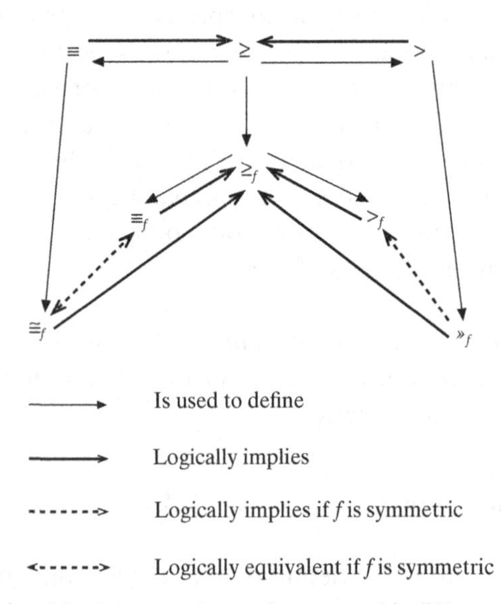

Figure 2. Relationships between the preference and indifference relations introduced in Sections 6.2–6.3.

6.4 CONSTRUCTING THE REPRESENTATION FUNCTION

We now know how representation functions can be used to derive combinative relations of preference and indifference. Before the logic of these derived relations can be investigated, we need to determine the construction of representation functions. As before, the intended interpretation is ceteris paribus preferences.

A recipe for this construction can be extracted from von Wright's pioneering book on preference logic. He defined ceteris paribus preferences as follows:

> [A]ny given total state of the world, which contains p but not q, is preferred to a total state of the world, which differs from the first in that it contains q but not p, but otherwise is identical with it.[13]

This recipe needs a few modifications before it can be put to use: (1) For reasons explained in Section 5.1, the set of alternatives should typically consist of much smaller units than possible worlds. (2) Where von Wright refers to "p but not q," that is, to $p\&\neg q$, we should instead refer to $p/_{\mathcal{A}}q$, as explained in Section 6.1. (3) Finally, von Wright's concept of "identity" is problematic. It is, to put it mildly, confusing to say of a round table and a square table that they are "identical" in all other respects than the shape of the table top. What can (at most) be said is that they are, given this difference, as similar as possible in all other respects.

With these modifications, the quoted passage can be rephrased as follows:

> Any given alternative that contains $p/_{\mathcal{A}}q$ is preferred to any alternative that differs from the first in that it contains $q/_{\mathcal{A}}p$, but is otherwise as similar as possible to it.

Before this recipe can be formalized, we need to operationalize "as similar as possible." In a follow-up article, von Wright attempted to solve this problem (but under another description) by means of an arithmetical count of differences in terms of logically independent atomic states of the world.[14] He assumed that there are n logically independent states of affairs $p_1, \ldots p_n$, and 2^n possible states of the world

[13] von Wright 1963a, p. 31. Cf. Quinn 1974, p. 124. Von Wright uses two other names for this type of preference, namely "unconditional preference" and "holistic preference" (von Wright 1963a, pp. 31–33, and 1972, pp. 140, 147).

[14] von Wright 1972, pp. 146–147.

$w_1, \ldots w_{2^n}$ that can be compared in terms of the n atomic states. If two states of affairs q and r are molecular combinations of m out of the n atomic states, then a ceteris paribus comparison of q and r keeps the other n–m states constant.

Unfortunately, this simple construction is not as promising as it might seem at first sight. Its major weakness is that the choice of atomic states can be made in different ways that give rise to different relations of similarity.[15] For an example of this, consider the following four sentential alternatives:

(1a) $\text{Cn}(\{p, q_1, q_2, q_3, q_4, q_5, q_6, q_7, q_8, q_9, q_{10}\})$
(1b) $\text{Cn}(\{\neg p, q_1, q_2, q_3, q_4, q_5, q_6, q_7, q_8, q_9, q_{10}\})$
(2a) $\text{Cn}(\{p, r_1, r_2, r_3, r_4, r_5, r_6, r_7, r_8, r_9, r_{10}\})$
(2b) $\text{Cn}(\{\neg p, \neg r_1, \neg r_2, \neg r_3, \neg r_4, \neg r_5, \neg r_6, \neg r_7, \neg r_8, \neg r_9, \neg r_{10}\})$

Intuitively, (1a) and (1b) seem to represent a ceteris paribus comparison between p and $\neg p$, whereas (2a) and (2b) do not. But suppose that $r_1, \ldots r_{10}$ are definable in terms of $p, q_1, \ldots q_{10}$ as follows:

$$r_1 \leftrightarrow (p \leftrightarrow q_1)$$
$$\ldots$$
$$r_{10} \leftrightarrow (p \leftrightarrow q_{10})$$

Then, in going from (1a) and (1b) to (2a) and (2b), we shift to another, expressively equivalent set of atomic sentences. Since there are no objectively given logical atoms, there is in general ample scope for choosing among sets of atomic sentences that are equivalent in terms of what can be expressed in the language, but not in terms of von Wright's similarity measure. Such assignments of atomicity can be viewed as a method to assign similarity and dissimilarity to pairs of states of affairs.[16]

It seems inescapable that a nontrivial explication of similarity will have to make use of more information than what is inherent in the logic. Probably the most transparent way to represent similarity is by means of a similarity relation, as follows:

Definition 6.12. *For any set Ψ, the four-place relation T is a similarity relation over Ψ if and only if, for all $U, V, W, X, Y, Z \in \Psi$:*

[15] This was discovered in connection to verisimilitude by Miller (1974) and Tichy (1974).
[16] Cf. Quinn 1974, pp. 124–125.

(T1) $T(W,X,Y,Z) \vee T(Y,Z,W,X)$ (completeness)
(T2) $T(U,V,W,X) \,\&\, T(W,X,Y,Z) \rightarrow T(U,V,Y,Z)$ (transitivity)
(T3) $T(X,X,Y,Z)$
(T4) $T(X,Y,Y,Y) \rightarrow X=Y$
(T5) $T(X,Y,Y,X)$ (symmetry)

The strict counterpart of T is defined as follows:

$$\hat{T}(W,X,Y,Z) \leftrightarrow T(W,X,Y,Z) \,\&\, \neg T(Y,Z,W,X)$$

$T(W,X,Y,Z)$ should be read "W is at least as similar to X as is Y to Z," and $\hat{T}(W,X,Y,Z)$ "W is more similar to X than is Y to Z." This axiomatization of the four-termed similarity relation was proposed by T. Williamson.[17] It is a generalization of a three-termed similarity relation that was introduced earlier by David Lewis.[18] Lewis's relation $S(X,Y,Z)$ should be read "X is at least as similar to Y as is Z." It can be defined from the four-termed relation through the relationship $S(X,Y,Z) \leftrightarrow T(X,Y,Z,Y)$.

(T1) and (T2) combine to say that similarity is a weak ordering (complete and transitive). (T3) and (T4) combine to say that maximal similarity obtains between two arguments if and only if they are identical, and (T5) states that the degree of similarity between two arguments does not depend on the order in which they are taken.

Given a similarity relation T over a sentential alternative set $\mathcal{A}$, we can derive a representation function that requires maximal similarity as measured by T:

Definition 6.13. *Let $\mathcal{A}$ be a set of sentential alternatives and T a similarity relation over $\mathcal{A}$. Then f is the* similarity-maximizing *representation function that is based on T, if and only if it is a representation function and, for all $x, y \in \cup\mathcal{A}$ and $A, B \in \mathcal{A}$:*

$\langle A,B \rangle \in f(\langle x,y \rangle)$ if and only if $x \in A$, $y \in B$, and $T(A,B,A',B')$ holds for all $A', B' \in \mathcal{A}$ such that $x \in A'$ and $y \in B'$.

[17] Williamson 1988. See also Hansson 1992b.

[18] Lewis 1973a, pp. 48ff; 1973b, p. 560; and 1981. He proposed the following postulates for S:

(S1) $S(X,W,Y) \vee S(Y,W,X)$
(S2) $S(X,W,X)$
(S3) $S(X,W,Y) \,\&\, S(Y,W,Z) \rightarrow S(X,W,Z)$
(S4) $S(X,W,W) \rightarrow X = W$

Furthermore, if $\geq$ *is a reflexive relation on* $\mathcal{A}$*, then* $\geq_f$ *is a* similarity-maximizing *preference relation if and only if it is based on a similarity-maximizing representation function.*

The two adequacy criteria introduced in Section 6.2 are satisfied by similarity-maximizing representation functions.

Observation 6.14. *Let* $\mathcal{A}$ *be a set of sentential alternatives and* f *a similarity-maximizing representation function over* $\mathcal{A}$*. Then* f *satisfies centring and symmetry.*

Definition 6.13 can be seen as a formalized version of the basic ideas behind von Wright's explication of ceteris paribus preferences, as quoted above. The most common criticism of von Wright's account is that it makes ceteris paribus a much stronger concept than the intuitive notion that it is intended to formalize. Rainer Trapp stated the argument forcefully in noting that, although it could sensibly be said to be better to have cholera than to have cancer, this comparison is not ceteris paribus in von Wright's sense since it would not hold in a world in which cholera was absolutely incurable whereas there existed safe remedies against all forms of cancer.[19]

This criticism is certainly valid, but only because von Wright's original explication of ceteris paribus preferences refers to total states of the world. With alternative sets consisting of *contextually* complete alternatives, the problem need not arise. When comparing cancer to cholera, your (implicit) alternative set does not contain alternatives in which the state of medical technology is different from what it is in the present state of the world. In general, the maximally similar but contextually irrelevant pairs of complete alternatives have been excluded when the alternative set was selected.

From another point of view, von Wright's approach to ceteris paribus preferences is too restricted in its choice of representations. When I said to my wife that a round table is better than a square one, I meant this to apply not only if the two tables are *maximally* similar in other respects, but also if they are, in a wider sense, *sufficiently* similar. Thus, I prefer the round table even if the square table has a somewhat different (and perhaps somewhat better) finish.

[19] Trapp 1985, p. 308. For two other instances of essentially the same criticism, see Danielsson 1968, pp. 58–59, and Hansson 1989a, p. 12.

One way to express this in the formal model is to let the representation function select not only the maximally similar representations but also other sufficiently similar representations of the two states of affairs.

Possibly the strongest argument in favour of an account in terms of sufficient, rather than maximal, similarity can be based on statements such as: "It would be better if this stick were shorter." This does not mean that it would be better if the stick were shorter but, given that, as similar as possible to how it is now. For that to be true, the difference in length would have to be as small as possible.[20] Instead, the speaker indicates that there is a range of lengths, shorter than that of the actual stick, such that it would be better if the length of the stick were within that range. Presumably, this is an interval with the actual length as its upper limit, so that if the actual length is l_0, and l_1 is in the interval, then any length strictly between l_0 and l_1 is also in the interval. However, although the quoted utterance indicates the existence of such an interval, it does not indicate its lower limit.

These informal examples indicate that it would be of interest to develop an account of ceteris paribus preferences in terms of *sufficiently* rather than *maximally* similar representations. A fairly general formal approach is to let the representation function satisfy the following property: If $\langle A,B \rangle$ is a sufficiently similar representation of $\langle p,q \rangle$, and $\langle A',B' \rangle$ is another representation of $\langle p,q \rangle$ whose arguments A' and B' are at least as similar to each other as A and B, then $\langle A',B' \rangle$ is also a sufficiently similar representation of $\langle p,q \rangle$.

Definition 6.15. *Let $\mathcal{A}$ be a set of sentential alternatives, f a representation function for $\mathcal{A}$, and T a similarity relation over $\mathcal{A}$. Then f is a* similarity-satisficing *representation function with respect to T if and only if the following holds for all $x, y \in \cup\mathcal{A}$ and $A, B, A', B' \in \mathcal{A}$:*

(1) If $\langle A,B \rangle \in f(\langle x,y \rangle)$, then $x \in A$ and $y \in B$.
(2) If $\langle A,B \rangle \in f(\langle x,y \rangle)$, $x \in A'$, $y \in B'$, and $T(A',B',A,B)$, then $\langle A',B' \rangle \in f(\langle x,y \rangle)$.

[20] The problem is of the same nature as the well-known problem with counterfactuals of the type "If this line were shorter, then . . .". Cf. Lewis 1973a, pp. 20–21. Under the (idealized) assumption that all rational numbers represent possible lengths of sticks, no maximally similar representation can be obtained in this case.

Furthermore, if $\geq$ is a reflexive relation on $\mathcal{A}$, then $\geq_f$ is a similarity-satisficing *preference relation if and only if it is based on a similarity-satisficing representation function.*

Neither of the two adequacy criteria referred to above hold in general for similarity-satisficing representation functions.

Observation 6.16. *Let $\mathcal{A}$ be a set of contextually complete alternatives and f a similarity-satisficing representation function for $\mathcal{A}$. Then:*

(1) It does not hold in general that f satisfies centring.
(2) It does not hold in general that f satisfies symmetry.

Similarity-maximizing representation functions are a proper subcase of similarity-satisficing representation functions:

Observation 6.17. *Let $\mathcal{A}$ be a set of sentential alternatives, and f a representation function for $\mathcal{A}$. Then:*

(1) If f is a similarity-maximizing representation function, based on some similarity relation T over $\mathcal{A}$, then it is also similarity-satisficing with respect to T.
(2) It does not hold in general (not even if $\mathcal{A}$ is contextually complete) that if f is similarity-satisficing, then it is also similarity-maximizing.

There is an important difference between how similarity-maximizing and similarity-satisficing representation functions are defined (Definitions 6.13 and 6.15). Whereas the former are defined in terms of a similarity relation, the latter are not fully determined by the similarity relation that is referred to in the definition. The reason for this is, intuitively speaking, that for similarity-satisficing representation functions we also need to know how similar is similar enough. In our definitions, this information is included in the representation function, but it might also have been given an independent representation.

The similarity-maximizing and similarity-satisficing constructions add further realism to the basic structure that was introduced in Section 6.2. They do this, however, at the price of introducing further variables into the formal apparatus. It is therefore worth investigating whether these complications can be dispensed with. In other words, can we construct a plausible representation function without introducing complicating extralogical components such as a similarity relation?

The following is an attempt at such a simplified construction:

Definition 6.18. *Let* $\mathcal{A}$ *be a sentential alternative set. The* maximal centred *representation function for* $\mathcal{A}$ *is the function f such that for all* $x, y \in \cup\mathcal{A}$:

(1) $f(\langle x,y\rangle) = \{\langle A,A\rangle \in \mathcal{A}\times\mathcal{A} \mid x, y \in A\}$ *if this set is nonempty*
(2) $f(\langle x,y\rangle) = \{\langle A,B\rangle \in \mathcal{A}\times\mathcal{A} \mid x \in A \;\&\; y \in B\}$ *otherwise*

Let $\geq$ *be a reflexive relation on* $\mathcal{A}$. *Then the f-extended preference relation* $\geq_f$ *is a* maximal centred preference relation.

This definition may be seen as a modification of von Wright's "absolute preferences." Von Wright applied these, however, to possible worlds. As he himself pointed out, this was a quite implausible construction.[21] For example, I cannot prefer (in this sense) having a cup of tea at this moment to having a cup of coffee without also preferring every possible world in which I have a cup of tea now to every possible world in which I have a cup of coffee now – even if, say, eternal peace begins today in the coffee world whereas World War Three breaks out in the tea world. Fortunately, this and similar counterexamples cannot be used against Definition 6.18, since it employs alternative sets that may be much smaller than possible worlds.

In order to make as much sense as possible of Definition 6.18, we should assume that it is applied to an alternative set that roughly corresponds to the contextual restrictions of informal discourse. In other words, the exclusion of too dissimilar representations, which was the task of the similarity relation, has already taken place when the alternative set was formed. Admittedly, this is a shaky defence of the construction. Nevertheless, this construction is so much simpler than the others that it is worth a careful investigation.

Maximal centred preference relations are similarity-maximizing.

Observation 6.19. *Let* $\mathcal{A}$ *be a set of sentential alternatives,* $\geq$ *a reflexive relation on* $\mathcal{A}$, *and f a representation function for* $\mathcal{A}$. *Then:*

(1) *If f is maximal centred, then there is some similarity relation* T *over* $\mathcal{A}$ *such that* $\geq_f$ *is similarity-maximizing with respect to* T.

[21] von Wright 1963a, pp. 29–30.

(2) *It does not hold in general (not even if $\mathcal{A}$ is contextually complete) that if $\geq_f$ is similarity-maximizing, based on some similarity relation T over $\mathcal{A}$, then $\geq_f$ is maximal centred.*

We have now defined three categories of pairwise combinative preference relations: similarity-satisficing (SS), similarity-maximizing (SM), and maximal centred (MC). They can be ordered as follows in terms of increasing generality:

$$\text{MC} \subset \text{SM} \subset \text{SS}$$

These three constructions are all based on the intuition that when comparing p to q we should look for pairs of alternatives $\langle A_1, A_2 \rangle$ that satisfy the following two conditions:

(1) *The representation condition*
A_1 is a representation of p, and A_2 is a representation of q.

(2) *The unfocused similarity condition*
A_1 and A_2 are maximally/sufficiently similar to each other, as compared to other pairs of alternatives that satisfy the representation condition.

If there is a privileged alternative A_0 that can serve as a reference point,[22] then the following is an interesting alternative to (2):

(2′) *The focused similarity condition*
A_1 is maximally/sufficiently similar to A_0, as compared to other alternatives that satisfy the representation condition with respect to p. In the same way, A_2 is maximally/sufficiently similar to A_0, as compared to other alternatives that satisfy the representation condition with respect to q.[23]

In the limiting case when the alternatives are possible worlds, A_0 in (2′) can be interpreted as the actual world. This is a construction with some tradition in the literature on preference logic.[24] In the present approach, which allows for smaller alternatives (cf. Section 5.1), it is often difficult to find a reasonable interpretation of the reference point. As one example of this, in the comparison between tables referred to

[22] This terminology was proposed by Wlodek Rabinowicz.

[23] Whereas a four-termed similarity relation is needed for defining the unfocused similarity condition, a three-termed similarity relation is sufficient for the focused condition.

[24] von Kutschera 1975; Trapp 1985; Hansson 1989a.

at the beginning of this chapter, it would not seem natural to regard one of the tables as a reference point for similarity judgments. Therefore, only the unfocused approach (represented by the SS, SM, and MC variants) will be investigated in detail here. However, before we leave the subject, a few words should be said about the formal relationship between the focused and the unfocused approach.

The similarity-maximizing preference relation introduced in Definition 6.13 may be called the *unfocused SM preference relation*. If we replace, in that definition, the condition

$T(A,B,A',B')$ holds for all $A', B' \in \mathcal{A}$ such that $x \in A'$ and $y \in B'$.

by the condition

$T(A,A_0,A',A_0)$ holds for all A' such that $x \in A' \in \mathcal{A}$,

and

$T(B,A_0,B',A_0)$ holds for all B' such that $y \in B' \in \mathcal{A}$,

then an alternative construction is obtained that we may call the *focused SM preference relation*. It turns out that if $\geq_f$ is a focused SM preference relation on the finite alternative set $\mathcal{A}$, based on a similarity relation T, then it is also an unfocused SM preference relation, based on another similarity relation T'.[25] The converse relationship does not hold.[26]

6.5 TRANSMITTED LOGICAL PROPERTIES

In this and the following section, logical properties of (unfocused) similarity-satisficing, similarity-maximizing, and maximal centred preference relations will be investigated.

[25] Let $\delta(X)$ be the number of elements Y in $\mathcal{A}$ such that $\hat{T}(Y,A_0,X,A_0)$. Let $T'(X,Y,Z,W)$ hold if and only if *either* $X = Y$ *or* $\delta(X) + \delta(Y) \leq \delta(Z) + \delta(W)$ and $Z \neq W$. Then T' satisfies conditions (T1)–(T5) of Definition 6.12. Furthermore, if x and y are $\mathcal{A}$-incompatible, then:

$T'(A,B,A',B')$ for all $x \in A' \in \mathcal{A}$ and $y \in B' \in \mathcal{A}$,
iff $\delta(A) + \delta(B) \leq \delta(A') + \delta(B')$ for all $x \in A' \in \mathcal{A}$ and $y \in B' \in \mathcal{A}$,
iff $\delta(A) \leq \delta(A')$ for all $x \in A' \in \mathcal{A}$ and $\delta(B) \leq \delta(B')$ for all $y \in B' \in \mathcal{A}$,
iff $T(A,A_0,A',A_0)$ for all $x \in A' \in \mathcal{A}$ and $T(B,A_0,B',A_0)$ for all $\in B' \in \mathcal{A}$.

It follows that T' gives rise to the same preference relation via Definition 6.13 as does T via the focused variant of that definition.

[26] This can be seen from the fact that focused SM preference relations satisfy transitivity for mutually exclusive relata, if the underlying exclusionary preference relation is transitive. This property does not hold for the unfocused variant. (See Observation 6.24.)

It is natural to ask to what extent various logical properties of the underlying exclusionary preference relation $\geq$ are reflected in the logic of the derived preference relation $\geq_f$. More precisely, a logical property is *transmitted* by f if and only if: If $\geq$ has this property, then so does $\geq_f$.[27]

Reflexivity is not transmitted by all representation functions, but it is transmitted by a wide range of representation functions, including those that are centred.

Observation 6.20. *Let $\geq$ be a reflexive relation on the sentential alternative set $\mathcal{A}$, and let f be a representation function for $\mathcal{A}$. Then $\geq_f$ is reflexive if and only if for all sentences x and all elements A_1 and A_2 of $\mathcal{A}$: If $\langle A_1,A_2\rangle \in f(\langle x,x\rangle)$, then $A_1 \geq A_2$.*

Corollary. *(1) If f satisfies weak centring, then $\geq_f$ is reflexive.*

(2) It does not hold in general (not even if $\mathcal{A}$ is contextually complete) that if f is similarity-satisficing, then $\geq_f$ is reflexive.

Reflexivity is a desirable property. Everything that we can compare – not only the elements of alternative sets – should be equal in value to itself.

Completeness of the exclusionary preference relation ($A \geq B \vee B \geq A$) is not in general transmitted to the f-extended preference relation:

Observation 6.21. *Let f be a representation function for the sentential alternative set $\mathcal{A}$, such that there are two elements p and q of $\cup\mathcal{A}$ and four pairwise distinct elements A_1, A_2, B_1, and B_2 of $\mathcal{A}$ such that $\langle A_1,B_1\rangle \in f(\langle p/_{\mathcal{A}}q, q/_{\mathcal{A}}p\rangle)$ and $\langle B_2,A_2\rangle \in f(\langle q/_{\mathcal{A}}p, p/_{\mathcal{A}}q\rangle)$. Then there is a complete relation $\geq$ over $\mathcal{A}$ such that $p \geq_f q \vee q \geq_f p$ does not hold.*

In other words, even if your preferences are sufficiently developed to cover all possible comparisons between (contextually) complete alternatives, they do not in general also cover all other possible comparisons. To see that this is plausible in the ceteris paribus interpretation, consider the four meals that can be composed out of the two dishes and the two drinks served at a small market stand. Suppose that you like each of the meals on the following list better than all those below it:

[27] This terminology was introduced in Hansson 1996a.

hamburger and beer
sandwich and coffee
sandwich and beer
hamburger and coffee

It would be wrong to say that you, in this context, prefer a meal with coffee to a meal with beer, or a meal with beer to a meal with coffee, or that you are indifferent between these two (incomplete) alternatives. You simply do not have a determinate ceteris paribus preference between the two. The ceteris paribus preference relation is incomplete, in spite of the fact that the underlying exclusionary preference relation is complete.[28]

Transitivity is not in general transmitted by representation functions, not even to maximal centred preference relations.

Observation 6.22. *Let $\geq$ be a complete and transitive relation on the contextually complete alternative set $\mathcal{A}$, and let f be a representation function on $\mathcal{A}$, such that $\geq_f$ is maximal centred. Then $p \geq_f q \;\&\; q \geq_f r \rightarrow p \geq_f r$ does not hold in general.*

If a maximal centred preference relation is restricted to a set of pairwise incompatible alternatives, then transitivity is transmitted from $\geq$ to $\geq_f$. Under the same restriction, transitivity-related properties are also transmitted to $\equiv_f$, $\gg_f$, and to a certain degree $>_f$.

Observation 6.23. *Let $\geq$ be a reflexive relation on the sentential alternative set $\mathcal{A}$, and let $\geq_f$ be the corresponding maximal centred preference relation. Furthermore, let p, q, and r be pairwise $\mathcal{A}$-incompatible elements of $\cup\mathcal{A}$. Then:*

(1) If $\geq$ is transitive, then $p \geq_f q \;\&\; q \geq_f r \rightarrow p \geq_f r$.
(2) If $\equiv$ is transitive, then $p \equiv_f q \;\&\; q \equiv_f r \rightarrow p \equiv_f r$.
(3) If $>$ is transitive, then $p \gg_f q \;\&\; q \gg_f r \rightarrow p \gg_f r$.
(4) If $\geq$ is IP-transitive, then $p \equiv_f q \;\&\; q \gg_f r \rightarrow p \gg_f r$.
(5) If $\geq$ is PI-transitive, then $p \gg_f q \;\&\; q \equiv_f r \rightarrow p \gg_f r$.

[28] In cases like these, no amount of additional specification of your exclusionary preferences can "resolve" the issue. Your preferences are fully specified, but they are this in a way that precludes (unconditional) preference or indifference between the two relata.

(6) If $\geq$ is complete and IP-transitive, then $p \equiv_f q$ & $q >_f r \rightarrow p >_f r$.
(7) If $\geq$ is complete and PI-transitive, then $p >_f q$ & $q \equiv_f r \rightarrow p >_f r$.

(By IP-transitivity is meant the property $A \equiv B$ & $B > C \rightarrow A > C$, and by PI-transitivity the property $A > B$ & $B \equiv C \rightarrow A > C$. If $\geq$ is complete, then it is IP-transitive if and only if it is PI-transitive.)

The corresponding properties do not hold for similarity-maximizing preferences in general:

Observation 6.24. *Let $\geq$ be a reflexive relation on the contextually complete alternative set $\mathcal{A}$, and let $\geq_f$ be a similarity-maximizing preference relation, based on $\geq$. Furthermore, let p, q, and r be pairwise $\mathcal{A}$-incompatible elements of $\cup\mathcal{A}$. Then:*

Neither of the properties (1)–(7) of Observation 6.23 holds in general.

One major transitivity property was missing in Observation 6.23: transitivity of $>_f$. This property turns out not to be transmitted even to maximal centred preference relations:

Observation 6.25. *Let $\geq$ be a complete relation on the contextually complete alternative set $\mathcal{A}$, such that the corresponding strict relation $>$ is transitive. Let f be maximal centred, and let p, q, and r be pairwise incompatible elements of $\cup\mathcal{A}$. Then it does not hold in general that $p >_f q$ & $q >_f r \rightarrow p >_f r$.*

When restricted to pairwise incompatible relata, acyclicity is transmitted to maximal centred $\gg_f$ (but not $>_f$). It is not transmitted to similarity-maximizing preferences.

Observation 6.26. *Let $\geq$ be a reflexive relation on the sentential alternative set $\mathcal{A}$, and let $\geq_f$ be the corresponding maximal centred preference relation. Furthermore, let $\{q_1, \ldots q_n\}$ be a finite set of pairwise $\mathcal{A}$-incompatible elements of $\cup\mathcal{A}$. Then:*

(1) If $>$ is acyclic, then so is the restriction of $\gg_f$ to $\{q_1, \ldots q_n\}$.
(2) It does not hold in general, not even if $\mathcal{A}$ is contextually complete and $\geq$ is complete, that if $>$ is acyclic, then so is the restriction of $>_f$ to $\{q_1, \ldots q_n\}$.

Observation 6.27. *Let* $\geq$ *be a complete relation on the contextually complete sentential alternative set* $\mathcal{A}$*, and let* $\geq_f$ *be a similarity-maximizing preference relation on* $\mathcal{A}$ *that is based on* $\geq$*. Furthermore, let* $\{q_1, \ldots q_n\}$ *be a finite set of pairwise* $\mathcal{A}$*-incompatible elements of* $\cup\mathcal{A}$*. Then:*

(1) It does not hold in general that if $>$ *is acyclic, then so is the restriction of* $\gg_f$ *to* $\{q_1, \ldots q_n\}$*.*
(2) It does not hold in general that if $>$ *is acyclic, then so is the restriction of* $>_f$ *to* $\{q_1, \ldots q_n\}$*.*

Hence, several transitivity-related properties are transmitted to maximal centred preferences but not in general to similarity-maximizing preferences. Since these properties apply only to pairwise incompatible relata, they are quite plausible, and it does not seem easy to find clear counterexamples against them.

6.6 SUPERSTRUCTURAL LOGICAL PROPERTIES

Some of the most interesting logical properties of combinative preferences cannot be transmitted from exclusionary preferences, for the simple reason that they are not defined for the latter. In particular, this applies to logical principles that refer to negated or disjunctive states of affairs. Properties of combinative preferences that are, due to their logical nature, undefined for exclusionary preferences, will be called *superstructural* properties. They are the subject of the present section.

Sören Halldén introduced the postulates $p{\equiv}q \rightarrow \neg q{\equiv}\neg p$ and $p{>}q \rightarrow \neg q{>}\neg p$.[29] Von Wright used the phrase "the principle of contraposition" for the latter of the two principles.[30] A similar postulate, $p{\geq}q \rightarrow \neg q{\geq}\neg p$, can be formed for weak preference. Here, 'contraposition' will be used as a common term for all postulates of this general form. Thus, $p{\geq_f}q \rightarrow \neg q{\geq_f}\neg p$ is contraposition of weak preference, $p{\equiv_f}q \rightarrow \neg q{\equiv_f}\neg p$ contraposition of indifference, and so on.

The principles of contraposition have a clear intuitive appeal. If you prefer playing the piano to playing football, then not playing the piano should be worse for you than not playing football. As the following observation shows, contraposition holds for $\geq_f$, $\equiv_f$, $>_f$, and $\gg_f$, except when one of the relata contextually implies the other. In particular,

[29] Halldén 1957, pp. 27–29, 36. [30] von Wright 1972, pp. 147–149.

contraposition holds in the important case when the relata are incompatible. This result holds for all preference relations that are based on a representation function in the manner of Definition 6.5.

Observation 6.28. *Let $\geq$ be a reflexive relation on the sentential alternative set $\mathcal{A}$ and f a representation function for $\mathcal{A}$. Furthermore, let p and q be elements of $\cup\mathcal{A}$ such that $p \nvDash_{\mathcal{A}} q$ and $q \nvDash_{\mathcal{A}} p$. Then:*

(1) $p \geq_f q \rightarrow \neg q \geq_f \neg p$,
(2) $p \equiv_f q \rightarrow \neg q \equiv_f \neg p$,
(3) $p >_f q \rightarrow \neg q >_f \neg p$, *and*
(4) $p \gg_f q \rightarrow \neg q \gg_f \neg p$.

That contraposition of strict preference should not hold if $p \vDash_{\mathcal{A}} q$ can be seen from an example proposed by Bengt Hansson (who did not, however, consider restricted versions of contraposition).[31] Let p denote that you win the first prize and q that you win some prize. Then $p >_f q$ may reasonably hold, but it does not hold that $\neg q >_f \neg p$. To the contrary, $\neg p$ is preferable to $\neg q$, since it leaves open the possibility of winning some other prize than the first prize. The same example can also be used against contraposition of weak preference ($p \geq_f q$ holds, but not $\neg q \geq_f \neg p$).

The following example can be used against contraposition of indifference if $p \vDash_{\mathcal{A}} q$: Let p denote that I have at least two copies of Rousseau's *Du contrat social* on my bookshelf and q that I have at least one copy of it. Since I need the book, but cannot use more than one copy, p and q are of equal value, that is, $p \equiv_f q$. However, it does not hold that $\neg q \equiv_f \neg p$. To the contrary, $\neg q$ is worse than $\neg p$, since it means that I am in the precarious situation of not having access to *Du contrat social.*

The most widely quoted argument against contraposition was provided by Roderick Chisholm and Ernest Sosa. They argued that "although that state of affairs consisting of there being happy egrets (p) is better than that one that consists of there being stones (q), that state of affairs that consists of there being no stones ($\neg q$) is no better, nor worse, than that state of affairs consisting of there being no happy egrets ($\neg p$)."[32]

[31] Bengt Hansson 1968, pp. 428–429.
[32] Chisholm and Sosa 1966a, p. 245.

Since stones and happy egrets can coexist, this is a comparison between compatible alternatives. Therefore, if it makes sense at all in a ceteris paribus setting, it does so according to Translation Procedure 3. In other words, when comparing the existence of happy egrets with that of stones, we should compare alternatives in which there are happy egrets but no stones to alternatives in which there are stones but no happy egrets, that is, $p\&\neg q$ to $q\&\neg p$. Next, let us compare $\neg q$ to $\neg p$. By the same argument, this should be a comparison between, on the one hand, there being no stones and not being no happy egrets and, on the other hand, there being no happy egrets and not being no stones. This is, hidden behind double negations, the same comparison between $p\&\neg q$ and $q\&\neg p$ that we have just made. Thus, from a logical point of view, it is unavoidable – once we have accepted Translation Procedure 3 – that $p>_f q$ holds if and only if $\neg q>_f\neg p$. What makes the example seem strange is that, although we can easily apply Translation Procedure 3 to p and q, unaided intuition halts before the negated statements and does not perform the same operation.

Halldén also introduced the two principles $p>q \leftrightarrow (p\&\neg q)>(q\&\neg p)$ and $p\equiv q \leftrightarrow (p\&\neg q)\equiv(q\&\neg p)$.[33] They have been accepted by von Wright.[34] The postulate $p>q \leftrightarrow (p\&\neg q)>(q\&\neg p)$ has been called "conjunctive expansion."[35] Here, that term will be used for all relationships of the same form. (Thus, $p\equiv q \leftrightarrow (p\&\neg q)\equiv(q\&\neg p)$ is conjunctive expansion of indifference, etc.)

Just like contraposition, conjunctive expansion holds for pairs of states of affairs, neither of which contextually implies the other. This result applies to all preference relations that are based on representation functions in the manner defined above.

Observation 6.29. *Let $\geq$ be a reflexive relation on the sentential alternative set $\mathcal{A}$, and f a representation function for $\mathcal{A}$. Furthermore, let p and q be elements of $\cup\mathcal{A}$ such that $p \nvDash_{\mathcal{A}} q$ and $q \nvDash_{\mathcal{A}} p$. Then:*

(1) $p\geq_f q \leftrightarrow (p\&\neg q)\geq_f(q\&\neg p)$,
(2) $p\equiv_f q \leftrightarrow (p\&\neg q)\equiv_f(q\&\neg p)$,
(3) $p>_f q \leftrightarrow (p\&\neg q)>_f(q\&\neg p)$, *and*
(4) $p\gg_f q \leftrightarrow (p\&\neg q)\gg_f(q\&\neg p)$.

[33] Halldén 1957, p. 28. [34] von Wright 1963a, pp. 24–25, 40, 60. [35] Jennings 1967.

Several authors, including Castañeda, Chisholm and Sosa, and Quinn have pointed out that conjunctive expansion cannot hold unrestrictedly since it would involve preferences with contradictory relata.[36] For concreteness, let p denote that a certain person is blind in her left eye and q that she is blind in two eyes. It is clearly worse to be blind in two eyes (q) than to be blind in the left eye (p). However, it does not follow that being blind only in the left eye ($p\&\neg q$) is better than contradiction ($q\&\neg p$).

Chisholm and Sosa chose to reject conjunctive expansion altogether, and so did Quinn.[37] As should be clear from the reasoning that led up to Translation Procedure 3, they threw the baby out with the bathwater. For pairs of states of affairs such that neither contextually implies the other, conjunctive expansion is a valid and illuminating principle.

Setsuo Saito came close to this conclusion, claiming that conjunctive expansion of indifference and strict preference hold "only when both $p\&\neg q$ and $\neg p\&q$ are logically possible, i.e., p and q do not imply each other."[38] However, logical possibility ($p \nvDash q$ and $q \nvDash p$) is not sufficient. We need the stronger notion of contextual possibility ($p \nvDash_{\mathcal{A}} q$ and $q \nvDash_{A} p$) that was used in the observation. This can be seen from the blindness example. Being blind in two eyes but not in the left eye ($q\&\neg p$) is not a logical contradiction, but in most discourses on human blindness it is a contextual impossibility, so that $q \vDash_{\mathcal{A}} p$ holds. (This is not true if being a three-eyed creature is among the alternatives discussed.) A plausible principle of conjunctive expansion must exclude cases of contextual, not only logical, impossibility.

Intuitively, we would expect $p \vee q$ to be intermediate in value between p and q. This turns out to hold if p and q are $\mathcal{A}$-incompatible. Otherwise, it does not hold in general.

Observation 6.30. *Let $\mathcal{A}$ be a sentential alternative set, $\geq$ a reflexive relation on $\mathcal{A}$, and f a representation function on $\mathcal{A}$. Let p and q be $\mathcal{A}$-incompatible elements of $\cup\mathcal{A}$. Then:*

(1) $p \geq_f (p \vee q) \leftrightarrow p \geq_f q$
(2) $(p \vee q) \geq_f p \leftrightarrow q \geq_f p$

[36] Castañeda 1958; Chisholm and Sosa 1966a, p. 245; Quinn 1974, p. 125.
[37] Chisholm and Sosa 1966a, p. 245; Quinn 1974, p. 125.
[38] Saito 1973, p. 388. Cf. Trapp 1985, p. 318.

(3) $p \geq_f q \rightarrow p \equiv_f (p \vee q)$ *does not hold in general, not even if* $\mathcal{A}$ *is contextually complete and f is maximal centred.*
(4) $p \geq_f q \rightarrow (p \vee q) \equiv_f q$ *does not hold in general, not even if* $\mathcal{A}$ *is contextually complete and f is maximal centred.*

Observation 6.31. *Let* $\geq$ *be a complete and transitive relation on the contextually complete alternative set* $\mathcal{A}$*. Let* $\geq_f$ *be the maximal centred extension of* $\geq$*. Then:*

(1) $p \geq_f q \rightarrow p \geq_f (p \vee q)$ *does not hold in general.*
(2) $p \geq_f q \rightarrow (p \vee q) \geq_f q$ *does not hold in general.*
(3) $p \equiv_f q \rightarrow p \equiv_f (p \vee q)$ *does not hold in general.*

The following unsurprising negative result will be referred to in a comparison in Section 7.7 of different combinative preference relations.

Observation 6.32. *Let* $\geq$ *be a complete and transitive relation on the contextually complete alternative set* $\mathcal{A}$*. Let* $\geq_f$ *be the maximal centred preference relation based on* $\geq$*. Then:*

(1) *It does not hold in general that if* $\models_{\mathcal{A}} p \rightarrow q$*, then* $p \geq_f q$*.*
(2) *It does not hold in general that if* $\models_{\mathcal{A}} p \rightarrow q$*, then* $q \geq_f p$*.*

Von Wright argued that "[d]isjunctive preferences are conjunctively distributive" in the sense that preferring $p \vee q$ to r is essentially the same as preferring p to r and also q to r.[39] It is indeed easy to find examples that conform with this pattern. To say that I prefer listening to Bach's or Handel's music to listening to Muzak seems to be just another way of saying both that I prefer listening to Bach's music to listening to Muzak and that I prefer listening to Händel's music to listening to Muzak. Further reflection will show, however, that this principle does not hold in general. The following argument is due to Castañeda: Let p, q, and r be three states of affairs such that q and r are equal in value, whereas p is valued higher than both q and r. Then it seems reasonable to claim that $p \vee q$ is strictly better than r, although q is not strictly better than r.[40] For concreteness, suppose that I am indifferent between spending my vacations in Liverpool (q) and spend-

[39] von Wright 1963a, p. 26. See also Bengt Hansson 1968, pp. 433–439.
[40] Castañeda 1969, pp. 258–259. Cf. Freeman 1973, p. 106.

ing them in Manchester (r), but prefer spending them on Hawaii (p) to both the other alternatives. I can then be assumed to prefer having vacations on Hawaii or in Liverpool to having vacations in Manchester ($(p \vee q) > r$), although I am indifferent between the two British alternatives ($q \equiv r$).[41]

There are, however, weakened versions of this distribution principle against which it seems to be much more difficult to find counterexamples. Several such weakened versions hold for maximal centred preference relations. For similarity-maximizing relations in general, only very watered down versions hold, and for similarity-satisficing relations in general, no results in this direction have been obtained.

Observation 6.33. *Let $\geq$ be a reflexive relation on the contextually complete alternative set $\mathcal{A}$, and $\geq_f$ the corresponding maximal centred preference relation. Furthermore, let p, q, and r be elements of $\cup\mathcal{A}$ that are pairwise $\mathcal{A}$-incompatible. Then:*

(1) $(p \vee q) \geq_f r \leftrightarrow p \geq_f r \;\&\; q \geq_f r$
(2) $p \geq_f (q \vee r) \leftrightarrow p \geq_f q \;\&\; p \geq_f r$
(3) $(p \vee q) \equiv_f r \leftrightarrow p \equiv_f r \;\&\; q \equiv_f r$
(4) $p \equiv_f (q \vee r) \leftrightarrow p \equiv_f q \;\&\; p \equiv_f r$
(5) $(p \vee q) \gg_f r \leftrightarrow p \gg_f r \;\&\; q \gg_f r$
(6) $p \gg_f (q \vee r) \leftrightarrow p \gg_f q \;\&\; p \gg_f r$
(7) $(p \vee q) >_f r \leftrightarrow p \geq_f r \;\&\; q \geq_f r \;\&\; (p >_f r \vee q >_f r)$
(8) $p >_f (q \vee r) \leftrightarrow p \geq_f q \;\&\; p \geq_f r \;\&\; (p >_f q \vee p >_f r)$
(9) It does not hold in general, not even if $\geq$ is complete and transitive, that $(p \vee q) >_f r \rightarrow p >_f r \;\&\; q >_f r$.
(10) It does not hold in general, not even if $\geq$ is complete and transitive, that $p >_f (q \vee r) \rightarrow p >_f q \;\&\; p >_f r$.

Observation 6.34. *Let $\geq$ be a complete and transitive relation on the contextually complete alternative set $\mathcal{A}$, and let $\geq_f$ be a similarity-maximizing preference relation based on $\geq$. Furthermore, let p, q, and r be elements of $\cup\mathcal{A}$ that are pairwise $\mathcal{A}$-incompatible. Then it holds for each of the conditions (1)–(8) of Observation 6.33 that it is not in general satisfied by $\geq_f$.*

[41] Cf. Section 6.3.

Observation 6.35. *Let* $\geq$ *be a complete and transitive relation on the contextually complete alternative set* $\mathcal{A}$. *Let* $p, q, r \in \cup\mathcal{A}$, *and let* $\geq_f$ *be a similarity-maximizing extension of* $\geq$. *Then:*

(1) $(p \vee q) \geq_f r \rightarrow p \geq_f r \vee q \geq_f r$ *holds if p, q, and r are pairwise* $\mathcal{A}$*-incompatible elements of* $\mathcal{A}$.
(2) $(p \vee q) \geq_f r \rightarrow p \geq_f r \vee q \geq_f r$ *does not hold in general, not even if f is maximal centred (unless p, q, and r are pairwise* $\mathcal{A}$*-incompatible).*
(3) $p \geq_f r \rightarrow (p \vee q) \geq_f r$ *does not hold in general, not even if p, q, and r are pairwise* $\mathcal{A}$*-incompatible and f is maximal centred.*
(4) $p \geq_f (q \vee r) \rightarrow p \geq_f q \vee p \geq_f r$ *holds if p, q, and r are pairwise* $\mathcal{A}$*-incompatible elements of* $\mathcal{A}$.
(5) $p \geq_f (q \vee r) \rightarrow p \geq_f q \vee p \geq_f r$ *does not hold in general, not even if f is maximal centred (unless p, q, and r are pairwise* $\mathcal{A}$*-incompatible).*
(6) $p \geq_f q \rightarrow p \geq_f (q \vee r)$ *does not hold in general, not even if p, q, and r are pairwise* $\mathcal{A}$*-incompatible and f is maximal centred.*

Observation 6.36. *Let* $\geq$ *be a complete and transitive relation on the contextually complete alternative set* $\mathcal{A}$, *and let p, q, and r be pairwise* $\mathcal{A}$*-incompatible elements of* $\mathcal{A}$. *Furthermore, let* $\geq_f$ *be a similarity-satisficing extension of* $\geq$. *Then:*

(1) $(p \vee q) \geq_f r \rightarrow p \geq_f r \vee q \geq_f r$ *does not hold in general.*
(2) $p \geq_f (q \vee r) \rightarrow p \geq_f q \vee p \geq_f r$ *does not hold in general.*

Summarizing this and the previous section, similarity-maximizing preference relations have a fairly weak logic. Some plausible logical properties are lacking for similarity-maximizing preferences in general, but hold for maximal centred preferences. Nevertheless, maximal centred preference relations should only be used with caution since they are based on a rather crude semantical construction.

7

Decision-Guiding Combinative Preferences

Following the plan outlined in Section 5.3, we now turn to combinative decision-guiding (choice-guiding) preferences. As will soon be seen, there are several ways in which a preference relation can guide decisions or choices among the elements of the alternative set. Therefore, several types of such relations need to be developed.

In Section 7.1, a distinction is made between two major approaches to decision making: the prognostic approach, which tries to predict future decisions, and the agnostic approach, which treats future decisions as completely open and undetermined. In Section 7.2, a combinative preference relation for the prognostic approach is investigated. In Sections 7.3 and 7.4, a general framework for the agnostic approach is constructed. In Section 7.5, several preference relations that fit into this general framework are introduced, and in Section 7.6 their properties are investigated. In Section 7.7, finally, the various combinative preference relations studied in this and the previous chapter are compared in terms of their logical properties.

7.1 TWO APPROACHES TO DECISION MAKING

The need for combinative decision-guiding preferences was explained in Section 5.3. As agents with limited cognitive capacities, we often have use for comprehensive comparisons between groups of alternatives. To express such comparisons in formal language, a preference relation is needed that can guide choices between subsets (not only elements) of the alternative set. Given the definitions in Section 5.2, it does not make much difference if the relata of this relation are taken to be sentences or sets of alternatives. Any sentence

p in $\cup\mathcal{A}$ can be taken to represent the set of all $X \in \mathcal{A}$ such that $p \in X$.[1]

Definition 7.1. *Let $p \in \mathcal{L}_{\mathcal{A}}$. Then:*

$$repr_{\mathcal{A}}(p) = \{X \in \mathcal{A} \mid p \in X\}$$

The index of $repr_{\mathcal{A}}$ is deleted whenever convenient.

A decision-guiding combinative preference relation can be defined as a relation $\geq'$ over $\mathcal{P}(\mathcal{A})\backslash\{\emptyset\}$ that extends the exclusionary preference relation $\geq$ over $\mathcal{A}$. We can use $p \geq' q$ as an abbreviated notation for $repr(p) \geq' repr(q)$.

To choose a subset of the alternative set is in general a less determinate choice than to choose an element of that same set. Provided that a subset with at least two elements was chosen, the (primary) choice leaves open a scope for a supplementary choice (or series of choices) to be made later. The secondary choice(s) can be treated in either of two ways, which correspond to two fundamentally different approaches to an as yet unmade decision. We can call these the *prognostic* and the *agnostic* approaches. When treating a future decision prognostically, I try to figure out how it will be made or how it will probably be made. When treating it agnostically, I treat it as completely undetermined and open. There is in most cases some clue as to how a future decision will be made. When applying the agnostic approach, we disregard this information.

The secondary choice can be made either by the agent herself or by someone else. In the latter case, the agnostic approach is adequate whenever the information needed for the prognostic approach is unavailable, so that the (primary) decision is made under uncertainty (ignorance) with respect to how the second decision will be made.

The most important applications of the agnostic approach refer to cases when it is assumed that the agent will herself make the secondary choice. Here, by adopting the agnostic approach we assign to the agent the role of being a free agent also in the secondary choice. As an example of this, if you decide to stop smoking today, then the

[1] In spite of the obvious connection, the two approaches have been developed more or less in isolation from each other. According to Kannai and Peleg (1984), the first investigation of preference relations over subsets (rather than elements) of an alternative set was Fishburn 1972.

probability that you will begin to smoke again very soon may very well be high. It can, however, be claimed that it would be counterproductive to take this probability as given and even to let it at all enter your deliberations. As a free agent, you should take control over your future actions.

The choice between a prognostic and an agnostic approach in cases like this is thus closely related to an issue that has been brought up by Wolfgang Spohn, namely whether or not a decision model should contain subjective probabilities for actions by the agent herself. Spohn emphatically recommends that no probabilities for "acts . . . which are under complete control of the decision maker" be included in the analysis.[2] It would seem to follow from this recommendation that the agnostic approach to agents' own decisions be universally adopted.

However, there are cases when a prognostic approach to one's own decisions is considered appropriate. Consider a nonsmoker's decision over whether or not to try a cigarette or a nonaddict's decision over whether or not to try an addictive drug. The probability that future (secondary) choices will lead to continued use of the drug is a major argument in the medical advice that is given in these cases. It can hardly be claimed that rationality requires the total exclusion of this type of probabilistic considerations. The reason for this is of course that the acts under discussion are not, in Spohn's words, under the complete control of the agent (at least not at the moment of the primary decision).

For another illustration of the distinction between a prognostic and an agnostic approach, consider the decision that I make one day when leaving work on how to spend the evening. For simplicity, we can assume that there are only three, mutually exclusive alternatives:

A I practise the piano.
B I watch a bad movie on TV at home.
C I watch a good movie at the cinema.

Since *A* and *B* take place at home and *C* takes place downtown, it is not unreasonable to first decide between {*A*,*B*} and {*C*} and in the first case decide later between *A* and *B*. If the first decision is made with a prognostic approach to the later decision, I should take into account

[2] Spohn 1977, p. 114. See also Jeffrey 1977 and Spohn 1999.

the probability that I will choose A, respectively B, if I first choose $\{A,B\}$.[3] According to the agnostic approach, I should instead treat the future decision between A and B as completely open.

In conclusion, both the prognostic and the agnostic approach are legitimate, and there are cases in which they both have considerable intuitive appeal. As we shall see, different types of formal constructions are needed for the two approaches.[4]

7.2 WEIGHTED PREFERENCES

One reasonable way to specify the prognostic approach to incomplete relata is to derive the values of these relata from those of complete alternatives, assigning weights to the latter according to their probabilities. In our example, suppose that practising the piano (A) has the value 4 on some value scale, watching a bad TV movie (B) the value 0, and watching a good film at the cinema (C) the value 2. Furthermore, suppose that the probability that I choose A in a later choice between A and B is 0.5. Then $\{A,B\}$ can be assigned the value 2. If, on the other hand, the probability of choosing A is only 0.1, then the value of $\{A,B\}$ is 0.4.

More generally speaking: Assign to each element of $\mathcal{A}$ a value and a weight. Let the value of each sentence in $\cup\mathcal{A}$ be equal to the weighted average of the values of the alternatives in which it holds. p is better than q if and only if the value computed for p is higher than that computed for q. If the weights are interpreted as probabilities, then this construction is an explication of "probably better."[5] The formal definition is straightforward enough:

Definition 7.2. *Let $\mathcal{A}$ be an alternative set. Any real-valued function v on $\mathcal{A}$ is a* value assignment *on $\mathcal{A}$.*

[3] It could perhaps be argued that if the agent in this example is incapable of carrying out a choice of A, then $\{B,C\}$ or $\{\{A,B\},C\}$ is a more suitable alternative set than $\{A,B,C\}$. However, choices not carried through are also choices and should be representable in the formal structure.

[4] There are more distinctions than this to be drawn with respect to the deciding agent's attitude to future decisions and actions. See Nozick 1969; Gibbard and Harper 1978; Price 1986; and Rabinowicz 1989.

[5] The same construction, with a definite interpretation of weights as probabilities, was called "expected utility preference" in Packard 1979. The special case when all weights are equal was called "averaging preference" by Packard. This special case has also been studied in Rescher 1968 and Fishburn 1972.

Any function w that assigns positive real numbers to the elements of $\mathcal{A}$ is a weight assignment *on $\mathcal{A}$.*[6]

A weight assignment w on a finite set $\mathcal{A}$ is calibrated *if and only if*

$$\sum_{X \in \mathcal{A}} w(X) = 1.$$

Definition 7.3. *Let $\mathcal{A}$ be a finite sentential alternative set, v a value assignment, and w a weight assignment on $\mathcal{A}$. The* weighted preference relation $\geq_{w,v}$ *on $\cup\mathcal{A}$, based on w and v, is the relation such that for all $p, q \in \cup\mathcal{A}$, $p \geq_{w,v} q$ if and only if*

$$\frac{\sum_{X \in repr(p)} w(X) \times v(X)}{\sum_{X \in repr(p)} w(X)} \geq \frac{\sum_{X \in repr(q)} w(X) \times v(X)}{\sum_{X \in repr(q)} w(X)}$$

Whenever convenient, $\geq_{w,v}$ is abbreviated $\geq_w$.

w can be interpreted as a probability measure, but the formal model does not exclude other interpretations.

A relation on the alternative set cannot replace v in the definition of weighted preferences; cardinal information is necessary. Therefore, weighted preferences cannot be derived from an underlying preference relation on the alternative set according to the general model proposed in Section 5.1. They are, however, consonant with the nonweighted preference relation over complete alternatives that can be defined from v in the obvious way. (This result should be compared to the corresponding result for $\geq_f$ in Observation 6.8.)

Definition 7.4. *Let $\mathcal{A}$ be a sentential alternative set and v a value assignment on $\mathcal{A}$. The* value-based preference relation *on $\mathcal{A}$, based on v, is the relation $\geq_v$ such that for all $X, Y \in \mathcal{A}$:*

$$X \geq_v Y \text{ if and only if } v(X) \geq v(Y).$$

Observation 7.5. *Let $\mathcal{A}$ be a finite sentential alternative set, v a value assignment, and w a weight assignment on $\mathcal{A}$. Let $\geq_{w,v}$ be the associated*

[6] It would be sufficient to require nonnegativity instead of positivity, but then $p \geq_w q$ would not be defined for all $p, q \in \cup\mathcal{A}$. Positivity has been adopted here in order to avoid unnecessary complications in Definition 7.3.

weighted preference relation and $\geq_v$ *the associated value-based preference relation. Furthermore, let A and B be elements of* $\mathcal{A}$*, and a and b sentences such that A is the only representation of a in* $\mathcal{A}$*, and B the only representation of b in* $\mathcal{A}$*. Then:*

$$a \geq_{w,v} b \text{ if and only if } A \geq_v B.$$

Some logical properties of weighted preferences are listed in the following observations. Several of these properties will be discussed in Section 7.6.

Observation 7.6. *Let* $\mathcal{A}$ *be a sentential alternative set and* $\geq_w$ *a weighted preference relation over* $\mathcal{A}$*. Furthermore, let* $p, q, r \in \cup\mathcal{A}$*. Then:*

(1) $p \geq_w q \vee q \geq_w p$ (completeness)
(2) $p \geq_w q \;\&\; q \geq_w r \rightarrow p \geq_w r$ (transitivity)

Observation 7.7. *Let* $\mathcal{A}$ *be a sentential alternative set and* $\geq_w$ *a weighted preference relation over* $\mathcal{A}$*. Furthermore, let* $p\&\neg q, q\&\neg p \in \cup\mathcal{A}$*. Then:*

(1) $p \geq_w q \rightarrow \neg q \geq_w \neg p$ *does not hold in general, not even if p and q are* $\mathcal{A}$*-incompatible.*
(2) $p >_w q \rightarrow \neg q >_w \neg p$ *does not hold in general, not even if p and q are* $\mathcal{A}$*-incompatible.*
(3) $p \equiv_w q \rightarrow \neg q \equiv_w \neg p$ *does not hold in general, not even if p and q are* $\mathcal{A}$*-incompatible.*
(4) $p \geq_w q \leftrightarrow (p\&\neg q) \geq_w (q\&\neg p)$ *holds if p and q are* $\mathcal{A}$*-incompatible.*
(5) $p \equiv_w q \leftrightarrow (p\&\neg q) \equiv_w (q\&\neg p)$ *holds if p and q are* $\mathcal{A}$*-incompatible.*
(6) $p >_w q \leftrightarrow (p\&\neg q) >_w (q\&\neg p)$ *holds if p and q are* $\mathcal{A}$*-incompatible.*
(7) $p >_w q \rightarrow (p\&\neg q) \geq_w (q\&\neg p)$ *does not hold in general.*
(8) $p \equiv_w q \rightarrow (p\&\neg q) \equiv_w (q\&\neg p)$ *does not hold in general.*

Observation 7.8. *Let* $\mathcal{A}$ *be a sentential alternative set and* $\geq_w$ *a weighted preference relation over* $\mathcal{A}$*. Furthermore, let* $p, q \in \cup\mathcal{A}$*. Then:*

(1) $p \geq_w q \leftrightarrow p \geq_w (p \vee q)$ *holds if p and q are* $\mathcal{A}$*-incompatible.*
(2) $p \geq_w q \rightarrow p \geq_w (p \vee q)$ *does not hold in general.*
(3) $p \equiv_w q \rightarrow p \equiv_w (p \vee q)$ *does not hold in general.*
(4) $p \geq_w q \leftrightarrow (p \vee q) \geq_w q$ *holds if p and q are* $\mathcal{A}$*-incompatible.*
(5) $p \geq_w q \rightarrow (p \vee q) \geq_w q$ *does not hold in general.*
(6) $p \equiv_w q \rightarrow (p \vee q) \geq_w q$ *does not hold in general.*

(7) $p \geq_w q \rightarrow p \equiv_w (p \vee q)$ *does not hold in general, not even if p and q are $\mathcal{A}$-incompatible.*
(8) $p \geq_w q \rightarrow (p \vee q) \equiv_w q$ *does not hold in general, not even if p and q are $\mathcal{A}$-incompatible.*

Observation 7.9. *Let $\mathcal{A}$ be a sentential alternative set and $\geq_w$ a weighted preference relation over $\mathcal{A}$. Furthermore, let $p, q, r \in \cup\mathcal{A}$. Then:*

(1) $(p \vee q) \geq_w r \rightarrow p \geq_w r \vee q \geq_w r$ *holds if p and q are $\mathcal{A}$-incompatible.*
(2) $(p \vee q) >_w r \rightarrow p \geq_w r \vee q \geq_w r$ *does not hold in general.*
(3) $(p \vee q) \equiv_w r \rightarrow p \geq_w r \vee q \geq_w r$ *does not hold in general.*
(4) $(p \vee q) >_w r \rightarrow p \geq_w r$ *does not hold in general, not even if p, q, and r are pairwise $\mathcal{A}$-incompatible.*
(5) $(p \vee q) \equiv_w r \rightarrow p \geq_w r$ *does not hold in general, not even if p, q, and r are pairwise $\mathcal{A}$-incompatible.*
(6) $p \geq_w (q \vee r) \rightarrow p \geq_w q \vee p \geq_w r$ *holds if q and r are $\mathcal{A}$-incompatible.*
(7) $p >_w (q \vee r) \rightarrow p \geq_w q \vee p \geq_w r$ *does not hold in general.*
(8) $p \equiv_w (q \vee r) \rightarrow p \geq_w q \vee p \geq_w r$ *does not hold in general.*
(9) $p >_w (q \vee r) \rightarrow p \geq_w q$ *does not hold in general, not even if p, q, and r are pairwise $\mathcal{A}$-incompatible.*
(10) $p \equiv_w (q \vee r) \rightarrow p \geq_w q$ *does not hold in general, not even if p, q, and r are pairwise $\mathcal{A}$-incompatible.*
(11) $p \geq_w r \rightarrow (p \vee q) \geq_w r$ *does not hold in general, not even if p, q, and r are pairwise $\mathcal{A}$-incompatible.*
(12) $p \geq_w q \rightarrow p \geq_w (q \vee r)$ *does not hold in general, not even if p, q, and r are pairwise $\mathcal{A}$-incompatible.*

Observation 7.10. *Let $\mathcal{A}$ be a sentential alternative set and $\geq_w$ a weighted preference relation over $\mathcal{A}$. Furthermore, let $p, q \in \cup\mathcal{A}$. Then:*

(1) *It does not hold in general that if $\vDash_{\mathcal{A}} p \rightarrow q$, then $p \geq_w q$.*
(2) *It does not hold in general that if $\vDash_{\mathcal{A}} p \rightarrow q$, then $q \geq_w p$.*

7.3 BASIC CRITERIA FOR AGNOSTIC PREFERENCES

Next, we are going to explore the agnostic approach to decision making. This means that all alternatives will be regarded as (equally) open and undetermined.

It should be clear that in the agnostic approach the value of *p* cannot be determined by the same method that we used in Section 7.2, namely by assigning different weights to the elements of *repr*(*p*). One might believe that the assignment of *equal* weights to the alternatives can be used, but it is easy to show with examples that this cannot be so. Let us consider the following modification of the example given in Section 7.1: There are two equally bad TV movies to choose between on different channels. We can then divide alternative B (watch a bad TV movie) into two variants: B_1 (watch bad TV movie 1) and B_2 (watch bad TV movie 2).[7] It is reasonable to require that $\{A,B_1,B_2\}$ and $\{A,B\}$ should be evaluated in the same way. The division of an alternative into two variants that are treated equally by the preference relation should not affect the decision outcome. This criterion cannot be satisfied in a model based on equal weights.

Another variation of the example serves to introduce yet another important property of the preference relation. Suppose that I replace A (in the original alternative set $\mathcal{A} = \{A, B, C\}$) by some other alternative A' that is equal in value to A. (In this case, perhaps: "I practise the flute.") Replacing A by the equally valued A' should have no effect, that is, $\{A',B\}$ and $\{A,B\}$ should be of equal value (according to the combinative preference relation) if A' and A are of equal value (according to the exclusionary preference relation).

Hence, we have two principles of invariance: one for the replacement of an alternative by two equally valued variants and the other for the replacement of an alternative by another, equally valued alternative. Given appropriate existence assumptions, these two principles sum up to the following postulate:

Positionality. *If for all $X \in \mathcal{B}_1$ there is some $Y \in \mathcal{B}_2$ such that $X \equiv Y$, and vice versa, then $\mathcal{B}_1 \equiv' \mathcal{B}_2$.*

Positionality ensures that alternatives that are equally ranked by $\geq$ can be interchanged, duplicated, or merged without affecting the evaluation in terms of $\geq'$. Hence, if $repr(p) = \{A,C,D\}$, $repr(q) = \{B,C,D\}$, and $A \equiv B$, then $p \equiv' q$. Similarly, if $repr(p_1) = \{A_1,B\}$, $repr(p_2) = \{A_1,A_2,B\}$, and $A_1 \equiv A_2$, then $p_1 \equiv' p_2$.

[7] Here, we disregard the possible value inherent in being able to choose between the films. More generally speaking, the present approach abstracts from the value that opportunities for choice may have *qua* opportunities. To express opportunity value, other types of models, which put value on variety among the elements of a subset of the alternative set, will have to be used. See Pattanaik and Xu 1990; Sen 1991; and Sugden 1998.

Another plausible requirement is that the effect of improving an alternative should not be negative.

Nonnegative response. *If* $\mathcal{B}_2 = (\mathcal{B}_1 \setminus \{X\}) \cup Y$ *for some* $X \in \mathcal{B}_1$*, then:*

(1) If $Y>X$*, then* $\mathcal{B}_2 \geq' \mathcal{B}_1$*.*
(2) If $X>Y$*, then* $\mathcal{B}_1 \geq' \mathcal{B}_2$*.*

7.4 EXTREMAL PREFERENCES

One interesting class of combinative preference relations is that which depends only on the maximal and minimal alternatives (according to the exclusionary preference relation $\geq$). To express them, we need some additional notation and terminology.

Definition 7.11. *Let* $\emptyset \neq \mathcal{B} \subseteq \mathcal{A}$*, let* $p \in \cup\mathcal{A}$*, and let* $\geq$ *be a relation on* $\mathcal{A}$*. Then:*

$$\mathfrak{max}(\mathcal{B}) = \{X \in \mathcal{B} \mid (\forall Y \in \mathcal{B})(X \geq Y)\}$$
$$\mathfrak{min}(\mathcal{B}) = \{X \in \mathcal{B} \mid (\forall Y \in \mathcal{B})(Y \geq X)\}$$

The elements of $\mathfrak{max}(\mathcal{B})$ *are the* ($\geq$-)maximal *elements of* $\mathcal{B}$*, those of* $\mathfrak{min}(\mathcal{B})$ *its* ($\geq$-)minimal *elements, and those of* $\mathfrak{max}(\mathcal{B}) \cup \mathfrak{min}(\mathcal{B})$ *its* ($\geq$-) extremal *elements.*

$\mathfrak{max}(p)$ *is an abbreviation of* $\mathfrak{max}(repr(p))$*, and* $\mathfrak{min}(p)$ is *an abbreviation of* $\mathfrak{min}(repr(p))$*. Furthermore:*

$\mathfrak{max}(\mathcal{B}) \geq \mathfrak{max}(\mathcal{D})$ *holds if and only if* $X \geq Y$ *for all* $X \in \mathfrak{max}(\mathcal{B})$ *and* $Y \in \mathfrak{max}(\mathcal{D})$.

$\mathfrak{min}(\mathcal{B}) \geq \mathfrak{min}(\mathcal{D})$ *holds if and only if* $X \geq Y$ *for all* $X \in \mathfrak{min}(\mathcal{B})$ *and* $Y \in \mathfrak{min}(\mathcal{D})$.[8]

Definition 7.11 leaves $\mathfrak{max}(p)$ and $\mathfrak{min}(p)$ undefined in the limiting case of sentences $p \in \mathcal{L}_{\mathcal{A}}$ such that $\vDash_{\mathcal{A}} \neg p$, and hence $repr(p) = \emptyset$. For most purposes, they can be left undefined. For some purposes, however, it is useful to have them defined by the introduction of two additional elements into the set of relata, with the following properties:

[8] The definitions for $\mathfrak{max}(\mathcal{B}) \geq \mathfrak{max}(\mathcal{D})$ and $\mathfrak{min}(\mathcal{B}) \geq \mathfrak{min}(\mathcal{D})$ are instances of a more general definition: for all $\mathcal{B}_1, \mathcal{B}_2 \subseteq \mathcal{A}$, $\mathcal{B}_1 \geq \mathcal{B}_2$ if and only if $X \geq Y$ for all $X \in \mathcal{B}_1$ and $Y \in \mathcal{B}_2$.

Definition 7.12. *Let* $\emptyset \neq \mathcal{B} \subseteq \mathcal{A}$. *Then* $\mathfrak{max}(\emptyset)$ *and* $\mathfrak{min}(\emptyset)$ *satisfy the following properties:*

(1) $\mathfrak{max}(\mathcal{B}) > \mathfrak{max}(\emptyset)$
(2) $\mathfrak{min}(\emptyset) > \mathfrak{min}(\mathcal{B})$
(3) $\mathfrak{max}(\emptyset) \equiv \mathfrak{max}(\emptyset)$
(4) $\mathfrak{min}(\emptyset) \equiv \mathfrak{min}(\emptyset)$

This definition has been chosen since it is convenient from a formal point of view (as will be seen in some of the proofs).[9] Under weak background assumptions, it has the counter-intuitive consequence that $\mathfrak{min}(\emptyset)>\mathfrak{max}(\emptyset)$.[10] Nevertheless, a partial intuitive motivation can be given as follows: For simplicity, let us assume that $\geq$ is complete and transitive. It holds for all α and β in $\cup\mathcal{A}$ that $\mathfrak{max}(\alpha\vee\beta)\geq\mathfrak{max}(\beta)$. We want this to hold as well if β is contradictory. In that case, $\mathfrak{max}(\alpha) \equiv \mathfrak{max}(\alpha\vee\beta)$. It follows from this and $\mathfrak{max}(\alpha\vee\beta)\geq\mathfrak{max}(\beta)$ that $\mathfrak{max}(\alpha)\geq \mathfrak{max}(\beta)$. Similarly, $\mathfrak{min}(\beta)\geq\mathfrak{min}(\alpha\vee\beta)$ and $\mathfrak{min}(\alpha\vee\beta)\equiv\mathfrak{min}(\alpha)$, so that $\mathfrak{min}(\beta)\geq\mathfrak{min}(\alpha)$.

After this *excursus* on $\mathfrak{max}(\emptyset)$ and $\mathfrak{min}(\emptyset)$, we can now return to the task of defining extremal preferences.

Definition 7.13. *Let* $\mathcal{A}$ *be a set of sentential alternatives and* $\geq$ *a relation on* $\mathcal{A}$. *A relation* $\geq'$ *on* $\mathcal{P}(\mathcal{A})\backslash\{\emptyset\}$ *is* extremal *with respect to* $\geq$ *if and only if, for all* $\mathcal{B}_1$, $\mathcal{B}_2 \subseteq \mathcal{A}$:

If $\mathfrak{min}(\mathcal{B}_1)\equiv\mathfrak{min}(\mathcal{B}_2)$ *and* $\mathfrak{max}(\mathcal{B}_1)\equiv\mathfrak{max}(\mathcal{B}_2)$, *then* $\mathcal{B}_1\equiv'\mathcal{B}_2$.

If we require transitivity of both $\geq$ and $\geq'$, then positionality and nonnegative response are sufficient to ensure that $\geq'$ is extremal.[11]

Theorem 7.14. *Let* $\geq$ *be a complete and transitive relation on a finite set* $\mathcal{A}$ *of sentential alternatives. Let* $\geq'$ *be a transitive relation on* $\mathcal{P}(\mathcal{A})\backslash\{\emptyset\}$ *that satisfies positionality and nonnegative response with respect to* $\geq$. *Then* $\geq'$ *is extremal.*

[9] See the proofs of Lemma 7.33 and Observation 10.37, in the Proofs.

[10] Let $\geq$ be transitive, and let $\mathcal{A}$ have two elements A_1 and A_2, such that $A_1>A_2$. Then we have $\mathfrak{min}(\emptyset)>\mathfrak{min}(\{A_1\})>\mathfrak{max}(\{A_2\})>\mathfrak{max}(\emptyset)$, hence $\mathfrak{min}(\emptyset)>\mathfrak{max}(\emptyset)$. This was pointed out by Wlodek Rabinowicz.

[11] Another set of conditions that is sufficient to guarantee that a preference relation is extremal was reported in Barbera, Barrett, and Pattanaik 1984.

In studies of extremal preference relations, the notation can be substantially simplified by letting each subset of $\mathcal{A}$ be represented by one of its *extremal pairs*, defined as follows:

Definition 7.15. *Let $\mathcal{A}$ be a set of sentential alternatives and $\geq$ a relation on $\mathcal{A}$. Let $\mathcal{B} \subseteq \mathcal{A}$. Then a pair $\langle X,Y \rangle$ is an* extremal pair *for $\mathcal{B}$ if and only if $X \in \mathfrak{min}(\mathcal{B})$ and $Y \in \mathfrak{max}(\mathcal{B})$.*

Let $\geq'$ be a relation on $\mathcal{P}(\mathcal{A}) \backslash \{\emptyset\}$, and let $\langle X_1, Y_1 \rangle$ and $\langle X_2, Y_2 \rangle$ be extremal pairs for subsets of $\mathcal{A}$. Then $\langle X_1, Y_1 \rangle \geq' \langle X_2, Y_2 \rangle$ is an alternative notation for $\{X_1, Y_1\} \geq' \{X_2, Y_2\}$.

Observation 7.16. *Let $\mathcal{A}$ be a set of sentential alternatives and $\geq$ a complete and transitive relation on $\mathcal{A}$. Let $\geq'$ be a transitive and extremal relation on $\mathcal{P}(\mathcal{A}) \backslash \{\emptyset\}$ that satisfies positionality with respect to $\geq$. Let $\mathcal{B}_1$, $\mathcal{B}_2 \subseteq \mathcal{A}$, and let $\langle X_1, Y_1 \rangle$ and $\langle X_2, Y_2 \rangle$ be extremal pairs for $\mathcal{B}_1$ and $\mathcal{B}_2$, respectively. Then:*

$$\mathcal{B}_1 \geq' \mathcal{B}_2 \text{ if and only if } \langle X_1, Y_1 \rangle \geq' \langle X_2, Y_2 \rangle.$$

7.5 FURTHER CONDITIONS ON EXTREMAL PREFERENCES

Given the new notation, we can now add further conditions on extremal preferences. One reasonable such condition is that an extremal preference relation should be sensitive to changes at either end; in other words, it should satisfy:

Sensitivity. *Let $\langle X_1, Y_1 \rangle$ and $\langle X_2, Y_2 \rangle$ be extremal pairs. Then:*

(1) If $X_1 \geq X_2$ and $Y_1 > Y_2$, then $\langle X_1, Y_1 \rangle >' \langle X_2, Y_2 \rangle$.
(2) If $X_1 > X_2$ and $Y_1 \geq Y_2$, then $\langle X_1, Y_1 \rangle >' \langle X_2, Y_2 \rangle$.

An important but less universal requirement is that decision making should be cautious.[12] A maximally cautious decision maker never chooses an alternative, the worst case of which is worse than the worst case of some other alternative:

Cautiousness. *Let $\langle X_1, Y_1 \rangle$ and $\langle X_2, Y_2 \rangle$ be extremal pairs. Then:*

$$\text{If } X_1 > X_2 \text{, then } \langle X_1, Y_1 \rangle >' \langle X_2, Y_2 \rangle.$$

[12] On cautious decision making, see Hansson 1997c, 1999a.

Exclusive cautiousness. *Let $\langle X_1, Y_1\rangle$ and $\langle X_2, Y_2\rangle$ be extremal pairs. Then:*

$$\langle X_1, Y_1\rangle >' \langle X_2, Y_2\rangle \textit{ if and only if } X_1 > X_2.$$

Two standard decision-making rules, namely the maximin and maximax rules, can be straightforwardly transferred to the present framework:

Definition 7.17. *Let $\geq$ be a relation on a set $\mathcal{A}$ of sentential alternatives. The* maximin preference relation *over extremal pairs in $\mathcal{A}$, based on $\geq$, is the relation $\geq_i$ such that:*

$$\langle X_1, Y_1\rangle \geq_i \langle X_2, Y_2\rangle \textit{ if and only if } X_1 \geq X_2.$$

Furthermore, the maximax preference relation *based on $\geq$ is the relation $\geq_x$ such that:*

$$\langle X_1, Y_1\rangle \geq_x \langle X_2, Y_2\rangle \textit{ if and only if } Y_1 \geq Y_2.$$

$p >_i q$ is an abbreviation of $p \geq_i q$ & $\neg(q \geq_i p)$, and $p \equiv_i q$ of $p \geq_i q$ & $q \geq_i p$. $p >_x q$ and $p \equiv_x q$ are defined analogously.

The maximin rule does not satisfy sensitivity, but it satisfies (and is fully characterized by) completeness and exclusive cautiousness. The latter is, arguably, a rather implausible property. It requires, for instance, that one be indifferent between owning a valueless piece of paper and owning a ticket in a two-ticket lottery in which the winner will receive €1,000,000 and the loser will receive nothing. The *interval maximin rule* is a modification of the maximin rule that avoids such extreme results. This rule maximizes both worst and best alternatives, but it gives maximization of the former absolute priority over maximization of the latter. In terms of postulates, it satisfies sensitivity and cautiousness, but not exclusive cautiousness.

Definition 7.18. *Let $\geq$ be a relation on the sentential alternative set $\mathcal{A}$. The* interval maximin preference relation *$\geq_{ix}$ based on $\geq$ is the extremal relation on $\mathcal{P}(\mathcal{A})\backslash\{\emptyset\}$ such that for all extremal pairs $\langle X_1, Y_1\rangle$ and $\langle X_2, Y_2\rangle$:*

(1) If $X_1 > X_2$, then $\langle X_1, Y_1\rangle >_{ix} \langle X_2, Y_2\rangle$.
(2) If $X_1 \equiv X_2$, then $\langle X_1, Y_1\rangle \geq_{ix} \langle X_2, Y_2\rangle$ if and only if $Y_1 \geq Y_2$.

$p>_{ix}q$ is an abbreviation of $p\geq_{ix}q$ & $\neg(q\geq_{ix}p)$, and $p\equiv_{ix}q$ is an abbreviation of $p\geq_{ix}q$ & $q\geq_{ix}p$.

Theorem 7.19. *Let $\mathcal{A}$ be a sentential alternative set and $\geq$ a relation over $\mathcal{A}$. Let $\geq'$ be a relation over $\mathcal{P}(\mathcal{A})\backslash\{\emptyset\}$ that is extremal with respect to $\geq$. Then $\geq'$ is the interval maximin extension of $\geq$ if and only if it satisfies sensitivity and cautiousness with respect to $\geq$.*

Similarly, *interval maximax* preference relations can be constructed that maximize both worst and best alternatives, but give maximization of the latter absolute priority over maximization of the former.

Definition 7.20. *Let $\geq$ be a relation on the sentential alternative set $\mathcal{A}$. The* interval maximax preference relation $\geq_{xi}$ *based on $\geq$ is the extremal relation on $\mathcal{P}(\mathcal{A})\backslash\{\emptyset\}$ such that for all extremal pairs $\langle X_1,Y_1\rangle$ and $\langle X_2,Y_2\rangle$:*

(1) If $Y_1>Y_2$, then $\langle X_1,Y_1\rangle >_{xi} \langle X_2,Y_2\rangle$.
(2) If $Y_1\equiv Y_2$, then $\langle X_1,Y_1\rangle \geq_{xi} \langle X_2,Y_2\rangle$ if and only if $X_1\geq X_2$.

$p>_{xi}q$ is an abbreviation of $p\geq_{xi}q$ & $\neg(q\geq_{xi}p)$, and $p\equiv_{xi}q$ of $p\geq_{xi}q$ & $q\geq_{xi}p$.

Perhaps the notation needs some explanation. In the indices, **x** refers to maximization of the ma**x**imum and **i** to maximization of the m**i**nimum. $\geq_x$ is a preference relation that only maximizes the maximum and $\geq_i$ only maximizes the minimum. $\geq_{ix}$ maximizes the minimum and after that the maximum, whereas $\geq_{xi}$ performs these maximizations in the reverse order of priority.

Although the interval maximin and interval maximax preference relations mitigate the rather strict principles of maximin and maximax preference relations, respectively, they do so only to a limited degree. It is therefore also of interest to study a wider category of extremal preferences that allows for all assignments of relative priorities to maximization of the best and worst alternatives. This can be done as follows:

Definition 7.21. *Let v be a value assignment on the sentential alternative set $\mathcal{A}$. Then $v_{MAX}(p)$ is the highest value of $v(A)$ for any A such that $p \in A \in \mathcal{A}$. Furthermore, $v_{MIN}(p)$ is the lowest value of $v(A)$ for any A such that $p \in A \in \mathcal{A}$.*

Definition 7.22. *Let $\mathcal{A}$ be an alternative set and v a value assignment on $\mathcal{A}$. Let δ be a number such that 0<δ<1. Then v_δ is the assignment of real numbers to the elements of $\cup\mathcal{A}$, such that for all $p \in \cup\mathcal{A}$:*

$$v_\delta(p) = \delta \cdot v_{MAX}(p) + (1-\delta) \cdot v_{MIN}(p).$$

A relation $\geq_E$ on $\cup\mathcal{A}$ is a max-min weighted preference relation *iff there is some v_δ satisfying these conditions and such that for all $p, q \in \cup\mathcal{A}$:*

$$p \geq_E q \text{ iff } v_\delta(p) \geq v_\delta(q).$$

7.6 PROPERTIES OF EXTREMAL PREFERENCES

We can now turn to an investigation of the logical properties of $\geq_i$, $\geq_x$, $\geq_{ix}$, $\geq_{xi}$, and $\geq_E$.

By definition, $\geq_E$ is complete and transitive. The combined property of being both complete and transitive is transmitted from an exclusionary preference relation $\geq$ to $\geq_i$, $\geq_x$, $\geq_{ix}$, and $\geq_{xi}$.

Observation 7.23. *Let $\mathcal{A}$ be a sentential alternative set, and let $\geq$ be a complete and transitive relation on $\mathcal{A}$. Then $\geq_i$, $\geq_x$, $\geq_{ix}$, and $\geq_{xi}$ are all complete and transitive.*

In Section 6.6 (Observations 6.28 and 6.29), the two negation-related properties of contraposition and conjunctive expansion were shown to hold for ceteris paribus preferences, unless one of the relata contextually implies the other. Contraposition does not hold for $\geq_i$, $\geq_x$, $\geq_{ix}$, $\geq_{xi}$, or $\geq_E$. For $\geq_i$ and $\geq_x$, conjunctive expansion holds only for strict preference, and for $\geq_{ix}$, $\geq_{xi}$, and $\geq_E$ it does not hold even in that case.

Observation 7.24. *Let $\geq$ be a complete and transitive relation on the contextually complete alternative set $\mathcal{A}$, and let $p,\neg p,q,\neg q \in \cup\mathcal{A}$. Then:*

(1a) $p \geq_i q \rightarrow \neg q \geq_i \neg p$ does not hold in general.
(1b) $p \equiv_i q \rightarrow \neg q \equiv_i \neg p$ does not hold in general.
(1c) $p >_i q \rightarrow \neg q >_i \neg p$ does not hold in general.
(2a) $p \geq_x q \rightarrow \neg q \geq_x \neg p$ does not hold in general.
(2b) $p \equiv_x q \rightarrow \neg q \equiv_x \neg p$ does not hold in general.
(2c) $p >_x q \rightarrow \neg q >_x \neg p$ does not hold in general.
(3a) $p \geq_{ix} q \rightarrow \neg q \geq_{ix} \neg p$ does not hold in general.
(3b) $p \equiv_{ix} q \rightarrow \neg q \equiv_{ix} \neg p$ does not hold in general.

(3c) $p>_{ix}q \rightarrow \neg q>_{ix}\neg p$ *does not hold in general.*
(4a) $p\geq_{xi}q \rightarrow \neg q\geq_{xi}\neg p$ *does not hold in general.*
(4b) $p\equiv_{xi}q \rightarrow \neg q\equiv_{xi}\neg p$ *does not hold in general.*
(4c) $p>_{xi}q \rightarrow \neg q>_{xi}\neg p$ *does not hold in general.*
(5a) $p\geq_{E}q \rightarrow \neg q\geq_{E}\neg p$ *does not hold in general.*
(5b) $p\equiv_{E}q \rightarrow \neg q\equiv_{E}\neg p$ *does not hold in general.*
(5c) $p>_{E}q \rightarrow \neg q>_{E}\neg p$ *does not hold in general.*

Observation 7.25. *Let* $\geq$ *be a complete and transitive relation on the contextually complete alternative set* $\mathcal{A}$, *and let* $p\&\neg q$, $q\&\neg p \in \cup\mathcal{A}$. *Then:*

(1a) $p\geq_{i}q \rightarrow (p\&\neg q)\geq_{i}q\&\neg p)$ *does not hold in general.*
(1b) $p\equiv_{i}q \rightarrow (p\&\neg q)\equiv_{i}q\&\neg p)$ *does not hold in general.*
(1c) $p>_{i}q \rightarrow (p\&\neg q)>_{i}(q\&\neg p)$
(2a) $p\geq_{x}q \rightarrow (p\&\neg q)\geq_{x}q\&\neg p)$ *does not hold in general.*
(2b) $p\equiv_{x}q \rightarrow (p\&\neg q)\equiv_{x}(q\&\neg p)$ *does not hold in general.*
(2c) $p>_{x}q \rightarrow (p\&\neg q)>_{x}(q\&\neg p)$
(3a) $p\geq_{ix}q \rightarrow (p\&\neg q)\geq_{ix}q\&\neg p)$ *does not hold in general.*
(3b) $p\equiv_{ix}q \rightarrow (p\&\neg q)\equiv_{ix}(q\&\neg p)$ *does not hold in general.*
(3c) $p>_{ix}q \rightarrow (p\&\neg q)>_{ix}(q\&\neg p)$ *does not hold in general.*
(4a) $p\geq_{xi}q \rightarrow (p\&\neg q)\geq_{ix}q\&\neg p)$ *does not hold in general.*
(4b) $p\equiv_{xi}q \rightarrow (p\&\neg q)\equiv_{xi}(q\&\neg p)$ *does not hold in general.*
(4c) $p>_{xi}q \rightarrow (p\&\neg q)>_{xi}(q\&\neg p)$ *does not hold in general.*
(5a) $p\geq_{E}q \rightarrow (p\&\neg q)\geq_{E}q\&\neg p)$ *does not hold in general.*
(5b) $p\equiv_{E}q \rightarrow (p\&\neg q)\equiv_{E}(q\&\neg p)$ *does not hold in general.*
(5c) $p>_{E}q \rightarrow (p\&\neg q)>_{E}(q\&\neg p)$ *does not hold in general.*

It should be obvious that the formulas listed in Observation 7.25 all hold if p and q are $\mathcal{A}$-incompatible. In that case, p and $p\&\neg q$ have the same representations in $\mathcal{A}$ and the same applies to q and $q\&\neg p$.

The principles of disjunctive distribution that were obtained in Section 6.6 for maximal centred preferences are not fully satisfied by any of $\geq_{i}$, $\geq_{x}$, $\geq_{ix}$, $\geq_{xi}$, or $\geq_{E}$, but the following weakened properties can be proved:

Observation 7.26. *Let* $\geq$ *be a complete and transitive relation over the contextually complete alternative set* $\mathcal{A}$. *Let* $p, q, r \in \cup\mathcal{A}$. *Then:*

(1a) $(p \vee q) \geq_i r \leftrightarrow p \geq_i r \;\&\; q \geq_i r$
(1b) $p >_i r \rightarrow (p \vee q) \geq_i r$ *does not hold in general, not even if p, q, and r are pairwise $\mathcal{A}$-incompatible.*
(2a) $(p \vee q) \geq_x r \leftrightarrow p \geq_x r \vee q \geq_x r$
(2b) $(p \vee q) >_x r \rightarrow p \geq_x r$ *does not hold in general, not even if p, q, and r are pairwise $\mathcal{A}$-incompatible.*
(3a) $(p \vee q) \geq_{ix} r \rightarrow p \geq_{ix} r \vee q \geq_{ix} r$
(3b) $(p \vee q) >_{ix} r \rightarrow p \geq_{ix} r$ *does not hold in general, not even if p, q, and r are pairwise $\mathcal{A}$-incompatible.*
(3c) $p >_{ix} r \rightarrow (p \vee q) \geq_{ix} r$ *does not hold in general, not even if p, q, and r are pairwise $\mathcal{A}$-incompatible.*
(4a) $(p \vee q) \geq_{xi} r \rightarrow p \geq_{xi} r \vee q \geq_{xi} r$
(4b) $(p \vee q) >_{xi} r \rightarrow p \geq_{xi} r$ *does not hold in general, not even if p, q, and r are pairwise $\mathcal{A}$-incompatible.*
(4c) $p >_{xi} r \rightarrow (p \vee q) \geq_{xi} r$ *does not hold in general, not even if p, q, and r are pairwise $\mathcal{A}$-incompatible.*
(5a) $(p \vee q) \geq_E r \rightarrow p \geq_E r \vee q \geq_E r$
(5b) $(p \vee q) >_E r \rightarrow p \geq_E r$ *does not hold in general, not even if p, q, and r are pairwise $\mathcal{A}$-incompatible.*
(5c) $p >_E r \rightarrow (p \vee q) \geq_E r$ *does not hold in general, not even if p, q, and r are pairwise $\mathcal{A}$-incompatible.*

Observation 7.27. *Let $\geq$ be a complete and transitive relation over the contextually complete alternative set $\mathcal{A}$. Let $p, q, r \in \cup\mathcal{A}$. Then:*

(1a) $p \geq_i (q \vee r) \leftrightarrow p \geq_i q \vee p \geq_i r$
(1b) $p >_i (q \vee r) \rightarrow p \geq_i q$ *does not hold in general, not even if p, q, and r are pairwise $\mathcal{A}$-incompatible.*
(2a) $p \geq_x (q \vee r) \leftrightarrow p \geq_x q \;\&\; p \geq_x r$
(2b) $p >_x q \rightarrow p \geq_x (q \vee r)$ *does not hold in general, not even if p, q, and r are pairwise $\mathcal{A}$-incompatible.*
(3a) $p \geq_{ix} (q \vee r) \rightarrow p \geq_{ix} q \vee p \geq_{ix} r$
(3b) $p >_{ix} (q \vee r) \rightarrow p \geq_{ix} q$ *does not hold in general, not even if p, q, and r are pairwise $\mathcal{A}$-incompatible.*
(3c) $p >_{ix} q \rightarrow p \geq_{ix} (q \vee r)$ *does not hold in general, not even if p, q, and r are pairwise $\mathcal{A}$-incompatible.*
(4a) $p \geq_{xi} (q \vee r) \rightarrow p \geq_{xi} q \vee p \geq_{xi} r$
(4b) $p >_{xi} (q \vee r) \rightarrow p \geq_{xi} q$ *does not hold in general, not even if p, q, and r are pairwise $\mathcal{A}$-incompatible.*

(4c) $p>_{xi}q \rightarrow p\geq_{xi}(q\vee r)$ *does not hold in general, not even if* p, q, *and* r *are pairwise* $\mathcal{A}$-*incompatible.*
(5a) $p\geq_{E}(q\vee r) \rightarrow p\geq_{E}q \vee p\geq_{E}r$
(5b) $p>_{E}(q\vee r) \rightarrow p\geq_{E}q$ *does not hold in general, not even if* p, q, *and* r *are pairwise* $\mathcal{A}$-*incompatible.*
(5c) $p>_{E}q \rightarrow p\geq_{E}(q\vee r)$ *does not hold in general, not even if* p, q, *and* r *are pairwise* $\mathcal{A}$-*incompatible.*

The following observation adds to the list of properties that distinguish the four types of preference relations from each other.

Observation 7.28. *Let* $\geq$ *be a relation on the sentential alternative set* $\mathcal{A}$, *and let* $p, q \in \cup\mathcal{A}$. *Then:*

(1) *If* $\vDash_{\mathcal{A}} p\rightarrow q$, *then* $p\geq_{i}q$.
(2) *If* $\vDash_{\mathcal{A}} p\rightarrow q$, *then* $q\geq_{x}p$.
(3a) *It does not hold in general that if* $\vDash_{\mathcal{A}} p\rightarrow q$, *then* $p\geq_{ix}q$.
(3b) *It does not hold in general that if* $\vDash_{\mathcal{A}} p\rightarrow q$, *then* $q\geq_{ix}p$.
(4a) *It does not hold in general that if* $\vDash_{\mathcal{A}} p\rightarrow q$, *then* $p\geq_{xi}q$.
(4b) *It does not hold in general that if* $\vDash_{\mathcal{A}} p\rightarrow q$, *then* $q\geq_{xi}p$.
(5a) *It does not hold in general that if* $\vDash_{\mathcal{A}} p\rightarrow q$, *then* $p\geq_{E}q$.
(5b) *It does not hold in general that if* $\vDash_{\mathcal{A}} p\rightarrow q$, *then* $q\geq_{E}p$.

Corollary. *Let* $\geq$ *be a relation on the sentential alternative set* $\mathcal{A}$, *and let* $p, q \in \cup\mathcal{A}$. *Then:*

(1) $p\geq_{i}(p\vee q)$.
(2) $(p\vee q)\geq_{x}p$.

The properties listed in the Corollary were used by Dennis Packard in axiomatic characterizations. Maximin preference is characterized by transitivity, completeness, $p\geq_{i}(p\vee q)$, and $p\geq_{i}r \;\&\; q\geq_{i}r \rightarrow (p\vee q)\geq_{i}r$. Similarly, maximax preference is characterized by transitivity, completeness, $(p\vee q)\geq_{x}p$, and $p\geq_{x}q \;\&\; p\geq_{x}r \rightarrow p\geq_{x}(q\vee r)$.[13]

Parts (1) and (2) of Observation 7.28 have a flavour of paradox. Part (1) may be called the "Nobel peace prize postulate." Let q denote that a certain statesman stops a war, and p that he first starts a war and then stops it. Let $\mathcal{A}$ be an alternative set that contains representations of p and q. Then $\vDash_{\mathcal{A}} p\rightarrow q$ is satisfied, and we can conclude that $p\geq_{i}q$, in other

[13] Packard 1979.

words, p is (in the maximin sense) at least as good a behaviour as q. This conclusion has been corroborated by the several cases in which a Nobel peace prize was awarded for behaviour like p. It is not difficult, either, to find examples that bring out the strangeness of part (2). We may, for instance, let q denote some violent action and p the same action, performed in self-defence. These properties of maximin and maximax preferences are strikingly counterintuitive and speak against their use as general-purpose decision tools.

Intuitively, one would expect $p \vee q$ to be intermediate in value between p and q, that is, the following postulate should hold:

$$p \geq' q \rightarrow p \geq' (p \vee q) \,\&\, (p \vee q) \geq' q \quad \textit{(disjunctive interpolation)}$$

In addition to its plausibility, this property will also turn out in Chapter 10 to be essential for the derivability of deontic logic from preference logic. It is satisfied by all our five preference relations for agnostic decision making:

Observation 7.29. *Let $\geq$ be a complete and transitive relation on the contextually complete alternative set $\mathcal{A}$, and let $p, q \in \cup\mathcal{A}$. Then:*

(1a) $p \geq_i q \rightarrow p \geq_i (p \vee q) \,\&\, (p \vee q) \geq_i q$

(1b) $p \geq_i q \rightarrow (p \vee q) \equiv_i q$

(2a) $p \geq_x q \rightarrow p \geq_x (p \vee q) \,\&\, (p \vee q) \geq_x q$

(2b) $p \geq_x q \rightarrow p \equiv_x (p \vee q)$

(3a) $p \geq_{ix} q \rightarrow p \geq_{ix} (p \vee q) \,\&\, (p \vee q) \geq_{ix} q$

(3b) $p \geq_{ix} q \rightarrow p \equiv_{ix} (p \vee q)$ *does not hold in general, not even if p and q are $\mathcal{A}$-incompatible.*

(3c) $p \geq_{ix} q \rightarrow (p \vee q) \equiv_{ix} q$ *does not hold in general, not even if p and q are $\mathcal{A}$-incompatible.*

(3d) $p \equiv_{ix} q \rightarrow p \equiv_{ix} (p \vee q)$

(4a) $p \geq_{xi} q \rightarrow p \geq_{xi} (p \vee q) \,\&\, (p \vee q) \geq_{xi} q$

(4b) $p \geq_{xi} q \rightarrow p \equiv_{xi} (p \vee q)$ *does not hold in general, not even if p and q are $\mathcal{A}$-incompatible.*

(4c) $p \geq_{xi} q \rightarrow (p \vee q) \equiv_{xi} q$ *does not hold in general, not even if p and q are $\mathcal{A}$-incompatible.*

(4d) $p \equiv_{xi} q \rightarrow p \equiv_{xi} (p \vee q)$

(5a) $p \geq_E q \rightarrow p \geq_E (p \vee q) \,\&\, (p \vee q) \geq_E q$

(5b) $p \geq_E q \rightarrow p \equiv_E (p \vee q)$ *does not hold in general, not even if p and q are $\mathcal{A}$-incompatible.*

Table 2. *Some properties of the various types of combinative preference relations introduced in Chapters 6 and 7.*

	$\geq_f$ (SS)	$\geq_f$ (SM)	$\geq_f$ (MC)	$\geq_w$	$\geq_i$	$\geq_x$	$\geq_{ix}$	$\geq_{xi}$	$\geq_E$
$p\geq'p$	–	+	+	+	+	+	+	+	+
$p\geq'q \vee q\geq'p$	–	–	–	+	+	+	+	+	+
$p\geq'q \;\&\; q\geq'r \rightarrow p\geq'r$	–	–	#	+	+	+	+	+	+
$(p\vee q)\geq'r \leftrightarrow p\geq'r \;\&\; q\geq'r$	–	–	#	–	+	–	–	–	–
$(p\vee q)\geq'r \leftrightarrow p\geq'r \vee q\geq'r$	–	–	–	–	–	+	–	–	–
$(p\vee q)\geq'r \rightarrow p\geq'r \vee q\geq'r$	–	#	#	#	+	+	+	+	+
$p\geq'(q\vee r) \leftrightarrow p\geq'q \;\&\; p\geq'r$	–	–	#	–	–	+	–	–	–
$p\geq'(q\vee r) \leftrightarrow p\geq'q \vee p\geq'r$	–	–	–	–	+	–	–	–	–
$p\geq'(q\vee r) \rightarrow p\geq'q \vee p\geq'r$	–	#	#	#	+	+	+	+	+
If $\models_{\mathcal{A}} p\rightarrow q$, then $p\geq'q$.	–	–	–	–	+	–	–	–	–
If $\models_{\mathcal{A}} p\rightarrow q$, then $q\geq'p$.	–	–	–	–	–	+	–	–	–
$p\geq'q \rightarrow p\geq'(p\vee q) \;\&\; (p\vee q)\geq'q$	#	#	#	#	+	+	+	+	+
$p\geq'q \rightarrow p\equiv'(p\vee q)$	–	–	–	–	–	+	–	–	–
$p\geq'q \rightarrow (p\vee q)\equiv'q$	–	–	–	–	+	–	–	–	–
$p\equiv'q \rightarrow p\equiv'(p\vee q)$	#	#	#	#	+	+	+	+	+

+ holds (unconditionally)
# does not hold in general, but holds if p, q, and r are pairwise $\mathcal{A}$-incompatible
– does not hold in general, not even if p, q, and r are pairwise $\mathcal{A}$-incompatible

(5c) $p\geq_E q \rightarrow (p\vee q)\equiv_E q$ *does not hold in general, not even if p and q are $\mathcal{A}$-incompatible.*

(5d) $p\equiv_E q \rightarrow p\equiv_E(p\vee q)$

Corollary. *Let $\geq$ be a complete and transitive relation on the contextually complete alternative set $\mathcal{A}$, and let $p\&q$, $p\&\neg q \in \cup\mathcal{A}$. Then:*

(1) $p\equiv_i(p\&q) \vee p\equiv_i(p\&\neg q)$

(2) $p\equiv_x(p\&q) \vee p\equiv_x(p\&\neg q)$

Finally, it should be noted that we have found several postulates that are satisfied by $\geq_i$ but not by $\geq_x$, $\geq_{ix}$, $\geq_{xi}$, or $\geq_E$. We have also found postulates that are satisfied by $\geq_x$ but not by $\geq_i$, $\geq_{ix}$, $\geq_{xi}$, or $\geq_E$. However, all postulates that we have found to be satisifed by $\geq_{ix}$ are also satisfied by $\geq_i$, $\geq_x$, $\geq_{xi}$, and $\geq_E$. Similarly, all postulates found to be satisifed by $\geq_{xi}$ are also satisfied by $\geq_i$, $\geq_x$, $\geq_{ix}$, and $\geq_E$, and all postulates found to be satisifed by $\geq_E$ are satisfied by $\geq_i$, $\geq_x$, $\geq_{ix}$, and $\geq_{xi}$. It is at present an open question whether any distinguishing postulates can be found for $\geq_{ix}$, $\geq_{xi}$, and $\geq_E$.

7.7 A RÉSUMÉ OF COMBINATIVE PREFERENCE RELATIONS

In Chapter 6, we investigated the logic of three types of combinative preference relatations that are intended to capture ceteris paribus preferences. In the present chapter, we have studied one preference relation that is intended for a prognostic approach to decisions ($\geq_w$) and five that are intended for an agnostic appoach ($\geq_i$, $\geq_x$, $\geq_{ix}$, $\geq_{xi}$, and $\geq_E$). Some of the logical properties of these relations are summarized in Table 2.[14] Although the picture is somewhat complicated, the general tendency is for the decision-theoretically oriented relations to satisfy stronger conditions.

[14] Two other tables comparing the properties of different preference relations can be found in Rescher 1968, pp. 306–307, and Packard 1979, p. 299.

8

Monadic Value Predicates

In addition to the comparative notions, 'better' and 'of equal value,' informal discourse on values contains monadic (one-place) value predicates, such as 'good,' 'best,' 'very bad,' 'fairly good,' and so on.

We predicate concepts such as goodness not only of complete alternatives but also – and more often – of particular states of affairs. Just like combinative preference relations, monadic predicates will be assumed to refer to states of affairs. They will be inserted into a structure that contains a combinative preference relation, so that their relations to the comparative notions can be studied. Throughout this chapter, $\geq'$ denotes a (weak) combinative preference relation that operates on the union $\cup\mathcal{A}$ of some contextually complete alternative set $\mathcal{A}$. As before, $>'$ and $\equiv'$ are its strict and symmetric parts. The more precise nature of $\geq'$ will be left open. Monadic predicates such as 'good' appear both in decision-guiding contexts where it would seem natural to connect them with a decision-guiding preference relation, and in other contexts in which a ceteris paribus preference relation may be more appropriate.

The standard reading of $p>'q$, "p is better than q," implies no commitment whatsoever with respect to the truth values of p and q. If p does not hold in the actual world, then "p would be better than q" is a more accurate paraphrase. A more precise general reading of $p>'q$ would be "p is or would be better than q," but in most contexts the shorter and more convenient phrase "p is better than q" will lead to no misunderstandings. Likewise, the monadic value predicates are noncommitted with respect to truth. The predicate for 'good,' G, means ". . . is good or would be good," but the shorter reading ". . . is good" is more convenient.[1]

[1] In this respect, I follow convention in preference logic. An exception is a system proposed by Lou Goble (1990), in which $Gp \rightarrow p$ is valid.

In Sections 8.1 and 8.2, some general classes of monadic predicates are identified. Sections 8.3 and 8.4 are devoted to the two principal predicates 'good' and 'bad.' In Section 8.5, the definability of monadic value concepts in terms of preferences is discussed in more general terms.

8.1 POSITIVE, NEGATIVE, AND CIRCUMSCRIPTIVE PREDICATES

What is better than something good is itself good. Many other value predicates – such as 'best,' 'not worst,' 'very good,' 'excellent,' 'not very bad,' 'acceptable' – have the same property. If one of these predicates holds for p, then it also holds for everything that is better than p or equal in value to p. This property will be called "$\geq'$-positivity," or (when there is no risk of confusion), simply "positivity."

Definition 8.1. *A monadic predicate H is* $\geq'$-positive *if and only if for all p and q:*

$$Hp \ \& \ q \geq' p \rightarrow Hq.$$

Similarly, 'bad' has the converse property that if p is bad, then whatever is worse than or equal in value to p is also bad. Other predicates that share this property are 'very bad,' 'worst,' and 'not best.' This property will be called "($\geq'$)-negativity."

Definition 8.2. *A monadic predicate H is* $\geq'$-negative *if and only if for all p and q:*

$$Hp \ \& \ p \geq' q \rightarrow Hq.$$

Intuitively, we expect the negation 'not good' of the positive predicate 'good' to be negative. Indeed, this can easily be shown to be a general pattern that holds for all positive and negative predicates.

Observation 8.3. *A monadic predicate H satisfies $\geq'$-positivity if and only if its negation $\neg H$ satisfies $\geq'$-negativity.*

An important class of positive predicates is that which represents 'best.' They are mirrored at the other end of the value scale by negative predicates that represent 'worst':

Definition 8.4. *Let $\geq'$ be a combinative preference relation. The following are monadic predicates defined from $\geq'$:*

$$Hp \leftrightarrow (\forall q)p \geq' q \qquad \text{(strongly best)}$$
$$Hp \leftrightarrow \neg(\exists q)q >' p \qquad \text{(weakly best)}$$
$$Hp \leftrightarrow (\forall q)q \geq' p \qquad \text{(strongly worst)}$$
$$Hp \leftrightarrow \neg(\exists q)p >' q \qquad \text{(weakly worst)}$$

The first two of these definitions correspond to the notions of strong and weak eligibility that were introduced in Section 2.4. As was shown there, with these definitions not all preference structures will have (strongly or weakly) best elements.[2]

In addition to the positive and negative predicates, ordinary language also contains a third category, namely those that are, intuitively speaking, bounded both upward and downward. 'Almost worst' and 'fairly good' are examples of this category. They will be called *circumscriptive* predicates. From a formal point of view, they can be defined as the combination of one positive and one negative predicate. Thus, "p is almost worst" may be defined as "p is very bad and p is not worst," employing the negative predicate 'very bad' and the positive predicate 'not worst.'

Definition 8.5. *A monadic predicate H is* $\geq'$-circumscriptive *if and only if there is a $\geq'$-positive predicate H^+ and a $\geq'$-negative predicate H^- such that for all p:*

$$Hp \leftrightarrow H^+p \;\&\; H^-p.$$

A $\geq'$-circumscriptive predicate is properly $\geq'$-circumscriptive *if and only if it is neither $\geq'$-positive nor $\geq'$-negative.*

Circumscriptive predicates satisfy the following condition:

[2] An alternative approach would be to insist that every preference structure should have 'best' – or rather 'optimal' – elements. One way to construct such a generalized notion of optimality is to treat it as a kind of stability. An element is optimal-stable if we cannot move away from it to some other element that is better. When there are no stable elements, we will have to settle for those that come as close as possible to being stable. In a finite structure, these are the retrievable elements. An element is retrievable if and only if every path (sequence of successive moves to better elements) away from it has a path back to it. For details of this notion of stability, see Hansson 1996b.

Definition 8.6. *A monadic predicate H is* $\geq'$-continuous *if and only if the following holds for all p, q, and r:*

$$\text{If } Hp, Hr, \text{ and } p\geq' q\geq' r, \text{ then } Hq.$$

It can be seen from the definitions that all $\geq'$-positive predicates are $\geq'$-continuous, and so are all $\geq'$-negative predicates.

Observation 8.7. *Let* $\geq'$ *be reflexive and transitive. Furthermore, let H be a* $\geq'$*-continuous predicate. Then H is either* $\geq'$*-positive,* $\geq'$*-negative, or properly* $\geq'$*-circumscriptive.*

8.2 NEGATION-COMPARING PREDICATES

As was noted in Section 6.5, we should not expect a ceteris paribus preference relation to be complete, even if it derives from an exclusionary preference relation that satisfies completeness. A much weaker property than full comparability between any two sentences ($p\geq' q \vee q\geq' p$) is comparability between any sentence and its negation ($p\geq'\neg p \vee \neg p\geq' p$). However, not even this weaker property can be expected to hold in general for ceteris paribus preferences. To see this, let p denote that you take a poison and q that you take an antidote. If you take the poison, then it is better for you to take the antidote than not to do so. Since the antidote is itself mildly toxic, if you do not take the poison, then it is better not to take the antidote. On the following list, each of the four alternatives is strictly better than all those below it.

$\text{Cn}(\{\neg p, \neg q\})$
$\text{Cn}(\{\neg p, q\})$
$\text{Cn}(\{p, q\})$
$\text{Cn}(\{p, \neg q\})$

If a ceteris paribus preference relation is applied to this set of alternatives, then we cannot expect either $q\geq'\neg q$ or $\neg q\geq' q$ to be validated. On the other hand, a decision-theoretically oriented preference relation may be expected to adjudicate in these types of cases. This will be useful, for instance, if you have to decide between q and $\neg q$ while the truth value of p is undetermined. In this particular example, we will have $q\geq_{i}\neg q$, $q\geq_{ix}\neg q$, $\neg q\geq_{x} q$, and $\neg q\geq_{xi} q$. We will also have either $q\geq_{w}\neg q$ or $\neg q\geq_{w} q$, and either $q\geq_{E}\neg q$ or $\neg q\geq_{E} q$, depending in both cases on the weights and values involved.

If, however, we settle in this case for a preference relation $\geq'$ such that neither $q\geq'\neg q$ nor $\neg q\geq' q$ holds, then the application of $\geq'$-related monadic value predicates to q and $\neg q$ is correspondingly problematic. It would seem strange to say that a state of affairs is good or bad if it is so deficient in value information that it cannot be compared to its negation. Therefore, predicates such as 'good,' 'bad,' and 'of neutral value' are not applicable to such states of affairs. In other words, these predicates have the following property:

Definition 8.8. *A monadic predicate H is* negation-comparing *with respect to $\geq'$ if and only if, for all p:*

$$Hp \rightarrow p\geq'\neg p \vee \neg p\geq' p.$$

8.3 DEFINING GOOD AND BAD

Definitions of 'good' and 'bad' in terms of a preference relation are a fairly common theme in the value-logical literature. There are two major traditions.[3] One of these defines 'good' as "better than its negation" and 'bad' as "worse than its negation." The first clear statement of this idea seems to be due to Albert Brogan.[4] It has been accepted by Mitchell, Halldén, von Wright, Åqvist, and others.[5]

The other tradition is to introduce, prior to 'good' and 'bad,' a set of neutral propositions. Goodness is predicated of everything that is better than some neutral proposition and badness of everything that is worse than some neutral proposition. The best-known variant of this approach was proposed by Chisholm and Sosa. According to these authors, a state of affairs is indifferent if and only if it is neither better nor worse than its negation. Then "a state of affairs is *good* provided it is better than some state of affairs that is indifferent, and . . . a state of affairs is *bad* provided some state of affairs that is indifferent is better than it."[6]

[3] A third type of definition is to equate 'good' with "better than most . . . ." See Langford 1942, esp. p. 336. For a critical comment, see Wallace 1972, p. 777.

[4] Brogan 1919.

[5] Mitchell 1950, pp. 103–105; Halldén 1957, p. 109; von Wright 1963a, p. 34, and 1972, p. 162; Åqvist 1968.

[6] Chisholm and Sosa 1966a, p. 246. They distinguish between indifference and neutrality. To be neutral means, in their terminology, to be equal in value to something that is indifferent.

Other variants of the same basic approach make use of tautologies or contradictions as neutral propositions.[7] However, it is far from clear how something can be compared in terms of value to a tautology or a contradiction. Admittedly, it may emerge as a technical result in a formal system that 'good' coincides with "better than tautology,"[8] but this should not be the formal definition through which 'good' is introduced. Its introduction must be based on intuitions relating to the comparisons that we can actually make, and these are comparisons between contingent states of affairs.

Perhaps the best way is to follow an axiomatic procedure. The first step is then to identify properties that 'good' and 'bad' should satisfy. These properties can be used as necessary and (if we are successful) sufficient conditions on a constructive definition.

'Good' and 'bad' are closely connected. Therefore, instead of developing postulates separately for each of them, we should direct our endeavours at a pair $\langle G,B\rangle$ of two interconnected monadic predicates, representing 'good' and 'bad,' respectively.

In Section 8.1, it was observed that G ('good') should satisfy positivity and that B ('bad') should satisfy negativity. In Section 8.2, it was argued that they should both be negation-comparing. In the following definition, these properties are defined for a pair $\langle G,B\rangle$ of predicates.

Definition 8.9. *Let $\langle G,B\rangle$ be a pair of monadic predicates.*

(1) It satisfies positivity/negativity *(PN) with respect to $\geq'$ if and only if G satisfies $\geq'$-positivity and B satisfies $\geq'$-negativity.*
(2) It satisfies negation-comparability *(NC) with respect to $\geq'$ if and only if both G and B are negation-comparing with respect to $\geq'$.*

It does not seem reasonable to say that something is (from the same point of view) both good and bad. If someone says, for example, that a particular novel is both good and bad, then this is perceived as paradoxical. We expect a "resolution" that typically distinguishes between different evaluation critieria, such as: "The plot is good, but the

[7] Tautologies were used in Danielsson 1968, p. 37. The use of contradictions was proposed by von Wright (1972, p. 164). The use of the empty set for similar purposes was proposed by Francis W. Irwin (1961).

[8] In a system proposed by Lou Goble (1989), the equivalences $Gp \leftrightarrow \top > \neg p$ and $Bp \leftrightarrow \neg p > \top$ (with $\top$ a tautology) emerge in this way. See also Observations 10.56 and 10.58.

language is bad." Due to our assumption of criterial constancy (Section 2.1), we can presume that goodness and badness are mutually exclusive:

Definition 8.10. *A pair* $\langle G,B \rangle$ *of monadic predicates satisfies* mutual exclusiveness *(ME) if and only if for all p:* $\neg(Gp \;\&\; Bp)$.

A related intuition is that a state of affairs and its negation are not (from the same point of view) both good or both bad. If you say "It is good to be married, and it is also good to be unmarried," then you typically mean that matrimony and bachelorhood are good in different respects or according to different criteria. Something similar can be said about the dismal pronouncement that "it is bad to be married, and it is also bad to be unmarried." As these examples show, uses of 'good' and 'bad' that allow both a sentence and its negation to represent something good (or both to represent something bad) are untypical and not without a certain flavour of paradox. In more typical usage, such patterns are excluded. This property will be called "nonduplicity."

Definition 8.11. *A monadic predicate H satisfies* nonduplicity *(ND) if and only if for all p:* $\neg(Hp \;\&\; H\neg p)$.

A pair $\langle G,B \rangle$ *of monadic predicates satisfies nonduplicity if and only if both G and B do so.*

In the presence of nonduplicity, NC can be reformulated for both positive and negative predicates:

Observation 8.12. *(1) If the monadic predicate H satisfies positivity and nonduplicity, then it satisfies negation-comparability (NC) if and only if it satisfies* $Hp \rightarrow p \geq' \neg p$.

(2) If the monadic predicate H satisfies negativity and nonduplicity, then it satisfies negation-comparability (NC) if and only if it satisfies $Hp \rightarrow \neg p \geq' p$.

We now have four *basic postulates for 'good' and 'bad,'* namely PN (positivity/negativity), NC (negation-comparability), ME (mutual exclusiveness), and ND (nonduplicity). As the following observation shows, ME is redundant in the presence of the other three:

Observation 8.13. *If* $\langle G,B \rangle$ *satisfies PN, ND, and NC, then it also satisfies ME.*

PN, ND, and NC (and the redundant ME) are satisfied by many pairs of predicates, including those that correspond to 'best' – 'worst' and 'very good'–'very bad.' We have not yet singled out the formal counterparts of 'good' and 'bad.'

'Good' and 'bad' are weak properties. For something to be (at all) good, it is only necessary that it satisfies the criteria of goodness to some very small degree. (A higher degree is required for 'very good.') Another way to express this is that 'good' and 'bad' should have *maximal coverage*. They should cover all the predicate pairs, such as 'best'–'worst,' 'very good'–'very bad,' etc., that satisfy the basic postulates. Given that $\langle G,B \rangle$ satisfies the basic postulates, there should be as few states of affairs as possible that are not covered by $\langle G,B \rangle$, that is, that are neither good nor bad.

This leads us to search for a "canonical" pair $\langle G_C,B_C \rangle$ that is maximal in the sense that if another pair $\langle G,B \rangle$ satisfies the basic postulates, then Gp implies G_Cp and Bp implies B_Cp. There is indeed such a pair, and it can be defined in a quite general fashion.

Definition 8.14.

$G_Cp \leftrightarrow (\forall q)(q{\geq}'^{*}p \rightarrow q{>}'\neg q)$ (canonical good)

$B_Cp \leftrightarrow (\forall q)(p{\geq}'^{*}q \rightarrow \neg q{>}'q)$ (canonical bad)

Theorem 8.15. *Let* $\geq'$ *be a relation that satisfies ancestral reflexivity* ($p{\geq}'^{*}p$). *Let* $\langle G_C,B_C \rangle$ *be as in Definition 8.14. Then:*

(1) $\langle G_C,B_C \rangle$ *satisfies PN, ND, and NC.*

(2) Let $\langle G,B \rangle$ *be a pair of monadic predicates that satisfies PN, ND, and NC. Then for all p:*

$$Gp \rightarrow G_Cp \text{ and } Bp \rightarrow B_Cp.$$

Ancestral reflexivity, the only condition on $\geq'$ that is needed in this theorem, is very weak indeed. It is satisfied by similarity-maximizing, but not in general by similarity-satisficing preferences. It is also satisfied by all the decision-guiding preference relations investigated in Chapter 7 ($\geq_w$, $\geq_i$, $\geq_x$, $\geq_{ix}$, $\geq_{xi}$, and $\geq_E$).

An intuition closely related to maximal coverage is that 'good' and 'bad' come so close to each other that they only have "neutral" values between them. One way to express this is that "things that are neither good nor bad are not among themselves better or worse." Another is that "if two things are of unequal value, then at least one of them must be good or at least one of them bad."[9] These two formulations can be shown to be equivalent for pairs of operators that satisfy PN.

Observation 8.16. *If $\langle G,B \rangle$ satisfies PN, then the following two conditions are equivalent:*

(a) $\neg Gp \mathbin{\&} \neg Bp \mathbin{\&} \neg Gq \mathbin{\&} \neg Bq \rightarrow \neg(p>'q) \mathbin{\&} \neg(q>'p)$
(b) $p>'q \rightarrow Gp \vee Bq$ (closeness)

As the following theorem shows, if there is any pair $\langle G,B \rangle$ that satisfies closeness in addition to the basic postulates, then so does $\langle G_C,B_C \rangle$.[10]

Theorem 8.17. *Let $\geq'$ satisfy ancestral reflexivity ($p \geq'^* p$). Then, if there is a pair $\langle G,B \rangle$ of predicates that satisfies PN, ND, NC, and closeness, then $\langle G_C,B_C \rangle$ satisfies (PN, ND, NC, and) closeness.*[11]

8.4 ALTERNATIVE ACCOUNTS OF GOOD AND BAD

Based on the results obtained in the previous section, G_C and B_C can be taken to represent 'good' and 'bad,' respectively. We can now return to the two traditional accounts of goodness and badness referred to above and compare them to G_C and B_C. The common definition of goodness as $p>'\neg p$ and badness as $\neg p>'p$ can be introduced as follows:

Definition 8.18.
$G_{N}p \leftrightarrow p>'\neg p$ (negation-related good)
$B_{N}p \leftrightarrow \neg p>'p$ (negation-related bad)

[9] Both quotations are from von Wright 1972, p. 161.
[10] On closeness, see Hansson 1990a.
[11] If $\geq'$ satisfies completeness, then $\langle G_C,B_C \rangle$ is also the only pair of predicates that satisfies the conditions given in the theorem. For a proof of uniqueness in this case and for other pairs of predicates that satisfy the condition in the general case, see Hansson 1990a.

It follows directly from the definition that $\langle G_N, B_N \rangle$ always satisfies ND, NC, and ME. Contrary to $\langle G_C, B_C \rangle$, however, it does not always satisfy PN. It does so, in fact, only when it coincides with $\langle G_C, B_C \rangle$.

Theorem 8.19. *Let $\geq'$ satisfy ancestral reflexivity ($p \geq'^* p$). Then $\langle G_N, B_N \rangle$ satisfies PN if and only if it coincides with $\langle G_C, B_C \rangle$.*

For major types of combinative preference relations, $\langle G_N, B_N \rangle$ does not in general satisfy PN.

Observation 8.20. *(1) There is a contextually complete alternative set $\mathcal{A}$ and a complete and transitive relation $\geq$ on $\mathcal{A}$ such that, if $\langle G_N, B_N \rangle$ is based on some extremal relation $\geq'$ based on $\geq$ that satisfies sensitivity, then G_N is not $\geq'$-positive and B_N is not $\geq'$-negative.*

(2) Let $\geq$ be a complete and transitive relation on the contextually complete alternative set $\mathcal{A}$. Let $\geq_i$ be the corresponding maximin preference relation. If $\langle G_N, B_N \rangle$ is based on $\geq_i$, then G_N is $\geq_i$-positive, but it does not hold in general that B_N is $\geq_i$-negative.

(3) Let $\geq$ be a complete and transitive relation on the contextually complete alternative set $\mathcal{A}$. Let $\geq_x$ be the corresponding maximax preference relation. If $\langle G_N, B_N \rangle$ is based on $\geq_x$, then B_N is $\geq_x$-negative, but it does not hold in general that G_N is $\geq_x$-positive.

(4) Let $\geq$ be a complete and transitive relation on the contextually complete alternative set $\mathcal{A}$. Let $\geq_f$ be a similarity-maximizing preference relation based on $\geq$, and let $\langle G_N, B_N \rangle$ be based on $\geq_f$. It does not hold in general that G_N is $\geq_f$-positive or that B_N is $\geq_f$-negative.

It follows from part (1) of Observation 8.20 that $\langle G_N, B_N \rangle$ does not in general satisfy PN if it is based on an interval maximin ($\geq_{ix}$), interval maximax ($\geq_{xi}$), or max-min weighted ($\geq_E$) preference relation. Since the positivity of 'good' and the negativity of 'bad' are indispensable properties of these predicates, $\langle G_N, B_N \rangle$ can be a plausible formalization of 'good' and 'bad' only if $\geq'$ is such that $\langle G_N, B_N \rangle$ satisfies PN. Therefore, Theorem 8.19 and Observation 8.20 speak in favour of using $\langle G_C, B_C \rangle$ rather than $\langle G_N, B_N \rangle$ as a general-purpose representation of 'good' and 'bad.'

Close connections can be obtained between $\langle G_N, B_N \rangle$ and other pairs of predicates that satisfy the basic postulates:

Observation 8.21. *Let* $\langle G,B\rangle$ *be any pair of predicates that satisfies PN, ND, and NC. Then:*

(1) $Gp \vee B\neg p \rightarrow G_N p$, *and*
(2) $Bp \vee G\neg p \rightarrow B_N p$.

Furthermore, if $\langle G,B\rangle$ *satisfies closeness, then:*

(3) $Gp \vee B\neg p \leftrightarrow G_N p$, *and*
(4) $Bp \vee G\neg p \leftrightarrow B_N p$.

The definition of 'good' and 'bad' proposed by Chisholm and Sosa can be introduced into the present formal framework as follows:

Definition 8.22.
$G_I p \leftrightarrow (\exists q)(p >' q \equiv' \neg q)$ (indifference-related good)
$B_I p \leftrightarrow (\exists q)(\neg q \equiv' q >' p)$ (indifference-related bad)

$\langle G_I,B_I\rangle$ does not in general satisfy any of the four basic postulates. For these to be satisfied, various conditions must be imposed on the preference relation.

For the definition of $\langle G_I,B_I\rangle$ to be at all useful, there should be at least one neutral element, that is, at least one q such that $q \equiv' \neg q$. Furthermore, it can be required that all neutral elements should be interchangeable, so that it does not matter which of them a sentence is compared to.[12] This amounts to the following requirement on the preference relation:

Definition 8.23. $\geq'$ *satisfies* calibration *if and only if:*

(1) There is some q *such that* $q \equiv' \neg q$, *and*
(2) If $q \equiv' \neg q$ *and* $s \equiv' \neg s$, *then for all* p: $p \geq' q \leftrightarrow p \geq' s$ *and* $q \geq' p \leftrightarrow s \geq' p$.

It follows from calibration that if $q \equiv' \neg q$ and $s \equiv' \neg s$, then $q \equiv' s$.

If calibration holds, then $\langle G_I,B_I\rangle$ satisfies ME. However, calibration is not sufficient to guarantee satisfaction of any of the other three basic postulates, PN, ND, and NC. Further conditions on $\geq'$ will have to be

[12] The definition of "neutral" proposed by Chisholm and Sosa, namely $q \equiv \neg q$, is not workable in all preference structures. A more general definition, in analogy with Definition 8.14, would be $Nq \leftrightarrow (\forall s)(q \equiv^* s \rightarrow s \equiv \neg s)$.

added in order to ensure that $\langle G_I,B_I \rangle$ satisfies these properties. (Transitivity is sufficient for PN, and completeness is sufficient for NC. ND does not follow even if calibration, transitivity, and completeness are all satisfied.)[13]

In summary, G_C and B_C have a special standing among the candidates for 'good' and 'bad.' They satisfy the four basic postulates, PN, ND, NC, and ME, and they include all other predicate pairs that satisfy these postulates. Furthermore, the pair $\langle G_C,B_C \rangle$ satisfies closeness whenever that property is compatible with the four basic postulates, and it coincides with $\langle G_N,B_N \rangle$ whenever G_N satisfies positivity and B_N negativity. $\langle G_C,B_C \rangle$ may be seen as a generalization of $\langle G_N,B_N \rangle$, covering cases in which the latter is not applicable but coinciding with it whenever it is applicable.

8.5 REDUCIBILITY THESES

Strong claims have been made in favour of the definability of monadic value predicates in terms of preference. Brogan claimed that "all value facts are facts about betterness."[14] Edwin Mitchell maintained that "[v]alue judgments . . . have the form '*A* is better than *B*,' or they can be reduced to this form."[15] I propose to call Mitchell's assertion the *reducibility thesis*. It comes in two forms:

Unrestricted reducibility thesis. *All monadic value predicates can be defined in terms of preference and indifference.*

Restricted reducibility theses. *The monadic value predicate . . . can be defined in terms of preference and indifference.*

Both Brogan and Mitchell expressly endorsed the unrestricted reducibility thesis.

The results of Sections 8.3 and 8.4 corroborate the reducibility theses for 'good' and 'bad.' These results should not be overinterpreted; definability does not imply conceptual precedence. That we can define 'good' and 'bad' in terms of 'better' (but not the other way around),

[13] For additional conditions, which can be used to ensure that ND holds, see Hansson 1990a.

[14] Brogan 1919, p. 96. [15] Mitchell 1950, p. 114.

once we possess the three concepts, does not necessarily mean that the latter concept is in any important sense simpler, more primitive, or more fundamental, than the other two.[16]

Furthermore, the validity of the restricted reducibility theses for 'good' and 'bad' is *not* sufficient to support the unrestricted reducibility thesis. Although two principal value predicates, 'good' and 'bad,' can be defined in terms of 'better' (as can, of course, 'best' and 'worst'), we have no evidence that all monadic value predicates can be so defined. Admittedly, interesting proposals have been put forward for definitions of some of these value predicates, such as:

very good = good among those that are good[17]
fairly good = good but not very good[18]
fairly good = good among those that are not very good[19]

Elegant as these proposals may be, they are only loosely connected to our intuitions about these value terms. The usage of most predicates of the type 'excellent,' 'very good,' 'quite good,' 'fairly good,' and so on, seems to be so vaguely delimited that they are resistant to solid logical reconstruction.[20] Obviously, a regimented language may be constructed in which all value judgments "have the form '*A* is better than *B*,' or they can be reduced to this form." However, such a regimented language will lack expressions for many of the commonly used value terms in ordinary language.

[16] Linguistic evidence points in the opposite direction. In a wide range of languages, the absolute form of adjectives ("tall") is the basic form, and the comparative form ("taller") is derived from the absolute. In a survey of 123 languages, no instance of the opposite relationship between the two forms was found (Ultan, cited in Klein 1980, p. 41).

[17] Wheeler 1972, esp. pp. 325–326. Wheeler's definition refers to attributives in general, of which 'good' is but an example (cf. p. 310).

[18] Wheeler 1972.

[19] Klein 1980, pp. 24–25.

[20] For a fuzzy definition of predicates of the type 'very good,' see Zadeh 1977, p. 384.

PART II

Norms

9

A Starting Point for Deontic Logic

This and the following four chapters are devoted to deontic logic, the logic of normative sentences. Informal normative notions are quite complex and have meaning components related to agency, possibility, commitment, conditionality, defeasibility, the application of rules, etc. The general strategy chosen here is to first identify instances in which as many as possible of these complicating factors are absent or can reasonably be abstracted from. This will be done in the present chapter. In Chapter 10, the logical investigation of these comparatively simple cases is carried out, and in Chapters 11–13 some of the complicating factors are reintroduced into the analysis.

9.1 WHAT NORMATIVE PREDICATES REFER TO

It is generally recognized that there are three major groups of normative expressions in ordinary language, namely prescriptive, prohibitive, and permissive expressions. In the formal language, they are represented by the corresponding three types of predicates. Here, prescriptive predicates will be denoted by "O" (with indices to differentiate between different such predicates), permissive predicates by "P," and prohibitive predicates by "W." (These are abbreviations of "ought", "permitted," and "wrong.")

We need to determine what type of arguments these normative predicates should be applied to. In nonformalized discourse on norms, words expressing prescriptions, prohibitions, and permissions refer in most cases to actions and other human behaviour. "Roberta is permitted to enter this building" refers to the action consisting of Roberta entering the building in question. Less commonly, normative expressions may refer to states of affairs that are not connected with actions, for example, "There ought to be no earthquakes."

For every action (or behaviour), there is a state of affairs consisting of that action (behaviour) taking place.[1] Therefore, we can take states of affairs as the more general case, under which the other case is subsumed.[2] Furthermore, as was mentioned in Section 5.2, states of affairs can in their turn be represented by sentences:

"Roberta is permitted to enter the building."

↓

"It is permitted that Roberta enter the building."

↓

Pq,

where P is a predicate expressing permission and q the sentence "Roberta enters the building."

This idealization in two steps leads to the representation of norms by predicates that take sentences as arguments. Furthermore, just like (dyadic and monadic) value predicates, these normative predicates will be assumed to allow for the substitution of logical equivalents. Hence, if p and p' are logically equivalent, then Op and Op' are equivalent, and so are Pp and Pp' and Wp and Wp'. As was observed in Section 1.4, this property is a useful simplifying assumption, but it sometimes leads to difficulties in terms of the intuitive interpretation. Its problematic nature is particularly acute in attempts to formalize free-choice permissions, that is, permissions to choose freely among a list of alternatives. In ordinary language, free-choice permission is usually expressed with "or": "You may have either coffee or tea." Some deontic logicians have attempted to formalize free-choice permission in terms of a predicate P that satisfies intersubstitutivity of logical equivalents and such that $P(p \vee q)$ expresses that the agent is free to choose between p and q. This predicate is then required to satisfy properties such as $P(p \vee q) \rightarrow Pp \,\&\, Pq$.[3]

[1] We can leave it open whether or not the action (behaviour) and the state of affairs are identical. To the extent that there is a distinction, it disappears in the present formalism since the two entities are represented by one and the same sentence.

[2] For the purposes of this presentation, it is not necessary to specify the structure of the action sentences. As an example, the language may contain a "do" operator E such that for any agent i, $E_i p$ denotes that i brings it about that p. See Section 13.1.

[3] See von Wright 1968, pp. 21–22; Føllesdal and Hilpinen 1970, pp. 22–23; Kamp 1973; Wolénski 1980; Stenius 1982; and Jennings 1985. It follows from $P(p \vee q) \rightarrow Pp \,\&\, Pq$, by substitution of $p \& q$ for p and $p \& \neg q$ for q, that $Pp \rightarrow P(p \& q)$, which is an extremely implausible property.

Unfortunately, in spite of its immediate appeal, this proposal is seriously defective. A free-choice permission to perform either of the actions *p* and *q* cannot be adequately represented as a property that holds for all sentences that are equivalent to $p \vee q$.[4] If it could, then for any logically equivalent disjunctions $p_1 \vee q_1$ and $p_2 \vee q_2$, it should be (free-choice) permitted to perform p_1 or q_1 if and only if it is (free-choice) permitted to perform p_2 or q_2. To see that this is not the case, let p_1 denote that I take an afternoon walk and carry an umbrella, q_1 that I take an afternoon walk and do not carry an umbrella, p_2 that I take an afternoon walk and shoot a policeman, and q_2 that I take an afternoon walk and do not shoot a policeman. Hopefully, I have a free choice between p_1 or q_1 but not between p_2 or q_2.

Hence, although predicates that satisfy the intersubstitutivity property will be used to represent normative notions, this is done only as a means to obtain a reasonably simple formal structure. Translations between the formal language and its informal counterpart must be performed with caution, always bearing in mind that some properties of informal normative statements may not be expressible in the formal language.

9.2 THE MULTIPLICITY OF NORMATIVE PREDICATES

In ordinary language, prescriptive expressions such as "ought to," "must," "should," "has a duty to," "has an obligation to" differ both in strength and in connotation. As an example of the latter, "obligations" typically derive from promises or agreements, whereas "duties" are associated with roles and offices in organizations and institutions.[5] It does not seem possible to do justice to such variations in connotation in a formal language without making it much too complex. Therefore, differences in connotation will be abstracted from. With respect to connotations, the prescriptive predicates of the formal language can be seen as corresponding to the common core of the various prescriptive expressions of natural language.

[4] A more promising approach is to represent free-choice permission to perform *p* or *q* as a property of the set $\{p,q\}$ rather than of the sentence $p \vee q$. David Makinson's (1984) analysis of disjunctive permission in terms of "checklist conditionals" satisfies this criterion. In this approach, a permission to "α or β" is expressed by the formula $(x)(x = \alpha \vee x = \beta \rightarrow Px)$. In classical first-order logic with identity, this formula is logically equivalent to $P\alpha \,\&\, P\beta$.

[5] Brandt 1965. Cf Mish'alani 1969; Forrester 1975.

The prescriptive words used in informal discourse differ in strength (stringency). It is instructive to compare them in this respect with words that express values. When you say that two things are both good, this does not necessarily mean that they have the same degree of goodness. One of them may be more good (better) than the other. Similarly, when you say that someone ought to do each of two things, it does not follow that the two requirements have equal strength. Moral requirements differ in stringency.[6] So do the words and phrases that we use to express them. "Must" is more stringent than "ought," and "ought" is more stringent than "should."[7]

This variability in the strength of both values and norms is an essential feature of any moral language that is at all suitable to treat the complexities of real life. We should be able to say: "That is a good action, but I would not say that it is *very* good." Similarly, we should be able to say: "This is something that you should do, but it would be wrong to say that you *must* do it."

It should be no surprise if a formal model that allows for only one degree of goodness is incapable of representing intuitive notions of value in a reasonable way. It would be a mistake to attribute any deficiencies found in such a model to the informal concept of goodness, unless they are robust to the introduction of degrees of goodness. An analogous argument applies to the deficiencies found in formal languages that allow for only one degree of moral requirement.

Unfortunately, most deontic logics, including standard deontic logic,[8] have exactly this all-or-nothing approach to moral requirements: An action is either prescribed with all the stringency that the system can muster, or it is not prescribed at all. A more realistic system of deontic logic should allow us to distinguish between the "ought" of "You ought to buy flowers for her" and the "must" of "You must throw out the lifebuoy to her."

[6] I take the meaning of this to be fairly obvious on an intuitive level. Every child knows the difference in stringency between the two requirements "do not speak with food in your mouth" and "do not erase the hard disk on mum's computer." There are various ways to explicate or codify the difference. The more stringent requirement is more important, its violation is worse, and it is less easily overridden (Chisholm 1964, p. 151).

[7] Guendling 1974. On differences in strength between moral requirements, see also Ladd 1957, p. 125; Sloman 1970, p. 391; Harman 1977, pp. 117–118; Jones and Pörn 1985; Meyer 1987, p. 87; Garcia 1989; Brown 1996.

[8] To be introduced in Section 10.1.

In accordance with this requirement, the formal language to be introduced here allows for prescriptive predicates of different strengths, distinguished by subscripts (O_x, O_y, etc.). Indices will be omitted whenever convenient. For similar reasons, prohibitive and permissive predicates will be assumed to differ in strength but not in connotation. Subscripts will be used in the same way to distinguish between different permissive and prohibitive predicates (P_x, P_y . . . , and W_x, W_y, etc.).

This multiplicity of predicates gives us reason to reconsider the relation between the three types of predicates. In traditional deontic logic, which has only one predicate of each type, the prohibitive predicate is assumed to correspond to the prescriptive predicate followed by a negation. Thus, "it is wrong to do *X*" is equated with "*X* ought not to be done."[9] Similarly, the permissive predicate is assumed to connect with the prescriptive predicate in such a way that an action may be performed (is permitted) if and only if it is not morally required not to perform it.[10]

This interrelation between prescription and permission relies on a further deviation from extraphilosophical usage. In ordinary language, "when saying that an action is permitted we mean that one is at liberty to perform it, that one may either perform the action or refrain from performing it."[11] We may call this *bilateral* permission.[12] In formal philosophy, however, "being permitted to perform an action is compatible with having to perform it."[13] Hence, Pq denotes that q (but not necessarily $\neg q$) is allowed. We may call this *unilateral* permission. The use of unilateral permission in regimented and formalized philosophical terminology is a convenient convention, since bilateral permission can be straightforwardly defined in terms of unilateral permission (as Pq & $P\neg q$), whereas no definition in the other direction is immediately available.[14] The unilateral interpretation of permission will be adopted here. With it, the interdefinability of prescription and permission is an

[9] Moore 1912, pp. 85–86; Stevenson 1944, pp. 99–100; von Wright 1951, p. 3.
[10] von Wright 1951, p. 3.
[11] Raz 1975, p. 161.
[12] It has also been called "facultative" permission (Alchourrón 1993, p. 55) and "optional" permission (Castañeda 1981, p. 76).
[13] Raz 1975, p. 161.
[14] Let P_1 denote unilateral and P_2 bilateral permission, and let O denote obligation. Then, as was pointed out to me by Wlodek Rabinowicz, we may define P_1p as $P_2p \vee Op$, but neither O nor P_1 is definable in terms of P_2 alone.

acceptable approximation. So is the interdefinability of prescription and prohibition.

It remains to transfer this interdefinability from the conventional models with only one predicate of each type to models that allow for several predicates of each type. The simplest and most principled solution is to require interdefinability at all levels of stringency. More precisely, for each prescriptive predicate O we will assume that there is a prohibitive predicate W and a permissive predicate P such that for any sentence q, Oq holds if and only if $W\neg q$, and it also holds if and only if $\neg P\neg q$. The corresponding interdefinability properties are assumed to hold for prohibitive and permissive predicates. For each W, there are O and P such that Wq holds if and only if $O\neg q$ and if and only if $\neg Pq$. For each P, there are O and W such that Pq holds if and only if $\neg O\neg q$ and if and only if $\neg Wq$.

9.3 DEONTIC ALTERNATIVES

The distinction between exclusionary and combinative preferences was essential in our development of preference logic. The relata of an exclusionary preference relation form a set of mutually exclusive alternatives, which is not the case for a combinative preference relation.

In general, normative discourse refers to compatible alternatives. As an example, consider the following series of admonitions to my eldest son a few days ago: "You must be back at six. You may invite your friend to have dinner with us. You are allowed to cross the street in front of the grocery store, but you are not allowed to cross the big street outside the park." The actions referred to in these normative expressions are independent in a logical sense and can be combined in various ways. As can easily be corroborated with other examples, this is typical for normative discourse. Therefore, deontic logic should be combinative.

In spite of this, mutually exclusive alternatives can be as useful in deontic logic as we have seen them to be in the logic of combinative preferences. Here as well, the formal treatment is substantially simplified if we assume that there is an underlying set of mutually exclusive alternatives, such that the normative appraisal of compatible alternatives should cohere with some reasonable appraisal of the elements of this alternative set.

As an example of this, when I allowed my son to cross the street in front of the grocery store, I did not mean that he was allowed to do this irrespective of his other actions. A better interpretation is that among the complete alternatives for his behaviour during the afternoon, there is at least one that is permitted and that includes him crossing the street in question.

Reconstructibility in terms of mutually exclusive alternatives can be seen as a consequence of the requirement of a reflective equilibrium (cf. Section 2.1). In ordinary deontic discourse, this requirement is not always satisfied. As was indicated by Aaron Sloman, most normative statements are uttered without a clear reference to an alternative set. However, "the less clear one is about which range of alternatives is under consideration when one is singled out as best, or what ought to be, the less clear it is what the implications are of what one is saying."[15] For the purposes of philosophical explication, it is expedient to regard normative sentences that are vague in this respect as ambiguous between different alternative sets.

In formalized deontic discourse, the alternative set represents that which is subject to normative appraisal. Hence, if both an action sentence and its negation are represented in the mutually exclusive deontic alternative set $\mathcal{A}$ (i.e., if they are both elements of $\cup\mathcal{A}$), then this means that the action represented by this sentence is treated as decision-theoretically unsettled. The contents of the alternative set will, of course, depend to a large extent on the factual and epistemic situation, that is, on what the facts are and what is known about them.[16] However, the situation (in this sense) does not unequivocally determine what states of affairs are taken into consideration in normative discourse. One and the same situation may be viewed from different *perspectives*. 'Perspective' will be used here as a technical term for that which determines what states of affairs are taken into consideration in the appraisal of a given situation.

One important source of the differentiation of perspectives is that a situation may be evaluated with respect to different types of events. It may be evaluated with respect to possible events in general (including actions). Alternatively, it may be evaluated exclusively with respect to actions, typically to actions by one (or several) specified agent(s). Perspectives that are restricted to actions may be called *agency*

[15] Sloman 1970, p. 387.

[16] A situation may be represented in formal language by a set of sentences. The introduction of this formalization is deferred to Section 12.3, since it is not needed until then.

perspectives. Since norms are typically action-guiding, these are the most common type of perspectives in normative discourse.[17] If a situation involves more than one possible agent, this gives rise to further differentiations. In an agency perspective of an individual *i*, the alternative set consists of sentences that represent actions by *i*. In a (collective) agency perspective of *i* and *j*, the alternative set consists of sentences that represent actions by *i* and/or *j*.[18] The normative status of a course of action may very well be different in different agency perspectives.[19]

Another major source of the differentiation of perspectives is that normative discourse is in most cases restricted to alternatives that are considered to be in some sense possible. Different criteria of possibility can be chosen by rational agents, and we often shift rapidly between such perspectives.[20] As one example, I can say of a drug addict that he ought to use sterile needles. When saying this, I take his drug abuse for granted. Soon afterward, I may say that he ought not to take drugs at all, thus shifting the discoursative boundary between settled and unsettled components of his behaviour. When I made the first statement, no alternative in which he refrained from injecting was included among the set of alternatives. In the second utterance, such an alternative was included.

An agent may legitimately exclude options from further consideration in order to economize in deliberation. Such cognitive decisions are often expressed in terms of possibility and give rise to "the frequent claim that one *cannot* do a certain thing because one has already decided to do something else, and not because one's will would not be

[17] The exceptions are "ideal" ought statements, which "are not prescriptive at all, either prudentially or morally, but express valuations. Such is 'Everybody ought to be happy.' This is not a prescription or command to anybody to act or to refrain." Robinson 1971, p. 195.

[18] "There seems to be a perfectly intelligible sense of 'alternative' in which one particular action may be an alternative to another even though they are not agent-identical. For example, two conspirators may both offer to shoot the president on a certain occasion, and they may be interested in finding out which one of these alternatives is to be preferred from a teleological point of view." Bergström 1966, p. 30.

[19] It is important to distinguish between agency perspectives and evaluator-relativity. Different agency perspectives give rise to different sets of alternatives, whereas evaluator-relativity concerns the existence of more than one legitimate evaluation of one and the same set of alternatives. On evaluator-relativity, see Sen 1982, 1983; Regan 1983; and Garcia 1986.

[20] Some such shifts are discussed in Chapter 11.

efficacious as regards that act."[21] The multiplicity of perspectives increases further when one deliberates after the fact on lost possibilities. As was pointed out by Lars Bergström, the alternatives to Brutus's murder of Caesar may be "those which Brutus actually considered, or those which he ought to have considered, or those which he could have considered."[22]

In summary, there is abundant scope for having different perspectives on the same situation and for appraising it in different ways. Mutually exclusive alternatives are useful as a formal device to specify and clarify some of these differences.

9.4 DELINEATING SITUATIONIST DEONTIC LOGIC

Many normative expressions compare or summarize what obtains in different situations or perspectives:

(1) "If your rich brother had at least paid your mother's hospital bills, then you would not have been obligated to send money to her."
(2) "If you borrow money, then you must pay it back."
(3) "You are not allowed to be cruel to animals."

Example (1) refers counterfactually to a situation that does not obtain at present and to a normative appraisal of (some suitable perspective on) that situation. Examples (2) and (3) are rules in the sense of applying to normative situations (and perspectives) in general, rather than to a particular (actual or counterfactual) situation.

It is essential to distinguish between norms that refer to situations in general [such as (3)] and norms that refer to a particular situation.[23] Unfortunately, this distinction is not obvious, since the English language (like many others) employs the same linguistic forms for both purposes. It is a common practice in deontic logic to follow natural language in this respect and to use the same symbolic form (Op) to express that something (p) is obligatory in the present situation and that it is so in general. Although common, this practice conceals

[21] Kapitan 1986, p. 246.

[22] Bergström 1966, p. 7.

[23] This distinction was made in Hansson 1988b. Similarly, Alchourrón (1993, p. 62) distinguished between "a norm for a single possible circumstance (which may be the actual circumstance)" and a norm for "*all possible circumstances.*" David Makinson (1999) distinguishes between norms "in all circumstances" and norms "in present circumstances."

important distinctions that should be highlighted rather than obscured in logical analysis.

Many if not most normative statements refer either to more than one situation or to some hypothetical situation that differs in specified ways from what obtains at present. Much of the complexity in deontic logic is connected with this variability in situations (and perspectives). As was indicated at the beginning of this chapter, we need to identify, as a starting point for deontic logic, a fragment of deontic discourse in which much of this complexity can be left out of consideration. This can be done by selecting those cases in which only one alternative set and one moral or legal appraisal of it need to be considered. More precisely, we will begin with *the deontic logic of that fraction of normative discourse that refers only to one moral appraisal of one situation, in one perspective.* In the intended interpretation, the situation referred to will be one that actually obtains. No changes in the situation or shifts in the perspective are allowed, and no general deontic statements (such as rules) will be represented. This restricted variant of deontic logic will be called *situationist* deontic logic.[24] It is the subject of Chapter 10. More complex, transsituational normative structures will be treated in Chapters 11–13.

[24] The term is not fully precise, since it only mentions the situation and not the perspective that is also kept constant.

10

Situationist Deontic Logic

The purpose of this chapter is to investigate the central fraction of deontic discourse that was dissected out in the previous chapter. In Section 10.1, the dominant approach to deontic logic, "standard deontic logic," is shown to suffer from deficiencies that are closely related to its basic semantic construction. This motivates the search for another semantic construction in which these deficiencies can be avoided. In Secion 10.2, such a construction is proposed. It provides us with a set of prescriptive predicates with differing stringency. In Section 10.3, the logic of this construction is investigated, and in Section 10.4 a series of representation theorems is presented. In Sections 10.5 and 10.6, some more specified deontic predicates are constructed within the same framework, and their properties are investigated.

10.1 STANDARD DEONTIC LOGIC (SDL)

Modern deontic logic began with a seminal paper by Georg Henrik von Wright in 1951.[1] With a minor modification,[2] his list of postulates has turned out to be characterizable by a simple semantical construction that has long dominated the subject: It is assumed that there is a subset of the set of possible worlds (the "ideal worlds") such that for any sentence p, Op holds if and only if p holds in all these worlds. Although there is some leeway in the meaning of the term *standard deontic logic* (SDL), the following definition seems to capture the gist of the matter:[3]

[1] von Wright 1951. On the origins of deontic logic, see Føllesdal and Hilpinen 1970 and von Wright 1998.

[2] Acceptance of the postulate $O(p \vee \neg p)$.

[3] According to Føllesdal and Hilpinen (1970, p. 34), the term 'standard deontic logic' was introduced by Bengt Hansson (1969).

Definition 10.1. *A model* $\langle \mathcal{A}, I \rangle$ for noniterative standard deontic logic *(noniterative SDL) consists of a set* $\mathcal{A}$ *of contextually complete alternatives and a nonempty subset* I *of* $\mathcal{A}$.

A noniterative deontic sentence *in* $\langle \mathcal{A}, I \rangle$ *is a truth-functional combination of sentences of the form* $O\alpha$*, with* $\alpha \in \mathcal{L}_{\mathcal{A}}$*. Such a sentence is* true *in* $\langle \mathcal{A}, I \rangle$ *if and only if it follows by classical truth-functional logic from the set* $\{O\alpha \mid \alpha \in \cap I\} \cup \{\neg O\alpha \mid \alpha \notin \cap I\}$*. It is* valid *if and only if it is true in all models.*

The valid sentences of noniterative SDL coincide with the theorems derivable from the following three axioms:

$$Op \rightarrow \neg O \neg p,$$
$$Op \,\&\, Oq \leftrightarrow O(p \,\&\, q), \text{ and}$$
$$O(p \vee \neg p).$$[4]

The term 'noniterative' refers to the fact that sentences containing iterations of the deontic predicate (such as OOp and $\neg O(Op \vee Oq)$) have been excluded. To cover them, modal semantics (with an accessibility relation) can be used.[5]

In many, probably most, accounts of standard deontic logic, the alternative set ($\mathcal{A}$) is assumed to consist of possible worlds. It is necessary to distinguish between those properties of SDL (or any other deontic logic) that depend on the choice of an alternative set and those that depend on the method used to assign normative status to the sentences represented in the chosen alternative set.

There is a crucial difference between possible worlds and the alternatives recommended in Section 9.3 for situationist deontic logic: The agent cannot choose between the possible worlds. This gives rise to several well-known drawbacks of SDL. One of these is that SDL cannot account for obligations that result from the neglect of other obligations. In ideal possible worlds, neither racism nor violence would exist, and therefore, in these worlds nobody would fight racism or help the victims of violence. It follows that there can be no duty to do either of these things.[6] Furthermore, SDL is unable to assign obligations to

[4] Føllesdal and Hilpinen 1970, p. 13.

[5] Ibid., pp. 15–19.

[6] Of course, a possible world may be counted as ideal in the technical sense of Definition 10.1 without being so in the usual sense of the word 'ideal.' Hence, violence, etc. may persist in some of the (technically) ideal worlds. Unfortunately, this move

individuals who would not exist in all the ideal worlds. Suppose that John was conceived as the result of a rape. Since the rape does not have obligatory status, there is at least one ideal world in which it has not been committed, and hence John does not exist in that world. Since there is nothing that John does (or refrains from doing) in all deontically perfect worlds, he is, according to SDL, subject to no obligations at all. He may do whatever he pleases.

These problems for SDL can be avoided by the choice of more realistic alternative sets. Instead of applying the logical structure of SDL to possible worlds, we can apply it to a set of alternatives that are open to the agent. The existence of racism and violence are circumstances beyond my control and should not be represented in an alternative set that is used to appraise various courses of action for me.[7] Indeed, in the best of the alternatives open to me, I fight racism and help victims of violence. Furthermore, in all the alternatives open to John, he exists,[8] so the argument that absolves him from all moral requirements does not go through.

Unfortunately, standard deontic logic also has other, serious deficiencies that are independent of the choice of an alternative set. It is an immediate consequence of the basic semantic idea of SDL – that of identifying obligatory status with presence in all elements of a certain subset of the alternative set – that the following property will hold:

$$\text{If } \vdash p \rightarrow q, \text{ then } \vdash Op \rightarrow Oq.$$

This property will be called *necessitation* since it says that whatever is necessitated by a moral requirement is itself a moral requirement.[9] As an example, suppose that I am morally required to take a boat without the consent of its owner and use it to rescue a drowning person. Let p denote this composite action that I am required to perform, and let q

does not help us. Let v denote that violence exists and h that I help victims of violence. If Oh holds, then h is true in all (technically) ideal worlds. Since h can only be true in worlds in which v is true, it follows that v holds in all (technically) ideal worlds, hence Ov holds. (Extensions of SDL can account for *conditional* norms of the type "if v then Oh." See Bengt Hansson 1969.)

[7] I.e., the corresponding sentences should not be elements of $\mathcal{L}_{\mathcal{A}}$.

[8] At least he does so at the time when the alternatives are open to him.

[9] It has many names, including "the inheritance principle" (Vermazen 1977, p. 14), "Becker's law" (McArthur 1981, p. 149), "transmission" (Routley and Plumwood 1984), "the consequence principle" (Hilpinen 1985, p. 191), and "entailment" (Jackson 1985, p. 178).

denote the part of it that consists in taking the boat without permission. Since q follows logically from p, I am logically necessitated to perform q in order to perform p. According to the postulate of necessitation, I then have an obligation to q. This is contestable, since I have no obligation to q in isolation.

Necessitation gives rise to most of the major deontic paradoxes. We may call these the *necessitation paradoxes*. Four of the most prominent are Ross's paradox, the paradox of commitment, the Good Samaritan, and the Knower. Ross's paradox is based on the instance $Op \rightarrow O(p \vee q)$ of necessitation. ("If you ought to mail the letter, then you ought to either mail or burn it.")[10] The paradox of commitment is based on the instance $O\neg p \rightarrow O(p \rightarrow q)$, which is interpreted as saying that if you do what is forbidden, then you are required to do anything whatsoever. ("If it is forbidden for you to steal this car, then if you steal it you ought to run over a pedestrian.")[11] The Good Samaritan operates on two sentences p and q, such that q denotes some atrocity and p some good act that can only take place if q has taken place. We then have $\vdash p \rightarrow q$, and it follows by necessitation that if Op then Oq. ("You ought to help the assaulted person. Therefore, there ought to be an assaulted person.")[12] Lennart Åqvist's Knower paradox makes use of the epistemic principle that only that which is true can be known. Here, q denotes some wrongful action, and p denotes that q is known by someone who is required to know it. Again, we have $\vdash p \rightarrow q$ and Op, and it follows by necessitation that Oq. ("If the police officer ought to know that Smith robbed Jones, then Smith ought to rob Jones.")[13]

The role of necessitation in the deontic paradoxes was summarized by von Wright when he concluded that "in a deontic logic which rejects the implication from left to right in the equivalence $O(p \& q) \leftrightarrow Op \ \& \ Oq$ while retaining the implication from right to left, the 'paradoxes' would not appear."[14] Given the intersubstitutivity of logically equivalent sentences, the left-to-right direction of that equivalence is logically equivalent with necessitation.

[10] Ross 1941, p. 62. Arguably, the "or" of the consequent is not truth-functional, but rather of the free-choice variant (see Section 9.1). Therefore, it is tempting to believe that this paradox depends only on the non–truth-functional properties of the "or" of English and other natural languages. However, as should be clear from the above example with the boat, the paradox persists if $p \vee q$ is replaced by a nondisjunctive logical consequence of p.

[11] Prior 1954. [12] Prior 1958, p. 144. [13] Åqvist 1967. [14] von Wright 1981, p. 7.

Deontic logic has on many occasions been at least as much concerned with paradoxes of its own creation as with issues that are relevant in the analysis of informal normative discourse. To avoid the paradoxes, the necessitation property must also be avoided. As we have seen, this in its turn requires that we give up the central semantic idea of standard deontic logic.

10.2 PREFERENCE-BASED DEONTIC LOGIC

The new semantic construction that we are searching for should allow for normative predicates of differing strengths (cf. Section 9.2). It is therefore natural to conceive them as connected with some underlying binary relation in about the same way as monadic value predicates are connected with an underlying preference relation (cf. Chapter 8).

Furthermore, in order to avoid unnecessary complications, the obvious first option is *not* to construct a new type of binary relation for the normative predicates, but instead to use a preference relation for them as well. In the rest of this chapter, I will attempt to show that such a preference-based deontic logic can provide a plausible account of normative discourse. This construction will also have the advantage of including representations of evaluative and normative notions in one and the same framework.

Before we embark upon this project, an obvious objection should be attended to: It may be argued that norms cannot be expressed in terms of values for the simple reason that a norm sentence and a value sentence cannot have the same meaning. For one thing, norms are typically action-guiding in a way that values are not. However, as was observed already by George Moore, a norm sentence and a value sentence may be equivalent in either an extensional or an intensional sense.[15] Extensional equivalence does not require identity of meaning and is therefore much more plausible than intensional equivalence.[16] Extensional equivalence is sufficient for our present semantical purposes, and nothing more will be sought for. With this proviso, attempts to explicate deontic logic in terms of preference logic need not be condemned beforehand, but should instead be judged according to how reasonable is the outcome of the analysis.

[15] Moore 1912, pp. 172–173. [16] Cf. Hansson 1991b.

There are two major proposals in the literature for defining normative predicates in value terms: (1) to equate 'ought' with 'best' and (2) to equate 'ought' with 'good.'[17] The former of these proposals is part and parcel of utilitarian moral philosophy. Since both 'best' and 'good' are positive predicates, in the sense explained in Section 8.1, a natural generalization of these proposals is to demand that *prescriptive predicates should satisfy positivity*. Another way of saying this is that whatever is better than or equal in value to something morally required is itself morally required.

Similarly, we can stipulate that whatever is at least as good as something permitted is itself permitted, or in other words that *permissive predicates should satisfy positivity*. For the third category of normative predicates, the converse requirement is the appropriate one: We can demand that *prohibitive predicates should satisfy negativity.*

At first sight, these three conditions on normative predicates may all seem reasonable. They are interrelated as follows:

Definition 10.2. *A (monadic) predicate H is* contranegative *with respect to a given relation $\geq'$ if and only if the following holds for all p and q:*

$$Hp \ \& \ (\neg p) \geq' (\neg q) \rightarrow Hq.$$

Observation 10.3. *Let O, P, and W be predicates with a common domain that is closed under negation, and such that for all p, Op if and only if $\neg P \neg p$, and Op if and only if $W\neg p$. Let $\geq'$ be a relation over this domain. Then the following three conditions are equivalent:*

(1) O satisfies $\geq'$-contranegativity,
(2) P satisfies $\geq'$-positivity, and
(3) W satisfies $\geq'$-negativity.

Hence, the conditions that *P* be positive and *W* negative can be combined into one single condition. We therefore now have two conditions on the connections between normative predicates and their underlying combinative preference relations. Focusing on the prescriptive predicates, they can be named as follows:

The positivity thesis. *Prescriptive predicates satisfy positivity.*

[17] On (1), see Moore 1903, p. 147; Bergström 1966, p. 91; and Sloman 1970, p. 388. On (2), see Gupta 1959 and von Kutschera 1975.

The contranegativity thesis. *Prescriptive predicates satisfy contranegativity (and consequently permissive predicates satisfy positivity and prohibitive predicates satisfy negativity).*[18]

Since both of these theses have intuitive appeal, one might wish them both to be complied with. Unfortunately, their combined effect has implausible consequences, as will be seen from Observations 10.4 and 10.6.

Observation 10.4. *Let O be a predicate with a domain that is closed under negation, and let $\geq'$ be a binary relation over that domain. Furthermore, let O be both $\geq'$-positive and $\geq'$-contranegative. Then: If $p >' q$ and $\neg p \geq' \neg q$, then Op if and only if Oq.*

For an example, let q denote that you give your hungry visitor something to eat and p that you serve her a gourmet meal. It is quite plausible to value p higher than q but nevertheless value $\neg p$ (which is compatible with serving her some food) at least as high as $\neg q$ (which means that you let her starve).[19] At any rate, this value assignment should not prevent us from maintaining that q is in some sense morally required without necessarily implying that p is also morally required (in that same sense). According to Observation 10.4, this is not possible if both the positivity and the contranegativity theses are required to hold.

We need an additional definition to prepare for our second observation on the combined effects of positivity and contranegativity:

Definition 10.5. *Let $\geq'$ be a relation over a domain that is closed under negation. Let $\mathcal{H}$ be a set of predicates in this same domain. Then $\mathcal{H}$ is a* fine-grained $\geq'$-positive set *if and only if (1) each element of $\mathcal{H}$ satisfies $\geq'$-positivity, and (2) for all elements p and q of the common domain, if $p >' q$, then there is some predicate $O \in \mathcal{H}$ such that Op and $\neg Oq$.*

Observation 10.6. *Let $\geq'$ be a complete relation over a domain that is closed under negation. Furthermore, let there be a fine-grained $\geq'$-positive set $\mathcal{H}$, all elements of which satisfy $\geq'$-contranegativity. Then $\geq'$ satisfies contraposition of strict preference (if $p >' q$, then $\neg q >' \neg p$).*

[18] Hansson 1991b.

[19] Cf. the counterexample to contraposition of strict preference in Section 6.6.

Admittedly, this observation is not strictly speaking a proof that positivity and contranegativity are incompatible. What it does show is, intuitively speaking, that we have to either (1) accept contraposition of strict preference, (2) give up fine-grainedness, (3) give up the positivity thesis, or (4) give up the contranegativity thesis. The first option is unsatisfactory in view of the implausibility of contraposition, as shown in Section 6.6 and in the above gourmet dinner example. The second option is equally unattractive. Something seems to be seriously wrong with a normative structure that does not allow for the introduction of additional degrees of stringency. We should therefore take a closer look at the positivity and contranegativity theses.

It turns out that the positivity thesis is not at all as plausible as it might first seem. Indeed, the gourmet dinner example provides a clear counterexample. Although serving a fine dinner is (under appropriate assumptions) better than just offering something to eat, the former need not be morally required if the latter is so. This is an instance of a general format for counterexamples against the positivity of prescriptive predicates: Let p represent something morally required and q a supererogatory variant of p. There is also another general format for such counterexamples: Let p represent something morally required and q a variant of p that is specified in some morally irrelevant way. As an example of the latter format, we can let p signify that I pay my debt to Adam and q that I pay this debt in banknotes that have odd serial numbers. For another example, let p denote that I visit my sick aunt and q that I do this, entering her flat with my left foot first. A reasonable deontic logic should allow us to have $Op, p \equiv' q$, and $\neg Oq$, which contradicts the positivity thesis.[20]

Examples such as these strongly suggest that prescriptive predicates should not in general be assumed to satisfy positivity. Should they satisfy contranegativity? An obvious approach to this question is to try to construct counterexamples against contranegativity according to patterns that parallel the two formats used above in counterexamples against positivity.

The first of these formats employs a morally better variant of a required action, such as the gourmet dinner in the example given above. A corresponding counterexample to the negativity of prohibi-

[20] This is in accordance with SDL. In this example, p will be an element of all the ideal alternatives, whereas q will be an element of only some of them. Hence, Op and $\neg Oq$.

tive predicates would have to make use of a morally inferior variant of a prohibited action. However, no convincing counterexample seems to be constructible in this way. Stealing is prohibited, and stealing from the poor is worse; that it is worse is certainly no reason why it should not be prohibited. In particular, there is no mirror image of supererogatory actions at the other end of the value scale.[21]

The second format refers to a prescribed action and some version of it that is specified in a morally irrelevant way. In order to try out the same construction on the contranegativity thesis, let W be a prohibitive predicate, p a sentence representing an action prohibited according to W, and q a variant of p that is specified in some morally irrelevant way. For concreteness, we can let p denote that I enter, in breeding season, a bird sanctuary to which I have no access, and q that I do this, entering with my left foot first. Then, with any reasonable moral assessment, if Wp then Wq.[22] No counterexample seems to be constructible in this way.

It is interesting to note in this context that SDL can be reconstructed in terms of a preference relation with respect to which the prescriptive predicate is contranegative but not positive.

Observation 10.7. *Let $\langle \mathcal{A}, I\rangle$ be a model for noniterative standard deontic logic and O its associated prescriptive predicate. Let $\geq$ be a complete and transitive relation over $\mathcal{A}$ that has I as its maximal element. Let $\geq_x$ be the maximax preference relation over $\mathcal{L}_{\mathcal{A}}$ that is based on $\geq$. Then:*

(1) O is $\geq_x$-contranegative, but
(2) it does not hold in general that O is $\geq_x$-positive.

The contranegativity thesis cannot be proved. It can, however, be corroborated by examples and by the lack of counterexamples. It can therefore be accepted on a preliminary basis, or at least adopted as a hypothesis to be tested. The ultimate criterion for its acceptability

[21] Arguably, it is worse to buy and burn a Stradivarius violin than to steal it and return it to its owner two months later. Nevertheless, the second but not the first of these two behaviours is prohibited. This may seem like a counterexample to the negativity of 'prohibited,' but it is not a true counterexample since it is based on an equivocation of moral and legal concepts. To buy and burn a Strad is prohibited by the moral standard according to which it is worse than to steal and return the instrument.

[22] SDL concurs with moral intuition in this example, since Wq follows from Wp and $\vdash q \rightarrow p$.

should of course be whether or not a plausible deontic logic can be based on it. In particular, can the well-known counterintuitive results in SDL be avoided in a deontic logic that still satisfies contranegativity? In order to answer this question, we need to investigate the logical properties of contranegative predicates. This will be done in the rest of this chapter. For the sake of completeness, the properties of positive predicates will be studied in parallel.

10.3 THE LOGIC OF POSITIVE AND CONTRANEGATIVE PREDICATES

There are many types of combinative preference relations and many types of contranegative (or positive) predicates with which they can be combined to obtain deontic logics with different properties. In this section, the focus is on some of the postulates that are well known from the literature on deontic logic. For each of these postulates, we are going to search for a property of the combinative preference relation $\geq'$ that is sufficient and necessary to guarantee that all $\geq'$-contranegative ($\geq'$-positive) predicates satisfy the postulate in question.

As was mentioned above, standard deontic logic has been axiomatized with the three axioms

$$Op \rightarrow \neg O \neg p,$$
$$Op \,\&\, Oq \leftrightarrow O(p \,\&\, q), \text{ and}$$
$$O(p \vee \neg p).$$

The left-to-right direction of the second axiom, $Op \,\&\, Oq \rightarrow O(p\&q)$, was called "agglomeration" by Bernard Williams.[23] A clarifying analysis of agglomeration was put forth by Walter Sinnott-Armstrong, who showed that it may hold for some but not all senses of 'ought.' Generally speaking, agglomeration is a more plausible property for 'ought' taken in the sense of "there is an overriding moral reason to" than for 'ought' taken in defeasible senses such as "there is a moral reason to."[24]

[23] Williams 1965, p. 118. It has also been called "aggregation" (Schotch and Jennings 1981, p. 152).

[24] Sinnott-Armstrong 1988, pp. 126–135. Essentially the same conclusion was reached, though after a less thorough analysis, in Hansson 1988b, pp. 344–345.

Agglomeration can be shown to hold for the contranegative predicates of a wide range of preference relations:

Theorem 10.8. *Let $\mathcal{A}$ be a set of contextually complete alternatives. The following are two conditions on a relation $\geq'$ in $\mathcal{L}_{\mathcal{A}}$:*

(1) $(p\geq'^*(p\vee q)) \vee (q\geq'^*(p\vee q))$

(2) Every $\geq'$-contranegative predicate O on $\mathcal{L}_{\mathcal{A}}$ satisfies agglomeration $(Op \,\&\, Oq \rightarrow O(p\&q))$.

If (1) holds, then so does (2). If $\geq'$ satisfies completeness, then (1) and (2) are equivalent.

Observation 10.9. *Let $\geq$ be a complete and transitive relation over the finite and contextually complete alternative set $\mathcal{A}$, and let $\geq'$ be a transitive relation on $\mathcal{L}_{\mathcal{A}}$ that satisfies positionality and nonnegative response*[25] *with respect to $\geq$. Furthermore, let O be a $\geq'$-contranegative predicate on $\mathcal{L}_{\mathcal{A}}$. Then O satisfies agglomeration.*

Observation 10.10. *Let $\geq$ be a complete and transitive relation over the finite and contextually complete alternative set $\mathcal{A}$. Furthermore, let O be a predicate on $\mathcal{L}_{\mathcal{A}}$ that is contranegative with respect to either $\geq_i$, $\geq_x$, $\geq_{ix}$, $\geq_{xi}$ or some max-min weighted preference relation $\geq_E$. Then O satisfies agglomeration.*

Agglomeration of positive predicates is associated with somewhat more demanding requirements on the preference relation:

Theorem 10.11. *Let $\mathcal{A}$ be a set of contextually complete alternatives. The following are two conditions on a relation $\geq'$ in $\mathcal{L}_{\mathcal{A}}$:*

(1) $((p\&q)\geq'^*p) \vee ((p\&q)\geq'^*q)$

(2) Every $\geq'$-positive predicate O in $\mathcal{L}_{\mathcal{A}}$ satisfies agglomeration.

If (1) holds, then so does (2). If $\geq'$ satisfies completeness, then (1) and (2) are equivalent.

Observation 10.12. *Let $\mathcal{A}$ be a contextually complete alternative set, and let $\geq$ be a complete and transitive relation on $\mathcal{A}$. Then:*

[25] Defined in Section 7.3.

(1) Every $\geq_i$-positive predicate on $\mathcal{L}_\mathcal{A}$ satisfies agglomeration.
(2) It does not hold in general that every $\geq_x$-positive predicate on $\mathcal{L}_\mathcal{A}$ satisfies agglomeration.
(3) It does not hold in general that every $\geq_{ix}$-positive predicate on $\mathcal{L}_\mathcal{A}$ satisfies agglomeration.
(4) It does not hold in general that every $\geq_{xi}$-positive predicate on $\mathcal{L}_\mathcal{A}$ satisfies agglomeration.
(5) It does not hold in general that every predicate on $\mathcal{L}_\mathcal{A}$ that is $\geq_E$-positive for some max-min weighted preference relation $\geq_E$ satisfies agglomeration.

The other direction of the second SDL axiom says, essentially:

$$O(p \& q) \rightarrow Op$$

It has been discussed in this form as a principle of "division of duties."[26] As was noted above, it is equivalent with the postulate of necessitation:

$$\text{If } \vdash p \rightarrow q, \text{ then } \vdash Op \rightarrow Oq$$

The following theorem refers, for technical reasons, to a somewhat stronger variant of necessitation in which $\vdash$ has been replaced by $\models_\mathcal{A}$:

Theorem 10.13. *Let $\mathcal{A}$ be a set of contextually complete alternatives. The following are three conditions on a relation $\geq'$ in $\mathcal{L}_\mathcal{A}$:*

(1) If $\models_\mathcal{A} q \rightarrow p$, then $p \geq'^* q$.
(2) Every $\geq'$-contranegative predicate O on $\mathcal{L}_\mathcal{A}$ satisfies strong necessitation (If $\models_\mathcal{A} p \rightarrow q$, then $Op \rightarrow Oq$).
(3) Every $\geq'$-positive predicate O on $\mathcal{L}_\mathcal{A}$ satisfies strong necessitation.

If (1) holds, then so do (2) and (3). If $\geq'$ satisfies ancestral reflexivity ($p \geq'^* p$), then (1), (2), and (3) are equivalent.

Observation 10.14. *Let $\mathcal{A}$ be a contextually complete alternative set, and let $\geq$ be a complete and transitive relation on $\mathcal{A}$. Then:*

(1) It does not hold in general that every $\geq_i$-contranegative or every $\geq_i$-positive predicate on $\mathcal{L}_\mathcal{A}$ satisfies necessitation.

[26] Hansson 1988b, p. 345.

(2a) *Every $\geq_x$-contranegative predicate on $\mathcal{L}_{\mathcal{A}}$ satisfies necessitation.*
(2b) *Every $\geq_x$-positive predicate on $\mathcal{L}_{\mathcal{A}}$ satisfies necessitation.*
(3) *It does not hold in general that every $\geq_{ix}$-contranegative or every $\geq_{ix}$-positive predicate on $\mathcal{L}_{\mathcal{A}}$ satisfies necessitation.*
(4) *It does not hold in general that every $\geq_{xi}$-contranegative or every $\geq_{xi}$-positive predicate on $\mathcal{L}_{\mathcal{A}}$ satisfies necessitation.*
(5) *It does not hold in general that every predicate on $\mathcal{L}_{\mathcal{A}}$ that is $\geq_E$-contranegative, or $\geq_E$-positive, for some max-min weighted preference relation $\geq_E$ satisfies necessitation.*

Hence, necessitation does not hold for contranegative (or positive) predicates other than under rather special conditions. It is also interesting to note that according to Theorem 10.13, necessitation applies equally to contranegative and positive predicates. The choice between contranegativity and positivity has no impact on whether necessitation holds or not. What matters is the choice of a preference relation. In particular, maximax preferences are unsuitable if we wish to avoid necessitation.

Judging by Observation 10.14, $\geq_i$ may seem to be a more plausible basis than $\geq_x$ for deontic logic. Unfortunately, though, $\geq_i$-contranegative and $\geq_i$-positive predicates satisfy an even worse property, *reverse necessitation*: If $\vdash p \rightarrow q$, then $Oq \rightarrow Op$. To exemplify the implausibility of reverse necessitation, let q denote that I give my youngest son a present on his birthday and p that I give him a present that I have stolen.

Theorem 10.15. *Let $\mathcal{A}$ be a set of contextually complete alternatives. The following are three conditions on a relation $\geq'$ in $\mathcal{L}_{\mathcal{A}}$:*

(1) *If $\vDash_{\mathcal{A}} p \rightarrow q$, then $p \geq'^* q$.*
(2) *Every $\geq'$-contranegative predicate O on $\mathcal{L}_{\mathcal{A}}$ satisfies strong reverse necessitation (If $\vDash_{\mathcal{A}} p \rightarrow q$, then $Oq \rightarrow Op$).*
(3) *Every $\geq'$-positive predicate O on $\mathcal{L}_{\mathcal{A}}$ satisfies strong reverse necessitation.*

If (1) holds, then so do (2) and (3). If $\geq'$ satisfies ancestral reflexivity ($p \geq'^* p$), then (1), (2), and (3) are equivalent.

Observation 10.16. *Let $\mathcal{A}$ be a contextually complete alternative set, and let $\geq$ be a complete and transitive relation on $\mathcal{A}$. Then:*

(1a) Every $\geq_i$-contranegative predicate on $\mathcal{L}_{\mathcal{A}}$ satisfies reverse necessitation.
(1b) Every $\geq_i$-positive predicate on $\mathcal{L}_{\mathcal{A}}$ satisfies reverse necessitation.
(2) It does not hold in general that every $\geq_x$-contranegative or every $\geq_x$-positive predicate on $\mathcal{L}_{\mathcal{A}}$ satisfies reverse necessitation.
(3) It does not hold in general that every $\geq_{ix}$-contranegative or every $\geq_{ix}$-positive predicate on $\mathcal{L}_{\mathcal{A}}$ satisfies reverse necessitation.
(4) It does not hold in general that every $\geq_{xi}$-contranegative or every $\geq_{xi}$-positive predicate on $\mathcal{L}_{\mathcal{A}}$ satisfies reverse necessitation.
(5) It does not hold in general that every $\geq_E$-contranegative or every $\geq_E$-positive predicate on $\mathcal{L}_{\mathcal{A}}$ satisfies reverse necessitation.

The following weakened form of necessitation has been called *disjunctive division* (of duties):[27]

$$O(p\&q) \rightarrow Op \vee Oq$$

Disjunctive division holds for the contranegative predicates associated with the major types of extremal preference relations. However, it does not hold for most of the corresponding positive predicates:

Theorem 10.17. *Let $\mathcal{A}$ be a set of contextually complete alternatives. The following are two conditions on a relation $\geq'$ in $\mathcal{L}_{\mathcal{A}}$:*

(1) $((p \vee q) \geq'^* p) \vee ((p \vee q) \geq'^* q)$
(2) Every $\geq'$-contranegative predicate O on $\mathcal{L}_{\mathcal{A}}$ satisfies disjunctive division ($O(p\&q) \rightarrow Op \vee Oq$).

If (1) holds, then so does (2). If $\geq'$ is complete, then (1) and (2) are equivalent.

Observation 10.18. *Let $\geq$ be a complete and transitive relation over the finite and contextually complete alternative set $\mathcal{A}$, and let $\geq'$ be a transitive relation on $\mathcal{L}_{\mathcal{A}}$ that satisfies positionality and nonnegative response with respect to $\geq$. Furthermore, let O be a $\geq'$-contranegative predicate on $\mathcal{L}_{\mathcal{A}}$. Then O satisfies disjunctive division.*

Observation 10.19. *Let $\geq$ be a complete and transitive relation over the contextually complete alternative set $\mathcal{A}$. Furthermore, let O be a predicate on $\mathcal{L}_{\mathcal{A}}$ that is contranegative with respect to either $\geq_i$, $\geq_x$, $\geq_{ix}$, $\geq_{xi}$,*

[27] Hansson 1988b.

or some max-min weighted relation $\geq_E$. Then O satisfies disjunctive division.

Theorem 10.20. *Let $\mathcal{A}$ be a set of contextually complete alternatives. The following are two conditions on a relation $\geq'$ in $\mathcal{L}_{\mathcal{A}}$:*

(1) $(p \geq'^* p\&q) \vee (q \geq'^* p\&q)$
(2) Every $\geq'$-positive predicate O on $\mathcal{L}_{\mathcal{A}}$ satisfies disjunctive division ($O(p\&q) \rightarrow Op \vee Oq$).

If (1) holds, then so does (2). If $\geq'$ is complete, then (1) and (2) are equivalent.

Observation 10.21. *Let $\mathcal{A}$ be a contextually complete alternative set, and let $\geq$ be a complete and transitive relation on $\mathcal{A}$. Then:*

(1) It does not hold in general that every $\geq_i$-positive predicate on $\mathcal{L}_{\mathcal{A}}$ satisfies disjunctive division.
(2) Every $\geq_x$-positive predicate on $\mathcal{L}_{\mathcal{A}}$ satisfies disjunctive division.
(3) It does not hold in general that every $\geq_{ix}$-positive predicate on $\mathcal{L}_{\mathcal{A}}$ satisfies disjunctive division.
(4) It does not hold in general that every $\geq_{xi}$-positive predicate on $\mathcal{L}_{\mathcal{A}}$ satisfies disjunctive division.
(5) It does not hold in general that every $\geq_E$-positive predicate on $\mathcal{L}_{\mathcal{A}}$ satisfies disjunctive division.

The following is another weakened version of necessitation:

$$Op \,\&\, Oq \rightarrow O(p \vee q) \text{ (disjunctive closure)}$$

It holds for the positive predicates associated with a wide range of preference relations, but in most cases not for the corresponding contranegative predicates.

Theorem 10.22. *Let $\mathcal{A}$ be a set of contextually complete alternatives. The following are two conditions on a relation $\geq'$ in $\mathcal{L}_{\mathcal{A}}$:*

(1) $(p \geq'^* p\&q) \vee (q \geq'^* p\&q)$
(2) Every $\geq'$-contranegative predicate O on $\mathcal{L}_{\mathcal{A}}$ satisfies disjunctive closure ($Op \,\&\, Oq \rightarrow O(p \vee q)$).

If (1) holds, then so does (2). If $\geq'$ is complete, then (1) and (2) are equivalent.

Observation 10.23. *Let $\mathcal{A}$ be a contextually complete alternative set, and let $\geq$ be a complete and transitive relation on $\mathcal{A}$. Then:*

(1) It does not hold in general that every $\geq_i$-contranegative predicate on $\mathcal{L}_{\mathcal{A}}$ satisfies disjunctive closure.
(2) Every $\geq_x$-contranegative predicate on $\mathcal{L}_{\mathcal{A}}$ satisfies disjunctive closure.
(3) It does not hold in general that every $\geq_{ix}$-contranegative predicate on $\mathcal{L}_{\mathcal{A}}$ satisfies disjunctive closure.
(4) It does not hold in general that every $\geq_{xi}$-contranegative predicate on $\mathcal{L}_{\mathcal{A}}$ satisfies disjunctive closure.
(5) It does not hold in general that every $\geq_E$-contranegative predicate on $\mathcal{L}_{\mathcal{A}}$ satisfies disjunctive closure.

Theorem 10.24. *Let $\mathcal{A}$ be a set of contextually complete alternatives. The following are two conditions on a relation $\geq'$ in $\mathcal{L}_{\mathcal{A}}$:*

(1) $(p \vee q \geq'^* p) \vee (p \vee q \geq'^* q)$
(2) Every $\geq'$-positive predicate O on $\mathcal{L}_{\mathcal{A}}$ satisfies disjunctive closure (Op & $Oq \rightarrow O(p \vee q)$).

If (1) holds, then so does (2). If $\geq'$ is complete, then (1) and (2) are equivalent.

Observation 10.25. *Let $\geq$ be a complete and transitive relation over the finite and contextually complete alternative set $\mathcal{A}$, and let $\geq'$ be a transitive relation on $\mathcal{L}_{\mathcal{A}}$ that satisfies positionality and nonnegative response with respect to $\geq$. Furthermore, let O be a $\geq'$-positive predicate on $\mathcal{L}_{\mathcal{A}}$. Then O satisfies disjunctive closure.*

Observation 10.26. *Let $\geq$ be a complete and transitive relation over the contextually complete alternative set $\mathcal{A}$. Furthermore, let O be a predicate on $\mathcal{L}_{\mathcal{A}}$ that is positive with respect to either $\geq_i$, $\geq_x$, $\geq_{ix}$, $\geq_{xi}$, or some max-min weighted relation $\geq_E$. Then O satisfies disjunctive closure.*

The negative results of Observation 10.23 may be somewhat surprising since disjunctive closure seems to be an almost self-evident property of a prescriptive predicate.[28] Some reflection will show, however, that it is not so self-evident after all. Consider its equivalent formulation for the corresponding permissive predicate:

[28] As I mistakenly took it to be in Hansson 1988b.

$$P(p\&q) \rightarrow Pp \vee Pq$$

Although this condition seems to hold in most cases, it does not hold when two mischiefs cancel each other. If Mary and Peter have a date at the local pub, it may be a permissible interference to both tell Mary that Peter will instead be waiting for her at the Italian restaurant downtown (p) and tell Peter that Mary will be waiting for him at that same restaurant (q). Nevertheless, it may be impermissible to redirect only one of them so that they miss the date, in other words, $P(p\&q)$ holds, but neither Pp nor Pq holds.

The following theorem of standard deontic logic,

$$Op \ \& \ O(p \rightarrow q) \rightarrow Oq,$$

is commonly referred to as *deontic detachment*.[29] Several counterexamples to this postulate have been given in the literature.[30] A counterexample can also be couched in terms of the following equivalent formulation of deontic detachment:

$$O(p \vee q) \& O\neg p \rightarrow Oq.$$

Suppose that for some reason you are morally required to come to a conference. You are also required not to come unannounced. Let p denote that you stay away from the conference and q that you give notice that you will come. Then $O(p \vee q)$ and $O\neg p$ both hold, but since you should not notify unless you come, Oq does not hold.

Deontic detachment can be connected to properties of the preference relation, as follows:

Theorem 10.27. *Let $\mathcal{A}$ be a set of contextually complete alternatives. The following are two conditions on a relation $\geq'$ in $\mathcal{L}_{\mathcal{A}}$:*

(1) $(p \geq'^{*} q) \vee ((\neg p \& q) \geq'^{*} q)$

(2) Every $\geq'$-contranegative predicate O on $\mathcal{L}_{\mathcal{A}}$ satisfies deontic detachment ($Op \ \& \ O(p \rightarrow q) \rightarrow Oq$).

If (1) holds, then so does (2). Furthermore, if $\geq'$ satisfies completeness, then (1) and (2) are equivalent.

Observation 10.28. *Let $\mathcal{A}$ be a set of contextually complete alternatives and $\geq$ a complete and transitive relation over $\mathcal{A}$. Then:*

[29] This name seems to have been introduced by P. S. Greenspan (1975, p. 260).
[30] McLaughlin 1955; Hansson 1988b.

(1) Every $\geq_i$-contranegative predicate on $\mathcal{L}_{\mathcal{A}}$ satisfies deontic detachment.
(2) Every $\geq_x$-contranegative predicate on $\mathcal{L}_{\mathcal{A}}$ satisfies deontic detachment.
(3) It does not hold in general that every $\geq_{ix}$-contranegative predicate on $\mathcal{L}_{\mathcal{A}}$ satisfies deontic detachment.
(4) It does not hold in general that every $\geq_{xi}$-contranegative predicate on $\mathcal{L}_{\mathcal{A}}$ satisfies deontic detachment.
(5) It does not hold in general that every $\geq_E$-contranegative predicate on $\mathcal{L}_{\mathcal{A}}$ satisfies deontic detachment.

Theorem 10.29. *Let $\mathcal{A}$ be a contextually complete alternative set. The following are two conditions on a relation $\geq'$ on $\mathcal{L}_{\mathcal{A}}$:*

(1) $(q\geq'^*p) \vee (q\geq'^*(\neg p \vee q))$
(2) Every $\geq'$-positive predicate O in $\mathcal{L}_{\mathcal{A}}$ satisfies deontic detachment.

If (1) holds, then so does (2). Furthermore, if $\geq'$ satisfies completeness, then (1) and (2) are equivalent.

Observation 10.30. *Let $\mathcal{A}$ be a contextually complete alternative set, and let $\geq$ be a complete and transitive relation on $\mathcal{A}$. Then:*

(1) Every $\geq_i$-positive predicate on $\mathcal{L}_{\mathcal{A}}$ satisfies deontic detachment.
(2) It does not hold in general that every $\geq_x$-positive predicate on $\mathcal{L}_{\mathcal{A}}$ satisfies deontic detachment.
(3) It does not hold in general that every $\geq_{ix}$-positive predicate on $\mathcal{L}_{\mathcal{A}}$ satisfies deontic detachment.
(4) It does not hold in general that every $\geq_{xi}$-positive predicate on $\mathcal{L}_{\mathcal{A}}$ satisfies deontic detachment.
(5) It does not hold in general that every $\geq_E$-positive predicate on $\mathcal{L}_{\mathcal{A}}$ satisfies deontic detachment.

Von Wright has proposed the following axiom for permissive predicates:[31]

$$P(p \& q) \& P(p \& \neg q) \rightarrow Pp$$

This is a plausible postulate, which does not seem to have been challenged in the literature. It does not have an established name, but may

[31] von Wright 1973, p. 44.

be called *permissive cancellation* since it allows for the cancellation of q and $\neg q$ from the two permissions. Permissive cancellation holds for a wide range of contranegative predicates. (Note that, according to Observation 10.3, a permissive predicate P is $\geq'$-positive if and only if the corresponding prescriptive predicate O is $\geq'$-contranegative.)

Theorem 10.31. *Let $\mathcal{A}$ be a set of contextually complete alternatives. The following are two conditions on a relation $\geq'$ in $\mathcal{L}_{\mathcal{A}}$:*

(1) $(p \geq'^* (p \& q)) \vee (p \geq'^* (p \& \neg q))$
(2) Every $\geq'$-positive predicate P on $\mathcal{L}_{\mathcal{A}}$ satisfies permissive cancellation $(P(p \& q) \ \& \ P(p \& \neg q) \rightarrow Pp)$.

If (1) holds, then so does (2). Furthermore, if $\geq'$ satisfies completeness, then (1) and (2) are equivalent.

Observation 10.32. *Let $\geq'$ be a relation on $\mathcal{L}_{\mathcal{A}}$. If it satisfies the following condition,*

$$((p \vee q) \geq'^* p) \vee ((p \vee q) \geq'^* q),$$

then all $\geq'$-positive predicates (P) satisfy permissive cancelling.

Observation 10.33. *Let $\geq$ be a complete and transitive relation on $\mathcal{A}$, and let P be a predicate on $\mathcal{L}_{\mathcal{A}}$. If P is either $\geq_i$-positive, $\geq_x$-positive, $\geq_{ix}$-positive, $\geq_{xi}$-positive, or $\geq_E$-positive, then it satisfies permissive cancelling.*

Theorem 10.34. *Let $\mathcal{A}$ be a set of contextually complete alternatives. The following are two conditions on a relation $\geq'$ in $\mathcal{L}_{\mathcal{A}}$:*

(1) $((p \vee q) \geq'^* p) \vee ((p \vee \neg q) \geq'^* p)$
(2) Every predicate P on $\mathcal{L}_{\mathcal{A}}$ such that for all p, $Pp \leftrightarrow \neg O \neg p$ for some $\geq'$-positive predicate O satisfies permissive cancellation.

If (1) holds, then so does (2). Furthermore, if $\geq'$ satisfies ancestral reflexivity ($p \geq'^* p$), then (1) and (2) are equivalent.

Observation 10.35. *Let $\geq$ be a complete and transitive relation on the set $\mathcal{A}$ of contextually complete alternatives. Let O and P be predicates on $\mathcal{L}_{\mathcal{A}}$ such that for all p, $Pp \leftrightarrow \neg O \neg p$. Then:*

(1) It does not hold in general that if O is $\geq_i$-positive, then P satisfies permissive cancellation.

(2) If O is $\geq_x$-positive, then P satisfies permissive cancellation.
(3) It does not hold in general that if O is $\geq_{ix}$-positive, then P satisfies permissive cancellation.
(4) It does not hold in general that if O is $\geq_{xi}$-positive, then P satisfies permissive cancellation.
(5) It does not hold in general that if O is $\geq_E$-positive, then P satisfies permissive cancellation.

In his 1951 paper, von Wright proposed a principle of "deontic contingency" for tautologies, namely: "A tautologous act is not necessarily obligatory," that is, $\neg O(p\vee\neg p)$.[32] Subsequent authors have in most cases accepted the opposite principle,

$$O(p\vee\neg p),$$

which may be called the *postulate of the empty duty*. As was reported above, this is one of the axioms of standard deontic logic. Nevertheless, it is clearly implausible. It is difficult to see why a sentence like "You are morally required to either commit or not commit mass murder" should be valid under all interpretations of moral requirement.[33]

The postulate of the empty duty does not hold in general for all contranegative (or all positive) predicates associated with any combinative preference relation. The reason for this is quite trivial: The empty predicate H, such that $\neg Hr$ holds for all arguments r in its domain, vacuously satisfies positivity and contranegativity with respect to any preference relation with an appropriate domain. In order to avoid this trivial case, the postulate of the empty duty should be reformulated as follows:

$$Op\rightarrow O(p\vee\neg p)$$

This form of the empty duty postulate is also a weakened form of necessitation. It can be shown to hold for some classes of contranegative and positive predicates:

Theorem 10.36. *Let $\mathcal{A}$ be a set of contextually complete alternatives. The following are two conditions on a relation $\geq'$ in $\mathcal{L}_{\mathcal{A}}$:*

(1) $p\geq'^{*}(p\&\neg p)$
(2) Every $\geq'$-contranegative predicate O satisfies $Op\rightarrow O(p\vee\neg p)$.

[32] von Wright 1951, p. 10.
[33] Cf. Jackson 1985, p. 191; Lenk 1978, p. 31.

If (1) holds, then so does (2). Furthermore, if $\geq'$ satisfies ancestral reflexivity ($p\geq'^{*}p$), then (1) and (2) are equivalent.

Observation 10.37. *Let $\mathcal{A}$ be a set of contextually complete alternatives, and let $\geq$ be a complete and transitive relation on $\mathcal{A}$. Then:*

(1) It does not hold in general that every $\geq_{i}$-contranegative predicate O on $\mathcal{L}_{\mathcal{A}}$ satisfies $Op\rightarrow O(p\vee\neg p)$.
(2) Every $\geq_{x}$-contranegative predicate O on $\mathcal{L}_{\mathcal{A}}$ satisfies $Op\rightarrow O(p\vee\neg p)$.
(3) It does not hold in general that every $\geq_{ix}$-contranegative predicate O on $\mathcal{L}_{\mathcal{A}}$ satisfies $Op\rightarrow O(p\vee\neg p)$.
(4) Every $\geq_{xi}$-contranegative predicate O on $\mathcal{L}_{\mathcal{A}}$ satisfies $Op\rightarrow O(p\vee\neg p)$.

Theorem 10.38. *Let $\mathcal{A}$ be a set of contextually complete alternatives. The following are two conditions on a relation $\geq'$ in $\mathcal{L}_{\mathcal{A}}$:*

(1) $(p\vee\neg p)\geq'^{*}p$
(2) Every $\geq'$-positive predicate O satisfies $Op\rightarrow O(p\vee\neg p)$.

If (1) holds, then so does (2). Furthermore, if $\geq'$ satisfies ancestral reflexivity ($p\geq'^{*}p$), then (1) and (2) are equivalent.

Observation 10.39. *Let $\mathcal{A}$ be a contextually complete alternative set, and let $\geq$ be a complete and transitive relation on $\mathcal{A}$. Then:*

(1) It does not hold in general that every $\geq_{i}$-positive predicate O on $\mathcal{L}_{\mathcal{A}}$ satisfies $Op\rightarrow O(p\vee\neg p)$.
(2) Every $\geq_{x}$-positive predicate O on $\mathcal{L}_{\mathcal{A}}$ satisfies $Op\rightarrow O(p\vee\neg p)$.
(3) It does not hold in general that every $\geq_{ix}$-positive predicate O on $\mathcal{L}_{\mathcal{A}}$ satisfies $Op\rightarrow O(p\vee\neg p)$.
(4) It does not hold in general that every $\geq_{xi}$-positive predicate O on $\mathcal{L}_{\mathcal{A}}$ satisfies $Op\rightarrow O(p\vee\neg p)$.
(5) It does not hold in general that every $\geq_{E}$-positive predicate O on $\mathcal{L}_{\mathcal{A}}$ satisfies $Op\rightarrow O(p\vee\neg p)$.

The *consistency axiom* of SDL, $Op\rightarrow\neg O\neg p$, differs from most other postulates of deontic logic in having the same formulation for a negative predicate and its corresponding contranegative predicate. Let W be the (negative) predicate such that, for all p, Wp holds if and only if $O\neg p$. It follows by substitution that $Op\rightarrow\neg O\neg p$ is valid if and only if $Wp\rightarrow\neg W\neg p$ is valid. There is no plausible way to construct a preference relation ≥′ such that all ≥′-contranegative (or all ≥′-positive) predicates satisfy the consistency postulate.

Observation 10.40. *Let ≥′ be a relation that satisfies ancestral reflexivity, and let there be an element p of the domain of ≥′ such that either p≥′(¬p) or (¬p)≥′p. Then:*

(1) There is some ≥′-contranegative predicate O that does not satisfy the consistency postulate (Op→¬O¬p).
(2) There is some ≥′-positive predicate O that does not satisfy the consistency postulate (Op→¬O¬p).

Some of the results of this section are summarized in Table 3. It is particularly interesting to note that agglomeration, disjunctive division, and permissive cancellation hold for a wide range of contranegative predicates, including the five types that are listed in the table.

Table 3. *Some properties of predicates that are contranegative or positive with respect to the major types of extremal preference relations.*

	contranegative, based on					positive, based on				
	$\geq_{\mathrm{i}}$	$\geq_{\mathrm{x}}$	$\geq_{\mathrm{ix}}$	$\geq_{\mathrm{xi}}$	$\geq_{\mathrm{E}}$	$\geq_{\mathrm{i}}$	$\geq_{\mathrm{x}}$	$\geq_{\mathrm{ix}}$	$\geq_{\mathrm{xi}}$	$\geq_{\mathrm{E}}$
Agglomeration *Op & Oq → O(p&q)*	+	+	+	+	+	+	−	−	−	−
Strong necessitation If $\vDash_{\mathcal{A}}$ *p→q*, then *Op→Oq*	−	+	−	−	−	−	+	−	−	−
Strong reverse necessitation If $\vDash_{\mathcal{A}}$ *p→q*, then *Oq→Op*	+	−	−	−	−	+	−	−	−	−
Disjunctive division *O(p&q) → Op ∨ Oq*	+	+	+	+	+	−	+	−	−	−
Disjunctive closure *Op & Oq → O(p∨q)*	−	+	−	−	−	+	+	+	+	+
Deontic detachment *Op & O(p→q) → Oq*	+	+	−	−	−	+	−	−	−	−
Permissive cancellation *P(p&q) & P(p&¬q) → Pp*	+	+	+	+	+	−	+	−	−	−
Consistency *Op → ¬O¬p*	−	−	−	−	−	−	−	−	−	−

10.4 REPRESENTATION THEOREMS

The previous section was devoted to classes of predicates that are contranegative (or positive) with respect to a given preference relation. Another useful approach is to specify particular prescriptive predicates that are contranegative with respect to a given preference relation. A fairly general method for this is to select some sentence to constitute the limit between the prohibited and the nonprohibited. Letting f be an element of $\mathcal{L}_{\mathfrak{A}}$, we can define a prohibitive predicate W such that for all p, Wp if and only if $f \geq' p$. We can base the corresponding prescriptive predicate on W in the usual way, or define it directly as follows:

Definition 10.41. *Let $\geq'$ be a transitive, combinative preference relation. The* sentence-limited contranegative predicate *based on $\geq'$ and an element f of its domain is the predicate O such that for all elements p of the domain of $\geq'$:*

$$Op \leftrightarrow f \geq' \neg p.$$ [34]

In the finite case, it can also be shown that all nonempty contranegative predicates are sentence-limited. (A predicate O is nonempty if and only if there is some p such that Op.)

Observation 10.42. *Let $\geq'$ be a complete and transitive relation over a set $\mathcal{L}_{\mathfrak{A}}$ of sentences that has a finite number of equivalence classes with respect to logical equivalence.*[35] *Furthermore, let O be a nonempty $\geq'$-contranegative predicate on $\mathcal{L}_{\mathfrak{A}}$. Then O is also a sentence-limited contranegative predicate based on $\geq'$ and some sentence f.*

It follows that under the background assumptions given in this observation, the logical properties of sentence-limited contranegative predicates are essentially those that were reported in Section 10.3 for contranegative predicates in general.

[34] By adjusting the definition to $Op \leftrightarrow f \geq'^{*} \neg p$, we can obtain contranegativity even if $\geq'$ is not transitive.

[35] This is a much weaker assumption than to require that $\mathcal{L}_{\mathfrak{A}}$ itself be finite. If a nonempty language is closed under truth-functional connectives, then it contains an infinite number of sentences: If it contains p, then it also contains $p\&(p \vee \neg p)$, $p\&(p \vee \neg p)\&(p \vee \neg p)$, ...

Using the sentence-limited construction, it has been possible to obtain axiomatic characterizations of predicates that are contranegative with respect to $\geq_i$, $\geq_x$, $\geq_{ix}$, $\geq_{xi}$, and $\geq_E$. Unfortunately, representation theorems have only been obtained under the restrictive assumption that $\mathcal{A}$ is a set of possible worlds. This is one of the limitations of SDL that were criticized in Section 9.3, and it is reintroduced here for purely technical reasons.[36]

Theorem 10.43. *Let $\mathcal{A}$ be a set of possible worlds. The following are equivalent conditions on a predicate O on $\mathcal{L}_{\mathcal{A}}$:*

(1) There is a complete and transitive relation $\geq$ on $\mathcal{A}$ and a sentence $f \in \cup\mathcal{A}$ such that:
O is the sentence-limited contranegative predicate with respect to f and the maximin preference relation $\geq_i$ that is based on $\geq$.
(2) O satisfies the postulates
(i) $O(p\&q) \to Op \vee Oq$,
(ii) If Oq and $\vdash p \to q$, then Op. (reverse necessitation), *and*
(iii) $O\bot$.

Theorem 10.44. *Let $\mathcal{A}$ be a set of possible worlds. The following are equivalent conditions on a predicate O on $\mathcal{L}_{\mathcal{A}}$:*

(1) There is a complete and transitive relation $\geq$ on $\mathcal{A}$ and a sentence $f \in \cup\mathcal{A}$ for which $\mathbf{max}(f)$ is nonmaximal, such that:
O is the sentence-limited contranegative predicate with respect to f and the maximax preference relation $\geq_x$ that is based on $\geq$.
(2) O satisfies the postulates
(i) $Op \,\&\, Oq \to O(p\&q)$ (agglomeration)
(ii) If Op and $\vdash p \to q$, then Oq. (necessitation)
(iii) $\neg O\bot$ (consistency)
(iv) There is some p such that Op. (nonemptiness)

Theorem 10.45. *Let $\mathcal{A}$ be a set of possible worlds. The following are equivalent conditions on a predicate O on $\mathcal{L}_{\mathcal{A}}$:*

(1) There is a complete and transitive relation $\geq$ on $\mathcal{A}$ and a sentence $f \in \cup\mathcal{A}$ with the property that either $\mathbf{min}(f)$ is nonminimal or $\mathbf{max}(f)$ is maximal, and such that:

[36] Consider, for instance, the step in the proof of Theorem 10.44 from $\mathbf{max}(f) \geq \mathbf{max}(\neg p)$ and $\mathbf{max}(f) \geq \mathbf{max}(\neg q)$ to $\mathbf{max}(f) \geq \mathbf{max}(\neg p \vee \neg q)$

O is the sentence-limited contranegative predicate with respect to f and the interval maximin preference relation $\geq_{ix}$ that is based on $\geq$.

(2) *O satisfies the postulates*

(i) $Op \,\&\, Oq \rightarrow O(p\&q)$ (agglomeration)

(ii) $O(p\&q) \rightarrow Op \vee Oq$

(iii) *If* $\vdash r \rightarrow s$, $\vdash s \rightarrow p$, *and* $\vdash p \rightarrow q$, *and* $\neg Or$, Os *and* $\neg Op$, *then* $\neg Oq$.

(iv) $O\bot$

Theorem 10.46. *Let $\mathcal{A}$ be a set of possible worlds. The following are equivalent conditions on a predicate O on $\mathcal{L}_{\mathcal{A}}$:*

(1) *There is a complete and transitive relation $\geq$ on $\mathcal{A}$ and a sentence $f \in \cup\mathcal{A}$ for which $\max(f)$ is non-maximal, and such that:*
O is the sentence-limited contranegative predicate with respect to f and the interval maximax preference relation $\geq_{xi}$ that is based on $\geq$.

(2) *O satisfies the postulates*

(i) $Op \,\&\, Oq \rightarrow O(p\&q)$ (agglomeration)

(ii) $O(p\&q) \rightarrow Op \vee Oq$

(iii) *If* $\vdash r \rightarrow s$, $\vdash s \rightarrow p$, *and* $\vdash p \rightarrow q$, *and* Or, $\neg Os$ *and* Op, *then* Oq.

(iv) $\neg O\bot$ (consistency)

(v) *There is some p such that Op.* (nonemptiness)

Definition 10.47.
$W^{+}p$ iff $W(p\&t)$ for all t such that $p\&t$ is consistent.
$P^{+}p$ iff $P(p\&t)$ for all t such that $p\&t$ is consistent.

Theorem 10.48. *Let $\mathcal{A}$ be a finite set of finitely representable*[37] *possible worlds, and let $0<\delta<1$. Then the following are equivalent conditions on a predicate O on $\mathcal{L}_{\mathcal{A}}$:*

(1) *O is a sentence-limited contranegative predicate with respect to a sentence $f \in \cup\mathcal{A}$ and a max-min weighted preference relation on $\mathcal{A}$ with δ as its index.*

(2) *O satisifes:*

(i) $Op \,\&\, Oq \rightarrow O(p\&q)$ (agglomeration)

(ii) $O(p\&q) \rightarrow Op \vee Oq$

[37] A set X is finitely representable if and only if there is a finite set X' such that $\mathrm{Cn}(X) = \mathrm{Cn}(X')$.

(*iii*) *If* W^+p, W^+q, $W(p \vee r)$, $\neg W(p \vee s)$, *and* $\neg W(q \vee r)$, *then* $\neg W(q \vee s)$.
(*iv*) *If* P^+p, P^+q, $P(p \vee r)$, $\neg P(p \vee s)$, *and* $\neg P(q \vee r)$, *then* $\neg P(q \vee s)$.

As can be seen from the postulates, the logic of the $\geq_x$-based prescriptive operator is a weakened version of SDL, and the logic of the $\geq_{xi}$-based operator is in turn weaker than that of the $\geq_x$-based operator. It can also be shown that the logic of the $\geq_E$-based operator is weaker than SDL:

Observation 10.49. *The SDL postulates imply the postulates of Theorem 10.48.*

The logic of the $\geq_{ix}$-based operator is weaker than that of the $\geq_i$-based operator, but since they both satisfy the postulate $O\bot$, neither of them is a weakened version of SDL.

The $\geq_x$-, $\geq_i$-, and $\geq_{ix}$-based operators all satisfy clearly implausible postulates (necessitation, reverse necessitation, and $O\bot$). The $\geq_{xi}$-based and, in particular, the $\geq_E$-based operator come out better than the others after the axiomatic analysis.

10.5 CANONICAL DEONTIC PREDICATES

A different approach to the construction of deontic predicates can be based on the definition of 'bad' that was introduced in Chapter 8. We can, in an extensional sense, equate 'wrong' with 'bad' (the bad–wrong connection). In other words, we can postulate that an action is wrong if and only if it is bad. 'Wrong' can in its turn, following mainstream traditions, be identified with 'ought not.'[38] Combining the two connections, we obtain a bad–ought connection: An action ought to be performed if and only if it is bad not to perform it.

Since 'bad' is a negative predicate, the bad–ought connection is a special case of the contranegativity thesis. Contrary to the latter more general notion, the bad–ought connection has a long history. In the tenth century, Abd Al-Jabbar defined an act as obligatory if one who omits it deserves blame for doing so.[39] Alexius Meinong claimed that 'bad' and 'permitted' are mutually exclusive.[40] In a discussion based on Meinong's ideas, Chisholm proposed that the obligatory is that which is "good to do and bad not to do."[41]

[38] See Section 9.2. [39] Hourani 1985, p. 102. [40] Meinong 1894, p. 92 (§29).
[41] Chisholm 1963, p. 10.

It must be emphasized that the bad–wrong and bad–ought connections are only very rough approximations. The words 'bad,' 'wrong,' and 'ought' do not necessarily have exactly the strengths necessary for interdefinability. Arguably, 'wrong' is mostly somewhat stronger than 'bad.' As was pointed out by Chisholm and Sosa, there are actions of "permissive ill-doing," that is, "minor acts of discourtesy which most of us feel we have a right to perform (e.g., taking too long in the restaurant when others are known to be waiting)."[42] Such acts may plausibly be said to be bad but not wrong.

With this important reservation, we can base a deontic logic on the assumption that 'ought' is coextensional with "bad if not."[43] Applying the bad–ought connection to the canonical predicate of badness from Section 8.3,

$$B_C p \leftrightarrow (\forall s)(p \geq'^* s \rightarrow \neg s >' s)$$

we obtain the following corresponding prescriptive predicate:

Definition 10.50. *Let $\geq'$ be a combinative preference relation. The* canonical contranegative predicate *based on $\geq'$ is the predicate O_C such that for all p in the domain of $\geq'$:*

$$O_C p \leftrightarrow (\forall s)(\neg p \geq'^* s \rightarrow \neg s >' s)$$

It follows directly from the definition that canonical 'ought' is contranegative and from Observation 10.42 that it is sentence-limited under plausible background conditions. Furthermore, it is the most inclusive contranegative predicate that satisfies the consistency postulate.

Observation 10.51. *Let $\geq'$ be a combinative preference relation and O_C the canonical contranegative predicate that is based on $\geq'$. Then:*

(1) If $\geq'$ satisfies ancestral reflexivity ($p \geq'^* p$), then O_C satisfies consistency ($O_C p \rightarrow \neg O_C \neg p$).
(2) If $\geq'$ satisfies completeness, then a $\geq'$-contranegative predicate O satisfies consistency ($Op \rightarrow \neg O \neg p$) if and only if it satisfies $Op \rightarrow O_C p$ for all p.

[42] Chisholm and Sosa 1966b, p. 326.

[43] The bad–ought connection is of particular interest in the analysis of the ideal ought (Seinsollen). That there ought to be more sunny days means, roughly, that it is a bad thing that there are not more sunny days.

10.6 MAXIMAL OBEYABLE AND MAXICONSISTENT PREDICATES

The consistency postulate referred to in Sections 10.3 and 10.5 is the postulate traditionally used in deontic logic:[44]

(1) $Op \rightarrow \neg O \neg p$, or equivalently $\neg(Op \& O \neg p)$.

This expression of deontic consistency is not general enough. Consider the following example:

You must pay at least 500 dollars. You are not allowed to pay more than 300 dollars.

Intuitively speaking, this is a clear case of deontic inconsistency. The inconsistency does not have the form $Op \& O \neg p$, but it can be expressed in the more general form $Op \& Oq$, where $\vdash \neg(p \& q)$. Hence, the consistency condition should generalized to:

(2) If $\vdash \neg(p \& q)$, then $\vdash \neg(Op \& Oq)$.

Deontic logicians have in general been content with (1) and have not replaced it with the more general (2). The reason for this is that most previously studied deontic logics, including SDL, satisfy necessitation. If neccessitation holds, then (1) and (2) are equivalent. (To see this, note that if $\vdash \neg(p \& q)$, then $O \neg p$ follows by necessitation from Oq.)

Not even (2) is general enough. Consider the following example:

You must report this incident either to the general or to the colonel. You are not allowed to report the incident to the general. You are not allowed to report the incident to the colonel.

This example can be formalized as $O(p \vee q)$, $O \neg p$, and $O \neg q$. Unless some other postulate for the O predicate is added (such as agglomeration), this set of three prescriptions does not violate (2). Nevertheless, this is clearly an inconsistent set of prescriptions.[45] Therefore, (2) should be generalized:

(3) $\{p \mid Op\}$ is consistent.

This can be taken as the most general form of the stipulation that prescriptions should be logically consistent. However, in order to ensure

[44] See, e.g., Føllesdal and Hilpinen 1970, p. 13.
[45] Cf. Royakkers 1996, p. 158.

that prescriptions can be complied with, consistency is not enough. Suppose that your only two options with respect to feeding your child are to steal food for the child or to let it starve. Then the following set of prescriptions is impossible to comply with, although it is consistent:

"You should not steal. You should not let your child starve."

Assuming that the alternative set contains the available options, we should therefore require not only that $\{p \mid Op\}$ be consistent but also that it be included in some element of the alternative set. This is our final definition of what it means (in a situationist setting) for a prescriptive predicate to be obeyable, or possible to comply with:[46]

Definition 10.52. *Let $\mathcal{A}$ be a set of contextually complete alternatives. A prescriptive predicate O in $\mathcal{L}_{\mathcal{A}}$ is* obeyable *if and only if there is some $A \in \mathcal{A}$ such that $\{p \mid Op\} \subseteq A$.*

Under plausible background conditions, the definition of obeyability can be simplified in an interesting way.

Observation 10.53. *Let $\mathcal{A}$ be a finite and contextually complete alternative set such that $\mathcal{L}_{\mathcal{A}}$ has a finite number of equivalence classes with respect to logical equivalence. Let $\geq$ be a complete and transitive relation over $\mathcal{A}$, and let $\geq'$ be a transitive relation on $\mathcal{L}_{\mathcal{A}}$ that satisfies positionality and nonnegative response with respect to $\geq$. Furthermore, let O be a $\geq'$-contranegative predicate on $\mathcal{L}_{\mathcal{A}}$. Then O is obeyable if and only if $\neg O\bot$.*

In Section 10.5, we identified O_C, the most inclusive contranegative predicate that satisfies the postulate $\neg(Op \;\&\; O\neg p)$. The most inclusive predicate that satisfies $\neg O\bot$ can be identified as follows:

Definition 10.54. *Let $\geq'$ be a combinative preference relation. The* maxiconsistent contranegative predicate *based on $\geq'$ is the predicate $O_{\top}$ such that for all p in the domain of $\geq'$:*

$$O_{\top}p \leftrightarrow (\forall s)(\neg p \geq'^{*} s \rightarrow \top >' s)$$

[46] On the consistency of normative rules, see Chapter 12.

Observation 10.55. *Let $\geq'$ be a combinative preference relation, and let O_T be the maxiconsistent contranegative predicate that is based on $\geq'$. Then:*

(1) If $\geq'$ satisfies ancestral reflexivity, then $\neg O_T\bot$.
(2) If $\geq'$ satisfies completeness, then every $\geq'$-contranegative predicate O satisfies $\neg O\bot$ if and only if it satisfies $Op \rightarrow O_T p$ for all p.

A simpler definition of O_T can be used if the combinative preference relation is complete and transitive:

Observation 10.56. *Let $\geq'$ be a complete combinative preference relation and O_T the maxiconsistent contranegative predicate that is based on $\geq'$. Then:*

(1) $O_T p \rightarrow \top >' \neg p$
(2) If $\geq'$ is transitive, then $O_T p \leftrightarrow \top >' \neg p$.

Note that the version of O_T introduced in Observation 10.56 corresponds to a bad–ought connection with another proposed definition of 'bad' than the one used in O_C: bad as "worse than a tautology."[47] The relation between O_T and O_C is clarified in the following observations:

Observation 10.57. *Let $\geq'$ be a combinative preference relation that satisfies completeness, transitivity, and the postulate $s \geq' \top \vee \neg s \geq' \top$. Let O_C be the canonical and O_T the maxiconsistent contranegative predicate that are based on $\geq'$. Then $O_T p \rightarrow O_C p$ holds for all p.*

Observation 10.58. *Let $\geq'$ be a complete and transitive combinative preference relation. Let O_C be the canonical and O_T the maxiconsistent contranegative predicate that are based on $\geq'$. Then the following three conditions are equivalent:*

(1) For all p, $O_C p \rightarrow O_T p$
(2) $\neg O_C \bot$.
(3) There is some s such that $\top \geq' s$ and $\top \geq' \neg s$.

The condition $s \geq' \top \vee \neg s \geq' \top$ that is used in Observation 10.57 follows from disjunctive interpolation. Furthermore, if disjunctive interpola-

[47] See Section 8.3.

tion holds, and there is an indifferent sentence (a sentence s such that $s \equiv' \neg s$),[48] then condition (3) of Observation 10.58 is satisfied. It follows that O_C and O_T coincide under the fairly plausible condition that $\geq'$ satisfies completeness, transitivity, disjunctive interpolation, and the existence of an indifferent sentence.

[48] See Section 8.3.

11

Conflicts and Counterfactuals

Situationist deontic logic, as delimited in Section 9.4 and outlined in Chapter 10, represents that fraction of deontic discourse that refers to only one (actual) situation and one normative appraisal of it. Deontic statements overstep these limitations primarily in two ways: First, they may refer to some situation that does not actually obtain. Second, they may refer to situations and alternative sets in general (deontic rules).

Ordinary language uses the same type of expressions, namely conditional sentences ("if"-sentences) to represent both hypothetical reasoning and reasoning by rules. As was indicated in Section 9.4, it is nevertheless essential to distinguish between these two types of normative expressions. Hypothetical deontic statements are treated in this chapter and deontic rules in Chapters 12 and 13.

Section 11.1 gives a formal account of a common move in informal normative discourse, namely to shift from a non–action-guiding to an action-guiding sense of a deontic word such as 'ought' or 'should.' In Section 11.2, it is shown how such shifts can be used for the (pragmatic) resolution of moral dilemmas. Section 11.3 outlines in general terms how the formal analysis from Chapter 10 has to be modified in order to deal with hypothetical deontic conditionals. In Section 11.4, this modification is performed for restrictive counterfactuals, a group of deontic conditionals that includes the contrary-to-duty conditionals.

11.1 SHIFTING TO ACTION-GUIDANCE

Chapter 10 resulted in a deontic logic that may contain both predicates that satisfy the "ought implies can" principle and predicates that violate this principle; in other words, both obeyable and nonobeyable predicates. This is an adequate feature of the logical model, since both types of predicates are needed to represent prescriptions made in nonformalized language. In natural languages, it is common for one and the same prescriptive word, such as 'ought' or 'should,' to be used in both obeyable and nonobeyable senses.

A good illustration of shifts between obeyable and nonobeyable senses can be derived from Michael Stocker's observation that "it would be at best a bad joke for me to suggest that if I have squandered my money, then I no longer ought to repay my debts."[1] There is an important sense in which, the morning after I lost all my money at the roulette table, I still ought to pay my debts. However, there is also an important sense of 'ought' that refers to my choice among the options open to me.

MORALIST: You have a large debt that is due today. You should pay it.
SPENDTHRIFT: It is impossible for me to do that. I do not have the money.
MORALIST: I know that.
SPRENDTHRIFT: Yes, and I already know what my obligations are. Please, as a moralist, tell me instead what I should *do*.
MORALIST: I have already told you. You should pay your debt.

Our Moralist is unhelpful, since she refuses to accept the shift in perspective demanded by Spendthrift when asking what she should *do*. With this phrase, Spendthrift called for action-guidance. The 'should' of "You should pay your debt" is not suitable for action-guidance, since it requires something that Spendthrift cannot do. As Sinnott-Armstrong has shown, when 'ought' is used for the purpose of advising, it implies 'can.'[2]

The shift in focus demanded by Spendthrift can be described as a shift from a morally adequate but nonobeyable prescriptive predicate O_M to a predicate O_A that is obeyable and therefore suitable for action-guidance. Let p designate that she pays off her debts. Then O_Mp and $\neg O_Ap$. Furthermore, let q denote that she pays her creditors at least as much as she can without losing her means of subsistence. Then, presumably, O_Mq and O_Aq.

What can be required of the logical relationship between O_A and O_M? Clearly, $O_M\alpha \rightarrow O_A\alpha$ does not hold for all α, since that would make

[1] Stocker 1987, p. 108.
[2] Sinnott-Armstrong 1988, p. 123. See also McConnell 1978. This refers, of course, to direct action-guidance and direct advice. Virtually any statement can be action-guiding in an indirect sense. Obligations that cannot be fulfilled may guide actions of apologizing or compensating, as clarified in Williams 1965.

O_A nonobeyable. It is much more reasonable to assume that in most cases O_A can be so constructed that the converse relationship, $O_A\alpha \rightarrow O_M\alpha$, will hold for all α, that is, that the action-guiding requirements are – intuitively speaking – a subset of the moral requirements.[3] Furthermore, since violations of moral obligations should be minimized, O_A should not be more lax than what is necessary to make it obeyable. If $O_A\alpha \rightarrow O_M\alpha$ holds, then we may therefore require $O_A\alpha$ to be a maximal obeyable restriction of O_M, in the following sense:

Definition 11.1. *Given a combinative preference relation $\geq'$, let O and O' be $\geq'$-contranegative predicates with the same domain. Then:*

O' is a restriction *of O if and only if: (1) for all p, if $O'p$, then Op, and (2) there is some p such that Op and $\neg O'p$.*

O' is a maximal obeyable restriction *of O if and only if it is an obeyable restriction of O and there is no obeyable restriction O'' of O such that O' is a restriction of O''.*

In general, a shift to action-guidance can be obtained by replacing any nonobeyable prescriptive predicate by an obeyable such predicate. In many cases, this obeyable predicate can be taken to be a maximal obeyable restriction of the original predicate. In the more orderly cases, there is only one such maximal obeyable restriction: As we saw in Section 10.6, under plausible background conditions, $O_\top$ is the only maximal obeyable restriction of any nonobeyable contranegative predicate.[4]

[3] This does not preclude that it may also, alternatively, be possible to construct some other action-guiding predicate O_A' that yields the same recommendations but does not satisfy $O_A'\alpha \rightarrow O_M\alpha$ for all α. In the above repayment example, we may let r denote that Spendthrift asks her creditors to wait until she can pay them. We then have O_Ar and $\neg O_Mr$. (This was pointed out to me by Wlodek Rabinowicz.) However, instead of r we may refer to $p\vee r$, her obligation to either pay her creditors or ask them to wait. Then we have $O_A(p\vee r)$ and $O_M(p\vee r)$. If the agent cannot realize p, then she realizes $p\vee r$ if and only if she realizes r. (This depends on $p, r \in \cup\mathcal{A}$ and the contextual completeness of $\mathcal{A}$.) Hence, a predicate O_A that validates $O_A(p\vee r)$ but not O_Ar may be used to obtain the desired action-guidance. This type of construction may be somewhat unintuitive in some cases. On the other hand, it leads to the satisfaction of the principle $O_A\alpha \rightarrow O_M\alpha$, which simplifies the formal structure.

[4] When there are several maximal obeyable restrictions, an alternative method is to use their intersection, so that $O_A\alpha$ holds if and only if $O'\alpha$ holds for all maximal obeyable restrictions O' of O_M.

In the literature on moral dilemmas in deontic logic, dilemmas have generally been represented in the form Op & $O\neg p$. For the reasons given in Section 10.6, this representation is not general enough. The moral requirements expressed by a deontic predicate O are logically in conflict whenever $\{p \mid Op\}$ is inconsistent. This may be the case even if there is no p such that Op & $O\neg p$. Furthermore, given an adequate choice of the alternative set $\mathcal{A}$, a (not necessarily logical) conflict of moral requirements occurs whenever $\{p \mid Op\}$ is not included in any element of $\mathcal{A}$, in other words whenever O is nonobeyable. For this conflict to qualify as a moral dilemma, O must be a predicate that adequately expresses moral requirements. In other words: *There is a moral dilemma whenever there is a morally adequate prescriptive predicate that is nonobeyable.* (The view that there are no moral dilemmas can be expressed as a view that all morally adequate prescriptive predicates are obeyable.)

In deontic logic, moral dilemmas give rise to a *noncompliance problem.*[5] This is essentially a formalized restatement of one of the most discussed problems in moral philosophy: In the presence of conflicting obligations, no acceptable course of action seems to be available, and the dictates of the O predicate cannot all be complied with.

Before dealing with the noncompliance problem in logical terms, let us consider the same problem in nonformalized deontic discourse. From an intuitive point of view, there is a sense in which we can act rationally in dilemmatic situations. What we then do is to choose an option that is not inferior to any other option. The student in Sartre's famous example had both an obligation to join the Forces Françaises Libres in Britain and a conflicting obligation to take care of his mother.[6] If he chose either to join the Free French Forces or to stay with his mother, then – given the moral appraisal of the options presupposed in the example – his choice would not be morally inferior to any other choice that he could have made. However, he may have had

[5] Agglomeration and necessitation combine to create another, much discussed problem in a deontic logic that allows for moral dilemmas: From Op and $O\neg p$ follows by agglomeration $O(p\&\neg p)$ and from this by necessitation Oq. Hence, if there is a moral dilemma, then everything is obligatory. This is absurd. As was observed by van Fraassen (1973), if a person is in a moral quandary, this does not mean that she should give up all moral distinctions. The derivation of this absurdity does not go through in deontic logic without necessitation.

[6] Sartre 1946, pp. 39–40.

other options that were prohibited in a stronger sense than any of these and that he was unequivocally required not to choose.

The choice of one of the horns of a dilemma resolves it from a pragmatic point of view, but in true moral dilemmas such a choice involves the breach of obligations that are not overridden in a moral sense. Therefore, the dilemma is not resolved from a *moral* point of view.[7]

Pragmatic resolutions of moral dilemmas depend on shifting strengths of moral requirement that can be expressed in deontic logic as distinctions between different prescriptive predicates. The obligations that Sartre's student has from a moral point of view can be represented by a nonobeyable prescriptive predicate, and those that he has from a pragmatic point of view by an obeyable predicate. Generally speaking, a pragmatic resolution of a moral dilemma can be formalized with the mechanism introduced in Section 11.1, namely a shift to an obeyable predicate. At least in many cases, this obeyable predicate can be taken to be a maximal obeyable restriction of the prescriptive predicate with which the dilemma is expressed.[8] Two examples will show how this works.

First, let us consider a situation in which I can either save A's life (p) or B's life (q), but not both. I am equally morally required to perform both of these incompatible actions, but my obligation to perform at least one of them ($p \vee q$) is still stronger. It can be argued that the deontic predicate that best reflects the moral features of this situation will be one that requires both that I save A and that I save B. Hence, $O_M p$ and $O_M q$ hold (and clearly so does $O_M(p \vee q)$). Since p and q cannot both be realized, O_M is not obeyable, and we have a moral dilemma in terms of O_M.

From the viewpoint of action-guidance, since O_M is nonobeyable, it should be replaced by its maximal obeyable restriction O_A. Since p and q are incompatible, $O_A p$ and $O_A q$ cannot both hold. By supposition, p

[7] See Hansson 1998a for an argument to the effect that a reasonable moral code cannot be so constructed that moral dilemmas will not arise and that moral dilemmas are an unavoidable component of a satisfactory moral life.

[8] Several authors have indicated that two types of prescriptive predicates are involved in the resolution of moral dilemmas. Williams (1965) noted that the 'ought' that occurs in statements of moral principle is not the same as that of the deliberative question "What ought I to do?". Lemmon (1962) distinguished between the logic of 'must' and that of 'ought,' maintaining that the former but not the latter satisfies the principle $\neg(Op \& O\neg p)$. Philippa Foot (1983) distinguished between 'type 1' and 'type 2' 'ought' statements, of which only the latter satisfy $\neg(Op \& O\neg p)$.

and q are morally required to exactly the same degree, so that O_A cannot distinguish between them. Therefore, $\neg O_A p$ and $\neg O_A q$. However, $p \vee q$ is morally required to a higher degree than either p or q; and since O_A is a *maximal* obeyable restriction of O_M, we have $O_A(p \vee q)$.[9] Hence, from the action-guiding point of view, it is obligatory to perform at least one of the two actions and permissible to perform either to the exclusion of the other.[10]

For our second example, suppose that I can save either A's life (p) or the lives of a dozen persons (r), but I cannot do both. Furthermore, suppose that I have a stronger moral requirement to do the latter than the former. A predicate O_M that represents moral requirement should be such that both $O_M p$ and $O_M r$ hold. For action-guidance, we need to replace O_M by its maximal obeyable restriction O_A. Since r is more strongly required than p, we then obtain $O_A r$ and $\neg O_A p$. In this case, the pragmatic resolution requires that one of the two horns of the dilemma be chosen in preference to the other.

Thus applied, the distinction between moral and action-guiding predicates provides a formal account of how moral dilemmas can exist, yet be susceptible to pragmatic resolutions. It should be observed that the moral 'ought' has not been eliminated, only supplemented. The nonobeyable moral 'ought' that accompanies the obeyable action-guiding 'ought' is the source of regret and other forms of moral residue. As was observed by Bernard Williams, theories that eliminate the 'ought' not acted upon have difficulties in accounting for moral residues.[11]

11.3 DEONTIC COUNTERFACTUALS

The following are deontic statements that refer to some situation that does not obtain at present:

(1) When I was a minor, I was not allowed to sell these assets.
(2) After she dies, her relatives are required to arrange for a decent funeral.

[9] The condition that $p \vee q$ is morally required to a higher degree than either p or q does not hold in general for any p and q, but it seems to hold in typical cases of moral conflicts between equally valued but incompatible courses of action.

[10] On how moral requirements can be overridden by a stronger disjunctive requirement, see also Sinnott-Armstrong 1988, p. 231.

[11] Williams 1965.

(3) If you had eaten your spinach, you would have been allowed to leave the table.
(4) If she had been more seriously ill, you would have been required to take her to the hospital.

In examples 3 and 4, the nonactual situation referred to is hypothetical. This is the most philosophically interesting case. Such examples can also be referred to as "deontic counterfactuals," counterfactual conditional sentences with deontic sentences as consequents. Letting $\Box\!\!\rightarrow$ denote the counterfactual conditional, we can write them in the form

$$p \Box\!\!\rightarrow \delta,$$

where p is a nonnormative and δ a normative sentence.

This notation is convenient, but it must be accompanied by a warning: It should not be taken for granted that all deontic counterfactuals can be subsumed under the same logical principles. It can, for instance, make a difference if the antecedent is an element of $\cup\mathcal{A}$ or a sentence representing some factual circumstance relevant for the choice of an appropriate alternative set.

As a starting point for the logical investigation of deontic counterfactuals, we can use the construction of situationist deontic logic that was carried through in some detail in Chapter 10. This construction has four major parts. Beginning with (1) an alternative set $\mathcal{A}$ and (2) an exclusionary preference relation $\geq$ on that set, we derived (3) a combinative preference relation $\geq'$ and (4) a deontic predicate O. Each part of this construction has to be reconsidered in the analysis of deontic counterfactuals.

First, the new alternative set has to be constructed. The following examples illustrate the major options:

(5) If it had been your mother's birthday, then you had been morally required to go and visit her.
(6) If your friends had arranged a party for you on that Friday, then you would have been morally required to go to that party.
(7) You are not allowed to use the company's phone for private calls. However, if the pay phone in the canteen is out of order, then you may use the company's phone for local private calls.
(8) If you marry Anthony, then you are not allowed to marry Bob.

(9) You should stay sober tonight. But if you drink, then you should not drive your car home.
(10) If you murder her, then you should murder her gently.[12]

Example (5) can be seen as constructed through the addition of the antecedent as a condition to the sentence "You are morally required to pay a visit to your mother," which (after linguistic adjustments) serves as the consequent of this conditional. The other examples can be seen as constructed in analogous ways.

In example 5, the alternative set seems to remain the same after the antecedent was added. The various activities that you could choose between are the same, one of them being to visit your mother. The fact that it is her birthday does not change the alternative set, but it changes our appraisal of its elements. In the other examples, the alternative set is changed. In example 6, a new alternative (going to the party) has been added. In examples 7–10, the antecedent removes one or several alternatives (using the pay phone, not marrying Anthony, staying sober, not murdering the victim). Examples 8–10 have the additional characteristic that this restriction of the alternative set depends on the agent's own actions. The antecedent is an element of $\cup\mathcal{A}$, that is, it represents an action considered to be open to the agent. Its function as an antecedent is to restrict the deontic discourse to those alternatives in which this action takes place, or in other words to those elements of $\mathcal{A}$ that contain the antecedent. Following Frank Jackson and Robert Pargetter, this type of conditionals will be called *restrictive* deontic conditionals.[13] They will be treated in Section 11.4.

Examples 9 and 10 are contrary-to-duty counterfactuals, that is, counterfactuals in which the antecedent represents the performance of a forbidden action. This is an important subclass of the restrictive counterfactuals.

For restrictive counterfactuals, the new alternative set should consist of those elements of the original alternative set that include the antecedent. For other types of deontic counterfactuals, no general logical rule seems to be available for constructing the new alternative set.

The second step is to find an exclusionary preference relation for the new alternative set. If the alternative set has been changed, then it

[12] Forrester 1984.
[13] Jackson and Pargetter 1987. A defining characteristic of restrictive conditionals, according to these authors, is that the antecedents of restrictive conditionals are "statements about the nature of the option chosen by the agent" (p. 80).

is obvious that a new preference relation is needed. Even if the alternative set is unchanged, the exclusionary preference relation may have to be modified. This can be seen from example 5, in which the antecedent leads to the assignment of a higher value to the alternative "go and visit mother." This corresponds to a preference revision of the type discussed in Chapter 4.

The third step is to construct a new combinative preference relation, based on the exclusionary preference relation. In general, we should expect this relation to be of the same type as that used in the original situation. If an interval maximin (max-min weighted, etc.) preference relation was used for the original alternative set, then an interval maximin (max-min weighted, etc.) preference relation should be applied to the new alternative set as well.

The fourth and final step is to construct a normative predicate that is based on the combinative preference relation obtained in the third step. If this preference relation differs from that used for the original (nonhypothetical) predicate, then the need for a new normative predicate is obvious. There are also deontic counterfactuals that require no changes at all in the first three steps, but only in the fourth. For instance, consider a normative appraisal of what I should do when I see that a child has fallen overboard into turbulent waters. There are three options: do nothing, try to save the child without risking my own life, and save the child at the risk of my own life. Arguably, it would be too much to require that I plunge into the sea to save the child at the risk of my own life. However:

(11) If I had caused the child to fall into the water, then I would have been morally required to plunge into the water and save her at the risk of my own life.

The set of alternatives and the exclusionary and combinative preference relations can be the same in the two cases. What differs is the stringency of the predicate of moral requirement.

In summary, the antecedent of a deontic counterfactual influences our appraisal of the consequent through three major mechanisms, which correspond to the first, second, and fourth steps discussed previously: It may change the alternative set, the exclusionary preference relation, and the stringency of the deontic predicate. The decision of how to perform these changes depends on a complex combination of factual and normative considerations. Therefore, we should not expect one single logical construction to cover all the possible cases. The best strategy in the logical analysis of deontic conditionals is to focus primarily on subcategories in which the full complexity does not materialize. One such case has already been indicated: restrictive deontic counterfactuals. It is the subject of the next section.

11.4 RESTRICTIVE CONDITIONALS

The drink-and-drive example (example 9 in the previous section) can be used to illustrate the analysis of restrictive counterfactuals. The first sentence of that example refers to an alternative set in which there are some alternatives such that you stay sober and some alternatives such that you drink. Among the latter, some but not all are such that you drive a car. In the second sentence, the antecedent restricts our attention to those alternatives in which you drink. In more general terms:

In the appraisal of the restrictive deontic counterfactual $p \mathrel{\Box\!\!\rightarrow} \delta$, the original alternative set $\mathcal{A}$ is replaced by $repr_{\mathcal{A}}(p)$.

The next step is to find an exclusionary preference relation for $repr_{\mathcal{A}}(p)$. In typical cases, the antecedent that made us restrict $\mathcal{A}$ gave us no reason to change our appraisal of the remaining elements.[14] Both before and after knowing that the person was going to drink, we considered drinking-and-driving worse than drinking-and-not-driving.

The obvious choice is therefore to restrict the relation $\geq$ that was used for $\mathcal{A}$ to the new domain $repr_{\mathcal{A}}(p)$:

In the appraisal of the restrictive deontic counterfactual $p \mathrel{\Box\!\!\rightarrow} \delta$, the original exclusionary preference relation $\geq$ is replaced by the relation $\geq_p = \geq \cap (repr_{\mathcal{A}}(p)) \times (repr_{\mathcal{A}}(p))$.

In the terminology of Chapter 4, $\geq_p$ is the outcome of a multiple subtraction of $\geq$ by $repr_{\mathcal{A}}(\neg p)$.

The third step, that of finding a new combinative preference relation, will depend on the construction of the original combinative preference relation. As was noted in the previous section, we should normally use the same type of preference relation for the restricted alternative set that was used for the original alternative set. In other words, $\geq_p'$ should have the same relation to $\geq_p$ as $\geq'$ to $\geq$.

[14] Positional choice may give rise to exceptions. Cf. Section 2.5.

If the original combinative preference relation $\geq'$ was extremal, then this principle has interesting consequences: $\geq_p'$ can be defined in terms of $\geq'$. The reason for this is that the minima and maxima in $\geq_p'$ of any sentence q coincide with the minima and maxima in $\geq'$ of $p\&q$. Hence:

In the appraisal of the restrictive counterfactual $p \;\Box\!\!\rightarrow \delta$, the original extremal combinative preference relation $\geq'$ is replaced by the relation $\geq_p'$, defined as follows:

$$q \geq_p' r \text{ if and only if } p\&q \geq' p\&r.$$

The last part of our construction consists in finding a new prescriptive predicate for the restricted alternative set. To simplify the discussion of this problem, we may assume that the consequent of the restrictive counterfactual has the form Oq, in other words that we are dealing with a sentence $p \;\Box\!\!\rightarrow Oq$. We can then use the conventional dyadic notation $O(q,p)$ for $p \;\Box\!\!\rightarrow Oq$. The words of caution from above should be repeated: The use of a unified notation is no proof that there is a unified logic.

The dyadic predicate should be contranegative:

In the appraisal of the restrictive deontic counterfactual $p \;\Box\!\!\rightarrow Oq$, the original prescriptive predicate O is replaced by a $\geq_p'$-contranegative predicate $O(\ ,p)$.

This is quite unspecific. Unfortunately, no universal recipe for the derivation of $O(\ ,\ )$ from O seems to be available. Intuitively, one may wish the transition from O to $O(\ ,\ )$ to preserve the degree of stringency of O so that $O(\ ,\ )$ is equally stringent as O. In general, however, there are no means to achieve this. Consider, for instance, the sentence-limited approach. If O is a prescriptive predicate based on the limit f, then we might wish to use either f or $p\&f$ as a limit for $O(\ ,p)$. However, the limit sentence f itself may have been reassessed in the process so that it can no longer be said to represent the same degree of stringency.

A somewhat more promising approach is to generalize the maxiconsistent predicate $O_{\top}$. If the original (monadic) predicate is maxiconsistent, then it would seem natural to construct the dyadic predicate in the same way. This can be done as follows:

Definition 11.2. *Let $\geq'$ be a combinative preference relation in $\mathcal{L}_{\mathcal{A}}$. The* dyadic maxiconsistent contranegative predicate *based on $\geq'$ is the predicate $O_{\top}(\ ,\ )$ such that:*

$$O_{\top}(q,p) \leftrightarrow (\forall s)\left(\neg q \geq_p{}'^* s \rightarrow \top >_p{}' s\right)$$

Observation 11.3. *Let $\mathcal{A}$ be a contextually complete alternative set, and let $\geq'$ be a transitive, complete, and extremal combinative preference relation in $\mathcal{L}_{\mathcal{A}}$. Then:*

(1) $O_{\top}(q,p) \leftrightarrow p >' p \& \neg q$
(2) $\neg O_{\top}(\perp,p)$
(3) $\neg O_{\top}(q,\perp)$
(4) $O_{\top}(q,\top) \leftrightarrow O_{\top}(q)$
(5) If $\geq'$ satisfies disjunctive interpolation, then:[15]

$$O_{\top}(q_1,p) \& O_{\top}(q_2,p) \rightarrow O_{\top}(q_1 \& q_2,p)$$

(6) If $\geq'$ is either $\geq_{x}$, $\geq_{ix}$, or $\geq_{xi}$, and $repr(p_1\&\neg q) \neq \emptyset \neq repr(p_2\&\neg q)$, then:[16]

$$O_{\top}(q,p_1) \& O_{\top}(q,p_2) \rightarrow O_{\top}(q,p_1 \vee p_2)$$

The dyadic operator $O_{\top}(,)$ is an extension of $O_{\top}$ to cover restrictive conditionals, in much the same way as Bengt Hansson's dyadic deontic operator extends SDL.[17] It must once again be emphasized that the dyadic $O_{\top}$ predicate has limited applicability and is decidedly *not* a general-purpose dyadic predicate for all types of "conditional obligations." The variations in meaning of "if"-sentences are so large that the construction of such a general-purpose predicate is not a realistic project.

[15] In dyadic logics that are extensions of SDL, the corresponding equivalence holds. $(O(q_1,p) \& O(q_2,p) \leftrightarrow O(q_1\&q_2,p))$. Bengt Hansson 1970a, p. 135; Spohn 1975, p. 239; Lewis 1974, pp. 11–12.

[16] This postulate holds in the dyadic deontic logics proposed by Bengt Hansson (1970a, p. 145) and David Lewis (1974, pp. 11–12). The reverse form $(O(q,p_1 \vee p_2))$ $\rightarrow O(q,p_1) \& O(q,p_2))$ was rejected by Bengt Hansson, who showed it to be implausible (1970a, p. 145).

[17] On how Bengt Hansson's dyadic logic extends SDL, see Bengt Hansson (1970a, p. 135). He interprets dyadic deontic logic as representing "secondary, reparative obligations, telling someone what he should do if he has violated (intentionally or not) a primary obligation" (p. 142). Hence, his logic is intended to represent restrictive conditionals, at least of the contrary-to-duty variant.

12

Rules and Normative Systems

For economy of thought, if for no other reason, moral norms are often regarded as emanating from general rules to be applied in various situations.[1]

There is an obvious and severe tension between the approach to norms introduced in Chapters 9–11 and the rule-based approach to be pursued in this chapter. In the previous chapters, the normative appraisal of each situation was based on the value appraisal of that same situation. Nothing prevented the uniqueness of each particular situation from being taken fully into account. In a rule-based approach, to the contrary, all appraisals of (the potentially infinite range of) possible situations have to be based on a (necessarily finite) set of predetermined general principles. Both approaches are common in informal moral discourse (but for obvious reasons the rule-bound approach dominates in legal contexts). They are not easy to reconcile, and since they reflect different patterns of actual normative discourse it is not at all evident that they can be reconciled in a nonmisleading way.

Section 12.1 introduces rules and Section 12.2 some basic principles for their application to factual situations. In Sections 12.3–12.5, three types of application are investigated. Finally, Sections 12.6–12.8 are devoted to changes in the moral code (promulgations and derogations).

12.1 RULES AND INSTANTIATIONS

Normative rules can be, roughly, of the forms expressed in the following sentences:

[1] See Shafer-Landau 1997 for a discussion of particularism, i.e., the view that rules cannot be used for moral guidance.

(1) "You are not allowed to be cruel to animals."
(2) "You are allowed to read all unclassified government documents."
(3) "If you have borrowed money, then you must pay it back."
(4) "If you have a ticket to the amusement park, then you may enter the park."

(1) and (2) are nonconditional rules, whereas (3) and (4) are conditional rules.[2]

Conditional normative rules must be distinguished from other types of if-sentences with a normative consequent, such as counterfactuals (see Sections 11.3 and 11.4). Similarly, nonconditional normative rules must be distinguished from norms that refer to a particular situation. The following sentence, said to someone who beats a cat,

(5) "You must stop beating Puss,"

differs from (1) in offering a norm for only one situation, namely the present one.

It is an important feature of rules that they are general. To express their generality, a formal language for rule-bound reasoning should include individual variables, and the language should be closed under instantiation of these variables. For a simple example, the norm

(6) If you borrow money from someone, then you ought to pay it back.

can be expressed as

$$b(i,j) \Rightarrow Op(i,j).$$

Here, b and p are predicates denoting borrowing and paying back, respectively, and i, j are individual variables.[3] The symbol $\Rightarrow$ is the *rule connective* that combines the antecedent and the consequent into a rule. It corresponds to a particular usage of "if . . . then" in informal

[2] I use the term "nonconditional" rather than the more common "unconditional," since the latter has too strong a connotation of absoluteness or indefeasibility. Note the distinction between conditionality and defeasibility. Both conditional and nonconditional rules can be defeasible.

[3] For simplicity, I leave out here the obvious specifications that are needed. In a less incomplete formalization of this example, the predicates should be specified with respect to sums, points in time, etc.

language. For the moment, its logical properties can be left open. (They will be investigated in Section 12.3.)

The rule of our example has the instantiation

$$b(\text{Smith, Jones}) \Rightarrow Op(\text{Smith, Jones})$$

that can be applied when Smith has borrowed money from Jones.

One way to express the generality of rules is to include universal quantifiers in the language. For simplicity, another method will be used here. Our formal language will include individual variables, which are implicitly understood as universally quantified whenever they occur in a rule. The following definition introduces the main components of a formal language suited for the analysis of normative rules and their application.

Definition 12.1. $I_C = \{1, 2, \ldots\}$ *is a set of (individual)* constants *that represent individuals.* $I_V = \{i, j \ldots\}$ *is a set of (individual)* variables *that range over* I_C.

The descriptive language $\mathcal{L}_D$ *consists of (1) atomic descriptive sentences, formed with n-ary predicates,* $n \geq 0$*, that take elements of* $I_C \cup I_V$ *as arguments, and (2) truth-functional combinations of atomic descriptive sentences. Elements of* $\mathcal{L}_D$ *are called* descriptive sentences.

The normative language $\mathcal{L}_N$ *consists of (1) atomic normative sentences that have the form* $O\alpha$*, with* $\alpha \in \mathcal{L}_D$*, and (2) truth-functional combinations of atomic normative sentences. Elements of* $\mathcal{L}_N$ *are called* normative sentences.

The variable-free fraction $\mathcal{L}_{DC}$ *of* $\mathcal{L}_D$ *consists of those elements of* $\mathcal{L}_D$ *that contain no instance of* I_V*. The variable-free fraction* $\mathcal{L}_{NC}$ *of* $\mathcal{L}_N$ *consists of those elements of* $\mathcal{L}_N$ *that contain no instance of* I_V.

The full language $\mathcal{L}$ *is the closure of* $\mathcal{L}_D \cup \mathcal{L}_N$ *under truth-functional operations and the rule connective* $\Rightarrow$.[4]

Cn *is a consequence operation on* $\mathcal{L}$*. It is supraclassical, compact, and satisfies the deduction property.*

Hence, in this chapter, the normative language will – for simplicity – be assumed to contain only one prescriptive predicate, O. The consequence operator Cn includes the logical principles adopted for this

[4] Full closure of the language under $\Rightarrow$ is assumed here only for the sake of simplicity. Formulas with repeated use of $\Rightarrow$, such as $(p \Rightarrow \delta) \Rightarrow q$, have no intended interpretation and will not be used.

predicate. As far as possible, full generality will be maintained with respect to which these logical principles are.

An instantiation is the outcome of a uniform substitution of individual constants for individual variables in an expression. We understand instantiation in the sense of full instantiation, that is, all variables are eliminated:

Definition 12.2. *Let* $\alpha \in \mathcal{L}$. *Then* $I(\alpha)$ *is the set of instantiations of* α, *that is,* $\beta \in I(\alpha)$ *if and only if* β *can be obtained from* α *through uniform substitution of an element of* I_C *for each element of* I_V *that appears in* α.

Furthermore, for any subset A of $\mathcal{L}$, $I(A) = \cup\{I(\alpha) \mid \alpha \in A\}$.

As was indicated above, conditional rules will be denoted $p \Rightarrow \delta$, where p is a descriptive sentence and δ a normative sentence. Nonconditional norms can be regarded as conditional norms with an empty, that is, tautological antecedent; hence, they can be denoted $\top \Rightarrow \delta$, where δ is a normative sentence.

Definition 12.3. *A* normative rule *is a sentence of the form* $p \Rightarrow \delta$, *with* $p \in \mathcal{L}_D$ *and* $\delta \in \mathcal{L}_N$.

Rules do not come alone, but as constituents of legal or moral codes. Such a code is a system consisting of a set of normative rules and procedures for applying them. The set N of normative rules will in typical cases contain both conditional and nonconditional rules.[5] Usually, the elements of a normative code will contain no individual constants, only individual variables, but this will not be mandatory.

12.2 TWO TYPES OF APPLICATION

The intended interpretation of normative rules should be obvious: A rule such as $p \Rightarrow Oq$ or $p \Rightarrow \neg Or$ is intended for application in all situations in which the antecedent (p) is true. However, the "all" of "all situations" should be taken with a grain of salt. The application of norms is not necessarily exceptionless.

[5] A legal code, in particular, also has other constituents such as legal definitions. They will not be treated here. (See Hansson and Makinson 1997, pp. 327–328, for a tentative attempt to include legal definitions in the formal model.)

In most legal systems, even strict rules such as that which prohibits homicide admit of exceptions in extreme cases. The same applies to any moral code that is action-guiding and preserves its action-guiding capacity when two rules conflict so that they cannot both be complied with. Indeed, as was noted by W. K. Frankena, it would be "very difficult to find a concrete rule which one would insist could never be rightly broken."[6] For the purposes of formal representation, there are essentially two ways to deal with this problem. According to what we may call the *defeasibility* account, rules are defeasible and can be rightly broken when they are defeated by other rules. According to the *exception clause* account, all rules have (implicit) conditions that exclude all cases in which they seem to be rightly broken.[7] As an example, let us consider the rule

(R1) Do not break a promise.

Assuming that promises can be rightly broken in order to save lives, the exception clause account sees (R1) as an abbreviated formulation of a rule such as

(R1′) Do not break a promise, unless keeping it would cost someone her life.

According to the defeasibility account, (R1) is the actual rule, but it can be defeated by other rules, such as

(R2) Do not cause the death of anyone.

The exception clause interpretation is counterintuitive, as we can see from the process whereby the supposed exception clauses can be discovered. They can hardly be detected by scrutiny of the rule in question alone or its motivating principles. Instead, they will be discovered when we consider examples or motivating principles that speak against it as a general rule and give rise to competing rules that are considered to have in some situations a stronger influence. Normal moral or legal reasoning takes the form of weighing (R1) and (R2) against each other, not of discovering an underlying rule (R1′) that is independent of (R2).

[6] Frankena 1952, p. 194.
[7] Cf. Al-Hibri 1980.

Furthermore, in order to capture the intricacies of actual moral and legal systems, exception clauses would have to be limitlessly complex. In particular, "weak" rules such as minor traffic rules would have to contain implicitly all the stronger rules of the system that could influence their validity under different circumstances. There could be no rule about stopping at red lights that did not contain stipulations about various emergencies such as the prevention of traffic accidents, the transportation of severely ill persons to the hospital, and the like. This has been pointed out by numerous authors. S. C. Coval and J. C. Smith noted that the circumstances that give rise to exceptions "are so varied and numerous that they often would not be thought of until they happen, so one could never get workable and complete definitions."[8] James Nickel observed that "a right containing sufficient qualifications and exceptions to avoid all possible conflicts would probably be too complex to be teachable to humans."[9] According to Frances Kamm, the view of duties as fully specified "is problematic, since it implies that, given that we never plot out all the exceptions in advance, we never know what our duties are. Also, it has difficulty accounting for compensation that we may owe for having failed to [fulfill an overridden duty to] keep our appointment, since it implies that we have not failed in any duty."[10]

In summary, the defeasibility account of rule-breaking is much more realistic than the exception clause account. By choosing this interpretation, we save ourselves from excessively complex structures of individual rules, but instead we have to deal with more complex notions of application.

The word "application" is ambiguous. There is a sense in which the rule "It is forbidden to kill another person" is applicable even in cases of justified self-defence when someone is, as a last resort, allowed to take the life of another. If such a killing is tried in court, the cited rule is one of those that the judges have to take into account. There is also another sense in which we can say that the rule was not applicable in this case. Application in the first sense will be called *unrestrained application* and in the second sense *restrained application*.[11] Generally speaking, unrestrained application is carried through in full even when

[8] Coval and Smith 1982, p. 455.

[9] Nickel 1982, p. 257. [10] Kamm 1985, pp. 121–122.

[11] This terminology was introduced in Hansson and Makinson 1997. Makinson (1999) also uses the terms "gross output" and "net output." The distinction is akin to that between prima facie and overall obligations. See Ross 1930, 1939.

it gives rise to contradiction. Restrained application is carried through as fully as is compatible with a principled avoidance of contradictions or of other consequences specified as undesirable.

Unrestrained application will be treated in Section 12.3 and restrained application in Sections 12.4 and 12.5.

12.3 UNRESTRAINED APPLICATION

Every application of a rule refers to a factual situation. Such a situation can be represented by the set of descriptive sentences that it assumes to be true. This set should be consistent and closed under logical consequence.[12] In typical cases, it will contain individual constants (elements of I_C), but since it refers to actual facts rather than general rules, it should not contain any individual variables (elements of I_V):

Definition 12.4. *A set S is a* situation *if and only if:*
(i) $S = \mathrm{Cn}(S) \cap \mathcal{L}_{DC}$, *and*
(ii) $\perp \notin S$.

A normative rule $p \Rightarrow \delta$ is intended to allow us to detach δ whenever p holds. Letting $a(N,S)$ denote the outcome of unrestrained application of N to S, we therefore have the following tentative principle for unrestrained application:

(1) If $p \in S$ and $p \Rightarrow \delta \in N$, then $\delta \in a(N,S)$.

Unfortunately, (1) is inaccurate since it does not take into account the distinction between individual constants and variables. S contains only individual constants, whereas in typical cases, N contains only individual variables. Let us again consider the repayment example from Section 12.1. Let

$$N = \{b(i,j) \Rightarrow Op(i,j)\}$$
$$S = \mathrm{Cn}(\{b(1,2)\})$$

Here, (1) is compatible with $a(N,S) = \emptyset$, which is counterintuitive; clearly, we want $Op(1,2) \in a(N,S)$ to follow by detachment in a case

[12] An alternative approach is to allow S not to be closed, but instead to construct operators of application in such a manner that if $\mathrm{Cn}(S) \cap \mathcal{L}_{DC} = \mathrm{Cn}(S') \cap \mathcal{L}_{DC}$, then the effect of applying any set of norms to S is the same as that of applying it to S'.

like this. There is an obvious way to rectify this error: Detachment should be performed on the set $I(N)$ of instantiations of N rather than on N itself. In this way, we obtain the following improved detachment principle for unrestrained application:

(2) If $p \in S$ and $p \Rightarrow \delta \in I(N)$, then $\delta \in a(N,S)$. (*simple detachment*)

In the example just given, $b(1,2) \Rightarrow Op(1,2) \in I(N)$ and thus $Op(1,2) \in a(N,S)$, as desired.

Simple detachment is a plausible principle, but it is not sufficient. To begin with, $a(N,S)$ should be closed under the logical principles that have been adopted for O. Suppose, for instance, that agglomeration holds for O. As an illustration, we may think of Cynthia, who is a civil servant at a county administration. She is responsible for responding to requests from the public for access to documents from the administration's files. In response to such a request, she may either hand out the requested document for reading or photocopying or refrain from doing so. Furthermore, she may either report her decision to her superior or refrain from doing so. Due to the agency's policy for openness in ecological matters, there is a rule that all decisions not to hand out a document concerning the environment must be reported to the superior officer. In other words, if the document concerns the environment (e), then Cynthia must either report (r) or hand out the document (h):

$$e \Rightarrow O(r \vee h)$$

According to another rule, if the requested document contains military information (m), then Cynthia must report to her superior unless she decides to refuse access to the document:

$$m \Rightarrow O(r \vee \neg h)$$

On one occasion, a document is requested that contains an analysis of the ecological effects of certain military activities. We then have (abstracting from any additional contents of N or S):

$$N = \{e \Rightarrow O(r \vee h), m \Rightarrow O(r \vee \neg h)\}$$
$$S = \text{Cn}(\{e,m\})$$

It follows by simple detachment that $O(r \vee h), O(r \vee \neg h) \in a(N,S)$. Since, by assumption, O satisfies agglomeration, we also expect Or to follow by unrestrained application in this case. This, however, does not follow from simple detachment. To avoid this unwanted limitation, we need to close $a(N,S)$ under Cn, or more precisely under those consequences under Cn that are eligible as actual norms, namely those that are variable-free normative sentences:

(3) $\mathrm{Cn}(a(N,S)) \cap \mathcal{L}_{\mathrm{NC}} \subseteq a(N,S)$ (*logical closure*).

The combination of simple detachment and logical closure yields the correct output in the cases that most easily spring to mind. However, there are cases in which it is insufficient. Consider the following case from a rule book for antidote treatment:

$N = \{p_1(j)\&m(i,j) \Rightarrow O(x(i,j)), p_2(j)\&m(i,j) \Rightarrow O(x(i,j))\}$
$S = \mathrm{Cn}(\{p_1(1) \vee p_2(1), m(2,1)\})$

where

$m(i,j)$ i is a medical doctor responsible for the treatment of j
$p_1(i)$ i has tributyl phosphate poisoning
$p_2(i)$ i has malathion poisoning
$x(i,j)$ i administers atropine to j

A reasonable operation of (restrained or unrestrained) application should in this case yield an output that contains $O(x(2,1))$. However, this result cannot be obtained by simple detachment and logical closure. One way to make it obtainable is to allow for the disjunctive combination of antecedents:

(4) If $p_1 \vee \ldots \vee p_n \in S$ and $p_1 \Rightarrow \delta, \ldots p_n \Rightarrow \delta \in I(N)$, then $\delta \in a(N,S)$.

There are also cases in which consequents may have to be combined disjunctively. Suppose that the penal code contains the following two rules: A person who has committed murder should be sentenced to at least eight years in prison; and a person guilty of complicity in murder should be sentenced to at least three years in prison. In a particular case, it can be proved that the defendant has committed one of these two crimes, but it remains unknown which of them he committed. We then have the following case (again, of course, leaving out the details):

$N = \{m(i)\&j(k,i) \Rightarrow Os_8(k,i)), c(i)\&j(k,i) \Rightarrow Os_3(k,i))\}$
$S = \mathrm{Cn}(\{m(1) \vee c(1), j(2,1)\})$

where

$m(i)$ i is guilty of murder
$c(i)$ i is guilty of complicity in murder

$j(k,i)$ k is a judge responsible for i's case
$s_3(k,i)$ k sentences i to at least three years in prison
$s_8(k,i)$ k sentences i to at least eight years in prison

Intuitively, it is quite clear that $O(s_8(2,1) \vee s_3(2,1)) \in a(N,S)$ should hold (and probably also that $Os_3(2,1)$ should follow from $O(s_8(2,1) \vee s_3(2,1)))$. However, (4) is compatible – even when combined with logical closure – with the unintuitive outcome $a(N,S) = \emptyset$ in this case. This can be remedied by strengthening (4) as follows:

(5) If $p_1 \vee \ldots \vee p_n \in S$ and $p_1 \Rightarrow \delta_1, \ldots p_n \Rightarrow \delta_n \in I(N)$, then $\delta_1 \vee \ldots \vee \delta_n \in a(N,S)$. (*disjunctive detachment*)

Since simple detachment is a special case of disjunctive detachment, we can summarize all of this as a requirement that $a(N,S)$ should satisfy the two principles of disjunctive detachment and logical closure.

Definition 12.5. *The operation of* standard unrestrained application *is the operation a such that for each set N of normative rules and each situation S, a(N,S) is the smallest set that satisfies the following two postulates:*

If $p_1 \vee \ldots \vee p_n \in \mathrm{S}$ *and* $p_1 \Rightarrow \delta_1, \ldots p_n \Rightarrow \delta_n \in I(N)$, *then* $\delta_1 \vee \ldots \vee \delta_n \in a(N,S)$. (disjunctive detachment)

$\mathrm{Cn}(a(N,S)) \cap \mathcal{L}_{NC} \subseteq a(N,S)$ (logical closure)

Under a weak condition of logical separability between $\mathcal{L}_D$ and $\mathcal{L}_N$, the rule connective ($\Rightarrow$) can, for the purposes of standard unrestrained application, be replaced by material implication ($\rightarrow$).

Definition 12.6. *Let* Cn *be a consequence operator on the language* $\mathcal{L}$ *and let* $\mathcal{L}_1$ *and* $\mathcal{L}_2$ *be subsets of* $\mathcal{L}$. *Then* $\mathcal{L}_2$ *is* separable *from* $\mathcal{L}_1$, *under* Cn, *if and only if it holds for each* $\alpha \in \mathcal{L}_1$ *and* $\beta \in \mathcal{L}_2$ *that:*

If $\beta \in \mathrm{Cn}(\{\alpha\})$, *then either* $\beta \in \mathrm{Cn}(\emptyset)$ *or* $\perp \in \mathrm{Cn}(\{\alpha\})$.

Theorem 12.7. *Let* $\mathcal{L}_N$ *be separable from* $\mathcal{L}_D$. *Let N be a finite set of normative rules, and let* $\mathcal{N}$ *be the result of replacing in N each instance of* $\Rightarrow$ *(the rule connective) by* $\rightarrow$ *(material implication). Then:*

$$a(N,S) = \mathrm{Cn}(I(\mathcal{N}) \cup S) \cap \mathcal{L}_{NC}$$

This result may be somewhat surprising, at least until it has been clarified that even if we use material implication as a rule connective, it does *not* follow that the set of normative rules is closed under logical consequence. If, for instance, $p_1 \rightarrow Oq$ is true, then so is $p_1 \& p_2 \rightarrow Oq$. However, from the fact that the normative rule $p_1 \rightarrow Oq$ is included in a certain legal or moral code, it does not follow that $p_1 \& p_2 \rightarrow Oq$ is included in that same code.

Similarly, it should be noted that the use of material implication as a rule connective does *not* imply that all true sentences are included in the (moral or legal) code. If $\neg p$ is true, then so is $p \rightarrow Oq$. (If it is true that "I did not drink coffee this morning," then it is also true that "If I drank coffee this morning, then I ought to kill the person who made the coffee.") However, even if the sentence $p \rightarrow Oq$ is true in this case, it does not follow that the normative rule expressible by this sentence is part of any legal or moral code.

If material implication is used as a rule connective, then the connective is redundant for nonconditional rules: Instead of $\mathsf{T} \rightarrow Op \in N$, we may equivalently write $Op \in N$. This means that, just as in natural languages (and in previous deontic logics), the same expression (Op) is used to denote an actual obligation and an obligation-creating rule.

In what follows, generality will be maintained with respect to the properties of unrestrained application. Therefore, the special symbol $\Rightarrow$ will be retained although it can in the standard case be replaced by material implication. The use of $\Rightarrow$ also has the important advantage of making a clear distinction between normative rules and norms applying to a particular situation. $\mathsf{T} \Rightarrow Op \in N$ cannot be reduced to $Op \in N$.

Standard unrestrained application has the following properties:[13]

Observation 12.8. *Let a denote standard unrestrained application, and let $N, N' \subseteq \mathcal{L}$ and $S, S' \subseteq \mathcal{L}_D$. Then:*

(1) If $N \subseteq N'$, then $a(N,S) \subseteq a(N',S)$ (left monotony)
(2) If $S \subseteq S'$, then $a(N,S) \subseteq a(N,S')$ (right monotony)
(3) $a(N,S) = a(a(N,S),S)$ (left idempotence)
(4) If $N \subseteq N' \subseteq N \cup a(N,S)$, then $a(N,S) = a(N',S)$ (left stability)
(5) If $a(N,S) \subseteq N' \subseteq N \cup a(N,S)$, then $a(N,S) = a(N',S)$ (strong left idempotence)

[13] Another way to study the logic of this operator is to focus on its left projection, a_S, with a constant situation S and such that for all N, $a_S(N)) = a(N,S)$. See Hansson and Makinson 1997, p. 317.

In restrained application, we require that the outcome be logically consistent. We may also want it to be obeyable, which is a stronger requirement than consistency. To begin with, let us treat the basic condition of consistency. Obeyable application is discussed in Section 12.5.

As noted in Section 12.1, the logical properties of O should be included in Cn. Since a set $A \subseteq \mathcal{L}$ is inconsistent if and only if it implies the inconsistent sentence $\perp$, the standard for consistency will depend on these properties. If, for instance, $\neg O\perp$ and agglomeration both hold, then any set containing both Op and $O\neg p$ is inconsistent, which need not be the case if these properties do not both hold.[14]

Consistent application, to be denoted $c(N,S)$, differs from unrestrained application in satisfying the requirement that $\perp \notin \text{Cn}(c(N,S))$. Furthermore, even when a set cannot be completely complied with, it should be complied with as far as possible. Therefore, consistent application should include a procedure for maximalization within the limits of consistency and selection from among the maximals – briefly, maximalization-and-choice.

Unrestrained application, as introduced in Section 12.3, consists of three suboperations: instantiation, detachment, and logical closure. We can construct consistent (or otherwise restrained) application by adding maximalization-and-choice as a fourth suboperation.

In order to combine the four suboperations, we first need to determine the order in which they should take place. Logical closure must be performed on detached norms, not on rules. Therefore, detachment must precede logical closure. Furthermore, instantiation must clearly take place before detachment.[15] This leaves us with four possibilities:

(i)
maximalization-and-choice
instantiation
detachment
logical closure

[14] A more substantial issue left open in this way is whether prescriptive consistency be required separately for each agent or jointly for all agents. Let $\mathcal{L}_D$ contain a "see to it that" operator E such that E_1p denotes that individual 1 sees to it that p is the case. Then clearly OE_1p & $OE_1\neg p$ is inconsistent, but should OE_1p & $OE_2\neg p$ be counted as inconsistent? (Cf. Hansson 1988b, pp. 343–344.)

[15] This can be seen from a trivial example such as $N = \{b(i) \Rightarrow \delta\}$ and $S = \text{Cn}(\{b(1)\})$. Here, $\delta \in a(N,S)$, as expected. If, instead, detachment is performed before instantiation, then the outcome will not contain δ.

(ii)
instantiation
maximalization-and-choice
detachment
logical closure

(iii)
instantiation
detachment
maximalization-and-choice
logical closure

(iv)
instantiation
detachment
logical closure
maximalization-and-choice

In option (i), maximalization-and-choice takes place on N, in option (ii) on $I(N)$, in option (iii) on the outcome of detachment on $I(N)$, and in option (iv) on $a(N,S)$.

Option (iv) does not make much sense. Maximalization-and-choice at this stage would consist of choosing a maximal consistent subset of $a(N,S)$. Since $a(N,S)$ satisfies the postulate of logical closure, if it is inconsistent, then it is equal to $\mathcal{L}_{NC}$ (the set of all variable-free normative sentences). Hence, the maximal consistent subset of $a(N,S)$ need have no relation at all to the particular code N and situation S that gave rise to the inconsistency.

Option (iii) is subject to a similar problem, as can be seen from the following example. We are concerned with the fire department in a small town, without sufficient resources to show up in two places simultaneously.

$N = \{f(i) \Rightarrow Od(i), c(i) \Rightarrow Od(i)\}$
$S_1 = \text{Cn}(\{f(1), c(2), \neg f(2)\})$
$S_2 = \text{Cn}(\{f(2), c(1), \neg f(1)\})$

where

$f(i)$ i's house is on fire
$c(i)$ i's cat has climbed a tree and refuses to come down
$d(i)$ the fire department drives immediately to i's house

According to option (iii), maximalization-and-choice has to be performed on the same set of (detached) norms in situation S_1 as in situation S_2. Therefore, the outcome of restrained (consistent) application of N to S_1 will be the same as that of restrained application of N to S_2. This is contrary to intuition; in the first situation, we want to have $Od(1)$ but not $Od(2)$, and conversely in the second situation. More generally speaking, option (iii) does not allow us to assign different priorities to two rules with the same consequent but different antecedents.

Option (i) means that maximalization-and-choice consists of finding a maximal subset N' of a given (uninstantiated) set N of rules such that $a(N',S)$ is consistent. This is a more plausible construction, but it also faces difficulties, as will be seen from the following example: A doctor in a remote country district is working under a normative code with three rules:

$$N = \{h(i) \Rightarrow Ov(i), s(i) \Rightarrow Ov(i), c(i) \Rightarrow Ov(i)\}$$
$$S = \mathrm{Cn}(\{h(1), s(2), s(3), c(4)\})$$

where

$h(i)$ i has a heart attack
$s(i)$ i has a severed hand
$c(i)$ i has a persistent cough
$v(i)$ the doctor visits i immediately

The four inhabitants 1, 2, 3, and 4 all live in the district and their houses are so located geographically that the doctor can visit any two of them immediately, but not more.[16] Let a denote standard unrestrained application. The only maximal subsets N' of N such that $a(N',S)$ is consistent are $\{h(i) \Rightarrow Ov(i), c(i) \Rightarrow Ov(i)\}$ and $\{s(i) \Rightarrow Ov(i)\}$. Their outputs include $\{Ov(1), Ov(4)\}$ and $\{Ov(2), Ov(3)\}$, respectively. But given the relative gravity of the three medical conditions involved, only $\{Ov(1), Ov(2)\}$, $\{Ov(1), Ov(3)\}$, and $\{Ov(1), Ov(2) \vee Ov(3)\}$ seem to be acceptable recommendations for the doctor.

Option (ii), finally, consists in maximalization-and-choice at the level of instantiations of normative rules. In other words, a maximal subset K of $I(N)$ is chosen, such that $a(K,S)$ is consistent, and we define $c(N,S) = a(K,S)$. As can easily be verified, this construction can yield

[16] To represent this, S may be constructed to include $\neg(v(1) \& v(2) \& v(3))$, $\neg(v(1) \& v(2) \& v(4))$, $\neg(v(1) \& v(3) \& v(4))$, and $\neg(v(2) \& v(3) \& v(4))$. See Section 12.5.

acceptable outputs in the examples that cause problems for (i), (iii), and (iv). It is by far the least problematic of the three options and will therefore be used here.

Before we are ready for the formal definition of consistent application, the precise nature of maximalization-and-choice on instantiations needs to be clarified. If there is exactly one subset of $I(N)$ that is maximal in the relevant sense, then that subset should be chosen. However, we need a method that can also be applied if there are several maximals. The problem of multiple maximals has already been extensively discussed in the literature on both belief change and nonmonotonic reasoning. The main possible responses are well known from these areas. If M is the collection of all maximal sets of a desired kind, then we use a selection function γ to choose a nonempty subset $\gamma(M)$ of M, and we intersect the elements of $\gamma(M)$, putting $\cap\gamma(M)$ as the desired output. This is the most general solution, called *partial meet* selection. It has two important limiting cases. In one of these, called *maxichoice* selection, γ always chooses a singleton, which amounts to choosing a single element of M. In the other limiting case, the *full meet* case, $\gamma(M)$ is always M itself, and hence the output is $\cap M$.[17]

In belief change, the most reasonable response is usually to employ the general procedure of partial meet selection, rather than any of its two limiting cases. Both limiting cases have turned out to be problematic in their epistemological interpretations.[18] In the application of normative rules, this is different. When I know that either p or q is true, but do not know which, it is in typical cases a reasonable response to believe in neither of them. When I have two obligations, Op and Oq, but cannot perform both, this is not a good reason to perform neither of them. Normative rules should be applied to as large an extent as possible. Therefore, maxichoice selection is an appropriate method for dealing with multiple maximals in the maximalization-and-choice suboperation of consistent application.[19] We can assume that

[17] Alchourrón, Gärdenfors, and Makinson 1985.

[18] Hansson 1999c, Sections 2.4, 3.3, 3.6, and 4.4.

[19] An argument to the contrary was offered by the late Carlos Alchourrón in a personal communication to David Makinson, reported in Hansson and Makinson 1997, p. 321: The code may contain the rule that whenever persons are in certain circumstances, they are obliged to perform a certain action. It may happen that more than one person is in those circumstances, but it is impossible that more than one carry out the action. One maximal set would oblige one person to act, another

consistent application is based on a partial meet operation, and that maxichoice is a plausible variant whereas full meet, the other limiting case, is not.

Definition 12.9. *γ is a* selection function *for the set $\mathcal{F}$ of sets if and only if for all sets $\mathcal{D} \in \mathcal{F}$:*

(1) $\gamma(\mathcal{D}) \subseteq \mathcal{D}$, and
(2) if $\mathcal{D} \neq \emptyset$, then $\gamma(\mathcal{D}) \neq \emptyset$.

Furthermore, γ is a maxichoice selection function *if and only if it also satisfies:*

(3) $\gamma(\mathcal{D})$ has at most one element.

Definition 12.10. *Let M be a set of normative rules and let $S \subseteq \mathcal{L}_D$.*

Then $cons(M,S)$, the set of maximal consistent S-applied subsets *of M, is the set such that $K \in cons(M,S)$ if and only if:*

(1) $K \subseteq M$
(2) $\bot \notin a(K,S)$
(3) If $K \subset K' \subseteq M$, then $\bot \in a(K',S)$

Definition 12.11. *Let a be an operator of unrestrained application. The operator c is an operator of* partial meet consistent application, *based on a, if and only if there is a selection function γ such that for all sets N of normative rules and all $S \subseteq \mathcal{L}_D$:*

$$c(N,S) = a(\cap\gamma(cons(I(N),S)),S)$$

maximal set would oblige another to act. Yet the judge may also be bound by a higher-order constraint: Never favour or prejudice one individual over another without sufficient reason. In the absence of sufficient reason, neither of the two individuals may legitimately be singled out for the duty, and so maxichoice selection is inappropriate.

However, in such an example neither full nor, more generally, partial meet selection improves matters. For by the symmetry of the case, if one individual is relieved of the duty, so should be the other, so that the desired act is not required of anybody. It would appear that in such a case the only rational procedure is to pass into a new dimension of "creative application": Either split the required action up into parts and oblige different individuals to perform different parts, or add to the rules available, or reinterpret their contents, or find additional facts, in order to render the code or the situation asymmetric, justifying selection of one individual over another. Here, such creative applications will not be considered. (They are presumably resistant to formalization.)

c(,) is an operator of maxiconsistent application *if and only if* γ *is a maxichoice selection function.*

If a is the standard operator and S is finite-based, then $c(N,S)$ can be simplified in a way that brings out similarities between partial meet consistent application and AGM revision. Let $\&S$ denote the conjunction of the elements of S. Then for any set K of normative rules (with $\rightarrow$ as the rule connective):

$$\begin{aligned}&\bot \in a(I(K),S)\\ &\text{if and only if } \bot \in \mathrm{Cn}(I(K) \cup S) \cap \mathcal{L}_{NC} \text{ (Theorem 12.7)}\\ &\text{if and only if } \bot \in \mathrm{Cn}(I(K) \cup S)\\ &\text{if and only if } \neg\&S \in \mathrm{Cn}(I(K))\end{aligned}$$

It follows from this that:

$$cons(M,S) = I(M)\bot\neg\&S$$

($X\bot y$ is the set of maximal subsets of X that do not imply y.) We can therefore express $c(N,S)$ as follows, assuming that $*$ is the partial meet revision operator on $I(N)$ associated with the selection function γ:

$$\begin{aligned}&c(N,S)\\ &= a(\cap\gamma(cons(I(N),S)),S)\\ &= a(\cap\gamma(I(N)\bot\neg\&S),S)\\ &= \mathrm{Cn}(\cap\gamma(I(N)\bot\neg\&S)\cup S)\cap\mathcal{L}_{NC} \text{ (Theorem 12.7)}\\ &= \mathrm{Cn}(\cap\gamma(I(N)\bot\neg\&S)\cup\{\&S\})\cap\mathcal{L}_{NC}\\ &= \mathrm{Cn}(I(N)*\&S)\cap\mathcal{L}_{NC}\end{aligned}$$

Hence, consistent application is more similar to AGM revision than it first seems. There are three major differences: (1) Contrary to belief sets, sets of normative rules have to undergo instantiation prior to receipt of the input; (2) Contrary to belief sets, neither N nor $I(N)$ is logically closed, but the outcome is nevertheless logically closed; (3) The output is restricted to a fragment ($\mathcal{L}_{NC}$) of the language.

Observation 12.12. *Let N be a set of normative rules, and let* $S \subseteq \mathcal{L}_D$. *Let c(,) be an operator of partial meet consistent application that is based on the standard unrestrained application operator. Then:*

(1) $c(N,S) = \mathrm{Cn}(c(N,S)) \cap \mathcal{L}_{NC}$ (closure)
(2) $c(N,S) \subseteq a(N,S)$ (restriction)
(3) $c(N,S) \nvdash \bot$ (consistency)

In the case of legal codes, a further specification of the mechanism for selection among elements of $I(N)$ is possible. Any well-organized legal code contains internal priorities. For example, principles of constitutional law have a stronger standing than those of nonconstitutional law, recently adopted laws have precedence over older ones of the same category, and so on. It would seem reasonable to base the ordering of subsets of $I(N)$ on this ordering of the elements of N. This can be done fairly easily under the following additional assumption:

The dominance principle. *Compliance with any number of lower norms can never outweigh noncompliance with a single higher norm.*

According to the dominance principle, ten decrees by government cannot outweigh a single law passed in parliament. This is a plausible principle in a legal context, but it cannot be expected that a moral code should be so structured as to satisfy the dominance principle.

Given the dominance principle, maximalization-and-choice on instantiations of rules in N should be broken up into steps according to a ranking of the rules themselves. Rigorously, in the finite case:

Definition 12.13. *A* dominance ordering *of a set N is a sequence $\langle N_1, \ldots N_n\rangle$, all elements of which are pairwise disjoint and such that $N = N_1 \cup \ldots \cup N_n$.*

The instantiation *of $\langle N_1, \ldots N_n\rangle$ is the sequence $I(\langle N_1, \ldots N_n\rangle) = \langle I(N_1), \ldots I(N_n)\rangle$*

Definition 12.14. *Let $\langle M_1, \ldots M_n\rangle$ be a dominance ordering and let $S \subseteq \mathcal{L}_D$. Then $cons(\langle M_1, \ldots M_n\rangle, S)$ is the set such that $K \in cons(\langle M_1, \ldots M_n\rangle, S)$ if and only if there are $K_1, \ldots, K_n$ such that $K = K_1 \cup \ldots \cup K_n$, and for each k with $1 \leq k \leq n$:*

(1) $K_k \subseteq M_k$

(2) $\left(\bigcup_{i<k} K_i\right) \cup K_k \cup S \nvdash \bot$

(3) If $K_k \subset K' \subseteq M_k$, *then* $\left(\bigcup_{i<k} K_i\right) \cup K' \cup S \vdash \bot$

It follows from the definition that if $\langle M_1, \ldots M_n\rangle$ is a dominance ordering of M, then $cons(\langle M_1, \ldots M_n\rangle, S) \subseteq cons(M, S)$.

Definition 12.15. *The operator $c(,)$ is an operator of* dominance-ordered partial meet consistent application *if and only if there is a* *selection function* γ *such that for all dominance orderings* $\langle N_1, \ldots N_n \rangle$ *of sets of normative rules and all* $S \subseteq \mathcal{L}_D$:

$$c(\langle N_1, \ldots N_n \rangle, S) = a(\cap\gamma(cons(I(\langle N_1, \ldots N_n \rangle), S)), S).$$

If γ *is a maxichoice selection function, then* $c(\ ,\ )$ *is an operator of* dominance-ordered maxiconsistent application.

12.5 OBEYABLE APPLICATION

The constructions introduced in Section 12.4 refer to *consistent* application. As was noted at the outset, consistency is only a minimal restraint in restrained application. For a legal or moral code to be action-guiding, the outputs from applying it should be obeyable.

The concept of obeyability was analyzed in Section 10.6. A prescriptive predicate O is obeyable if and only if there is some alternative (element A of $\mathcal{A}$) in which it is completely complied with (i.e., $\{p \mid Op\} \subseteq A$). The alternative set is presumed to represent the relevant sense of possibility; in other words, it should consist of exactly those courses of action that are open to the agent.

This representation of obeyability is not immediately available in the framework developed in the previous sections of this chapter, since alternative sets have up to now been dispensed with. In order to represent obeyability, we must either add alternative sets (or make some other addition that represents obeyability) or find some way of interpreting the apparatus of partial meet consistent application so that it incorporates obeyability. The latter option means that either Cn, S, or N has to include a representation of obeyability.

To begin with Cn, it is possible to modify Cn to ensure that nonobeyable injunctions are inconsistent. If both $\neg O\bot$ and the agglomeration postulate are satisfied for Cn, then this can be done simply by including $\cap\mathcal{A}$ in Cn($\emptyset$). More precisely, in order to ensure that consistent application complies with obeyability according to a certain alternative set $\mathcal{A}$, we can replace the original consequence operator Cn by a new operator Cn′, such that for all X, Cn′(X) = Cn(($\cap\mathcal{A}$)$\cup X$). It is easy to show that the standard properties of consequence operators are satisfied by Cn′ if they are satisfied by Cn.[20]

[20] Cf. Definition 3.3 and Observation 3.4. See also Hansson 1992c, p. 102.

To see how this works, let $\mathcal{L}_{\mathcal{A}}$ contain all the action-describing sentences under consideration, and let $p\&q$ be such a sentence. Let $p\&q \notin A$ for all $A \in \mathcal{A}$. Provided that $\mathcal{A}$ is contextually complete, it follows from this that $\neg(p\&q) \in A$ for all $A \in \mathcal{A}$, hence $\neg p \vee \neg q \in \cap\mathcal{A}$. Suppose that Op & $Oq \in c(N,S)$. Then, since $c(N,S)$ is Cn′-closed and agglomeration is included in Cn′, $O(p\&q) \in c(N,S)$, hence since $p\&q$ and $\bot$ are logically equivalent under Cn′, $O\bot \in c(N,S)$, contrary to $\neg O\bot \in \text{Cn}(\emptyset)$ and the consistency of $c(N,S)$. We can conclude that Op & $Oq \notin c(N,S)$.

This method is appealingly simple. Unfortunately, it has the unintuitive effect that the logic has to be changed if the alternative set is changed. In order to avoid this strange feature, we can instead include $\cap\mathcal{A}$ in S, in other words let the specification of the situation contain a complete specification of the standard for obeyability. The reason this works is that due to Theorem 12.7, we can write the definition of $c(\ ,\ )$ in such a way that Cn only appears in subformulas of the form $\text{Cn}(\Xi \cup S)$. Therefore, including $\cap\mathcal{A}$ in S has the same effect as including it in Cn – but with the important difference that we now allow the standard of obeyability to be different for different situations, while the logic remains the same.

However, this construction is not without its problems. It requires that we give up the possibility of combining one and the same situation with different alternative sets (cf. Section 9.3). Furthermore, it gives rise to the following monotonicity property:

Monotonicity of infeasibility. *If p is infeasible (not included in any alternative) in the alternative set associated with a situation S, and $S \subseteq S'$, then p is infeasible in the alternative set associated with S'.*

(This follows from the simple fact that if $\cap\mathcal{A} \subseteq S$, and $S \subseteq S'$, then $\cap\mathcal{A} \subseteq S'$.) It is not difficult to construct cases in which this monotonicity property is implausible. For one example, let S be a situation such that you are alone on a desert island 1 mile from the mainland, with no boat or aircraft available. It is then reasonable to assume that the alternative set associated with S contains no alternative in which you immediately set off for the mainland. Hence, if m expresses that you go immediately to the mainland, then $\neg m \in S$.

Next, let t denote that there is a tunnel through which you can walk to the mainland. Let $S' = \text{Cn}(S \cup \{t\})$. Then, since Cn satisfies monotony, we have $\neg m \in S'$, contrary to intuition.[21] Hence, although letting S include information about the alternative set is a useful approximation in many cases, it is not a general solution to our problem.

The third option mentioned above, namely to let N include information about the alternative set, is worth no further investigation since N should contain no elements of $\mathcal{L}_D$ and since it is much more credible that the set of available possibilities is a function of the situation than that it is a function of the set of normative rules.

This leaves us with one more option to investigate, namely the introduction of an explicit representation of the alternative set. The formal development of this proposal is straightforward: The set N of rules is then applied, not to a situation S but to a pair $\langle S,\mathcal{A}\rangle$ of a situation and an alternative set.

Definition 12.16. *The operator* $ç(\ ,\ ,\ )$ *is an operator of* partial meet obeyable application *if and only if there is an operator* $c(\ ,\ )$ *of partial meet consistent application, such that for all sets N of normative rules, all* $S \subseteq \mathcal{L}_D$, *and all alternative sets* $\mathcal{A}$:

$$ç(N,S,\mathcal{A}) = c(N,S\cup(\cap\mathcal{A}))$$

If $c(\ ,\ )$ *is maxiconsistent, then* $ç(\ ,\ ,\ )$ *is an operator of* maximal obeyable application.

Hence, we have two passable models for obeyable application: one in which standards of obeyability are included in the representation of situations (S) and another in which they are separately represented ($\mathcal{A}$). Although the former model has advantages in terms of notational economy, the latter has the important advantages of being both more general and more realistic.

[21] One possible response to this problem is to restrict the definition of a situation (Definition 12.4) so that only some, not all, closed, consistent subsets of $\mathcal{L}_{DC}$ qualify as situations. Then, although S in this example is a situation, S' need not be. (The outcome of incorporating t into S will have to be represented by some other set, which does not include $\neg m$.) It may indeed be the case that there are few if any situations S and S' such that $\varnothing \subset S \subset S'$. Then monotonicity of infeasibility holds, but it is innocuous since its antecedent is not satisfied in the potentially problematic cases. In Rott 1989 and Hansson 1992d, a similar solution was chosen for a related problem in the theory of conditionals.

Changes in sets of normative rules must be distinguished from changes in norms that refer to a particular situation. One of these types of change may occur without the other. A rule that has no application can be abolished or introduced without influencing any actual norms. On the other hand, actual norms can be changed due to changes in the factual situation, although the normative system is unchanged.

$c(N,S)$ can be changed due to changes in either N or S, and $ç(N,S,\mathcal{A})$ due to changes in either N, S, or $\mathcal{A}$. A complete account of the dynamics of actual norms will therefore have to be a fairly complex theory that allows for changes both in the normative system and in the factual situation. No such all-embracing theory will be attempted here. Instead, the focus will be on changes in sets of normative rules, that is, changes in legal or moral codes.

Changes in the set of rules (normative change) must be distinguished from applications. Application, restrained or not, always refers to a particular case requiring resolution. When performing a consistent application, we shelve certain of the rules (or rather, certain instantiations of the rules), not using them for the case at hand. However, these rules are not deleted, and when dealing with another case we return to the full set of rules as our starting point. Normative change, on the other hand, is performed on the rules themselves and is not directly concerned with individual cases. In normative change, certain rules may be deleted, after which they cannot be used unless reinstated. Application is the task of judges, normative change the task of legislatures. Operations of normative change can be divided into two major classes. Operations that add a normative rule to the code are called promulgations (or amendments; the former term will be used here). Operations that remove a normative rule from the legal or moral code are called derogations.

12.7 PROMULGATION

Two major forms of promulgation have been proposed in the literature. In what may be called *formal promulgation*, a new rule is added to the code, without anything else being changed.[22] In what may be

[22] This is the only type of promulgation recognized by Bulygin and Alchourrón (1977, esp. p. 27).

called *material promulgation*, previous parts of the code are deleted whenever that is needed to give room to the new piece of legislation.[23]

Formal promulgation corresponds much better than material promulgation to the way new statutes are added to actual legal codes. Suppose that Parliament, when issuing a new law, fails to proclaim that a certain old law is thereby superseded. Editors of statute books are not then entitled to remove the old law from the books. Even if it will be defeated by the newer law in all potential cases of application, it is still part of the legal code (and may become effective again if the newer law is derogated).

Formal promulgation leaves us with more conflicts in the legal code than does material promulgation. Much of the task that has been assigned to (material) promulgation, namely the task of selecting among conflicting rules, is in actual practice performed in the process of (restrained) application. This means that the resolution of conflicts between legal rules takes place in a more situation-dependent and case-by-case manner than has often been assumed in logic-oriented studies of legal change.

Since changes in moral codes are less well organized than changes in legal codes, the structure of promulgation in moral contexts is more difficult to pin down. My tentative suggestion is that the acquisition of new moral principles does not typically cause previously accepted conflicting principles to be given up immediately and once and for all. To the contrary, the previous principles are retained unless subjected to separate operations of derogation. To the extent that this suggestion is correct, formal rather than material promulgation will be the most adequate representation of promulgations in moral as well as legal codes.

The most obvious way to represent formal promulgation is set-theoretical addition. The addition of a normative rule δ to a set N of normative rules would result in the set $N \cup \{\delta\}$ of normative rules. Unfortunately, this simple solution does not in general provide us with a workable legal code. As we saw in Section 12.4, the rules included in a legal code are ordered in terms of their priority. A new rule must be assigned a place in that ordering.

[23] This type of promulgation is discussed in Alchourrón and Makinson 1981, esp. p. 130.

As a practical illustration of this, suppose a new statute is issued that prohibits the use of sexist language. It makes a big difference whether the new statute is, say, a government edict or an amendment of the constitution. In the former case, it will have a much weaker position in relation to constitutional guarantees of freedom of speech. In order to specify the effects of a promulgation, it is necessary to establish the position of the new rule in the system of priorities.

In order to express this formally, let the state of the legal code prior to promulgation be represented by the dominance ordering $\langle N_1, \dots N_n \rangle$. Single-sentence inputs can be given the form $\langle \emptyset, \emptyset, \dots \{\delta\} \dots, \emptyset \rangle$, where δ has the position intended for it in the new code. It is easy to generalize to multiple inputs of the form $\langle M_1, \dots M_n \rangle$, and promulgation can then be represented by the simple formula:

$$\langle N_1, \dots N_n \rangle \oplus \langle M_1, \dots M_n \rangle = \langle N_1 \cup M_1, \dots N_n \cup M_n \rangle$$

12.8 DEROGATION

There are two commonly distinguished types of derogation. *Formal derogation* of a normative rule means to "merely *drop* it from the code," leaving the rest of the code intact even if it implies the removed element.[24] *Material derogation* means to reject not only the particular normative rule, but also whatever other parts of the code are needed to prevent the remainder from implying it. The difference has also been described as that between rejecting a formulation of a norm and rejecting the norm itself.[25]

For reasons similar to those given above for promulgation, the simpler operation of formal derogation seems to be the one that best reflects how actual regulations are removed from actual legal codes. In any well-ordered jurisdiction, if old statutes are removed, this is done by explicitly listing the discarded parts of the legislation, not by proclaiming that any regulation with a certain effect is from now on not a valid regulation.

[24] Alchourrón and Makinson 1981, p. 130. These authors reserve the term "derogation" for what is here called material derogation and use "abrogation" for formal derogation.

[25] Bulygin and Alchourrón 1977, p. 27.

Formal derogation differs from promulgation in not requiring that new priorities be set. To remove a norm δ from $\langle N_1, \ldots N_k \rangle$ means to remove it from whatever level at which it was included, hence:

$$\langle N_1, \ldots N_k \rangle \ominus \delta = \langle N_1 \setminus \{\delta\}, \ldots N_k \setminus \{\delta\} \rangle$$

and more generally, for derogation of a set M of normative rules:

$$\langle N_1, \ldots N_k \rangle \ominus M = \langle N_1 \setminus M, \ldots N_k \setminus M \rangle$$

Although lawgivers and other agents who change the law do not typically perform operations that have the *form* of material derogation, they sometimes perform operations that have the *purpose* that material derogation has – namely, to see to it that a certain normative rule is no longer a part of or an effect of the legal system. Something should be said about how this purpose can be achieved. It will be helpful to treat permissions and obligations separately.

The purpose of "materially derogating" a nonconditional permission $\top \Rightarrow Pq$ from N is to see to it that q is no longer permitted – that it is forbidden. This cannot in general be achieved by deleting elements from N. Instead, the natural way to do it is to add $\top \Rightarrow O\neg q$ to N and give it sufficient priority to defeat any rule that contradicts it. Similarly, to "materially derogate" $p \Rightarrow Pq$, $p \Rightarrow O\neg q$ is added to N and given sufficient priority to prevail over all conflicting rules.[26]

This account accords with actual legal practice, as can be seen from a few simple examples. Suppose the lawgiver decides that smoking in public places should no longer be permitted. This decision is normally effectuated by prohibiting smoking in public places. Similarly, suppose that the lawgiver also wants to ensure that it is no longer permitted to drive on highways if one has poor eyesight. This is done by issuing the corresponding conditional prohibition.

The purpose of a "material derogation" of an obligation $\top \Rightarrow Oq$ is to see to it that q is no longer required, that is, to permit $\neg q$. This can be done in at least two different ways. First, a lawgiver can promulgate $\top \Rightarrow P\neg q$, giving it sufficient priority to defeat $\top \Rightarrow Oq$. Second, she may delete $\top \Rightarrow Oq$ together with whatever elements of N have to be given

[26] Hence, "material derogation" of permissions requires the addition of new rules to the set of normative rules. There is an interesting parallel here with contraction by possibility statements in belief revision. In order to remove a sentence such as "it is possible that p" from a belief set, some other sentence (such as $\neg p$) may have to be added to the belief set. See Levi 1988; Rott 1989; Fuhrmann 1991b; Hansson 1999c, Section 5.1.

up in order to ensure that $\mathsf{T} \Rightarrow Oq$ is not a logical consequence of the remainder of N. Similarly, there are two ways to "derogate materially" $p \Rightarrow Oq$: to promulgate $p \Rightarrow P \neg q$ and to delete $p \Rightarrow Oq$. In both cases, the second option (deletion of $\mathsf{T} \Rightarrow Oq$, respectively $p \Rightarrow Oq$) can be achieved with a partial meet construction. This type of operation may be useful as part of the lawgiver's preparatory deliberations, but the official decision to change the law will take the form of one or several formal derogations.

13

Legal Relations

One of the most interesting applications of deontic logic is to use it for the characterization and classification of legal relations such as rights, claims, and powers. In Section 13.1, the formal framework from Chapter 12 is supplemented in order to express the dependence of norms on symbolic actions recognized by law. In Section 13.2, a typology of legal relations is introduced, and in Section 13.3 it is used in an analysis of the concept of a right. Section 13.4 is devoted to a comparison of the present framework with Wesley Hohfeld's classic typology of legal relations.

13.1 POTESTATIVE RULES

Several authors, notably Stig Kanger, Frederick Fitch, and Lars Lindahl, have used deontic logic to characterize and classify legal relations.[1] The basic framework, first developed by Stig Kanger, makes use of two operators: a prescriptive predicate O and a dyadic action predicate E, such that E_ip means "i sees to it that p."[2] The atomic sentences on which the classification is built have the form $[\neg]O[\neg]E_i[\neg]p$, where $[\neg]$ is a placeholder that can either be deleted or replaced by a negation sign. (Hence, $OE_i\neg p$ and $\neg O\neg E_ip$ are atomic sentence types in this language, and there are six others.)

These systems are clarifying in important respects, but there are non-negligible features of actual legal relations that they do not cover. Perhaps foremost among these is that many norms are activated by

[1] Kanger 1957; Kanger and Kanger 1966; Fitch 1967; Lindahl 1977, 1994. For an overview, see Herrestad 1996. I will use the term 'legal relation' to cover normative states assigned by the legal system. (It would perhaps be more precise to reserve 'legal relation' for normative states that involve at least two persons and use 'legal position' for normative states that only refer to one person.)

[2] On this operator, see Kanger 1972; Pörn 1970, p. 11; Hansson 1986a.

symbolic actions of different kinds, typically (but not exclusively) by speech acts such as those of claiming, ordering, and permitting. The jailor must discharge the prisoner (only) if the judge orders him to do so. I am permitted to enter your home (only) if you permit me to do so. Some loans are payable (only) on demand, and so on. In Hart's words, it is "characteristic of those laws that confer rights (as distinguished from those that only impose obligations) that the obligation to perform the corresponding duty is made by law to depend on the choice of the individual who is said to have the right or the choice of some person authorized to act on his behalf."[3]

Hence, important legal relations are constituted by legal rules, according to which the normative status of actions depends on symbolic actions, which in most cases take the form of claims made and permissions granted by the concerned individuals. The examples already given can help us determine the features of such symbolic actions that are essential for their formal representation. There are various (verbal and nonverbal) ways in which you can allow me to enter your home. What they all have in common is that, in the given legal system, they count as declarations by you to the effect that I may enter your house. The symbolic contents of your action can be expressed by specifying (i) a normative sentence ("you may enter this house") and (ii) the agent (in this case you) by whom or on whose behalf a symbolic assertion of this norm is made.

Similarly, consider a case in which I must pay back my loan immediately if my creditor requires me to do so. Here, the symbolic action of the creditor can be seen as a declaration to the effect that I must pay back the loan.

The pattern common to these two examples is appealingly simple, and it seems to be important in actual legal systems. It may not exhaust the forms that symbolic action can take in legal contexts, but it is a basic pattern that should be explored before other, more complex structures are introduced.

Most of the building blocks needed to express this pattern were developed in Chapter 12. The only substantial addition needed is a representation of symbolic action. For that purpose, we can use the operator Dc, such that $Dc_i\delta$ denotes that "an action has been performed that counts as a declaration by i that δ."[4]

[3] Hart 1954, p. 49. [4] It was introduced in Hansson 1990b.

Definition 13.1. *Dc is the operator of declarations. An* atomic declarative sentence *is an expression $Dc_i\delta$ such that $i \in I_C \cup I_V$ and $\delta \in \mathcal{L}_N$. The subscript i is called the* declarant.

A declarative sentence *is a truth-functional combination of atomic declarative sentences. The set of declarative sentences is denoted $\mathcal{L}_{Dc}$.*

The symbolic action referred to in $Dc_i\delta$ will in most cases be an action by *i*, but it may also be an action by someone who represents *i*, such as a parent, a guardian, or a solicitor. It may be a speech act or some other symbolic action such as a gesture or the signing of a contract.

A legal relation that depends on symbolic action will be represented by a normative rule with a declarative antecedent:[5]

Definition 13.2. *A* potestative rule *is a sentence of the form $\alpha \Rightarrow \delta$, where $\alpha \in \mathcal{L}_{Dc}$ and $\delta \in \mathcal{L}_N$.*

This definition leaves open the representation of action and agency. In a potestative rule such as $Dc_iOp \Rightarrow Op$, we expect *p* to represent an action. However, in order to focus on the role of declarations, the structure of actions will be left unspecified.

A potestative rule confers what may be called legal "power" or "capacity," or – with the better terms that are available in German and Latin – a *Befugnis* or *potestas*. The definition provides us only with a very general framework for such powers that needs further specification in several respects. To begin with, it allows for potestative rules that make norms depend on symbolic actions with quite unrelated normative contents, such as:

$$Dc_i\delta_1 \Rightarrow \delta_2$$

where, say, δ_1 represents that John ought to pay his debt to *i* and δ_2 that he ought to go on holiday with his wife. This is not how the potestative rules of a well-organized legal system should be constructed. In such a system, those symbolic actions whose performance or nonperformance determine the status of a norm δ should count as declarations about δ. Hence, they should satisfy the following property:

[5] It is not, strictly speaking, a normative rule in the sense of Definition 12.3, since $\mathcal{L}_{Dc}$ replaces $\mathcal{L}_D$.

Definition 13.3. *A potestative rule* $\alpha \Rightarrow \delta$ *satisfies* pertinence *if and only if* α *is a truth-functional combination of sentences of the forms* $Dc_i\delta$ *or* $Dc_i\neg\delta$ *(for various* $i \in I_C \cup I_V$*).*

Pertinence excludes some of the most obviously inadequate types of potestative rules, and it allows for the most obviously reasonable ones, such as $Dc_i\delta \Rightarrow \delta$ and $\neg Dc_i\neg\delta \Rightarrow \delta$. However, there are inadequate potestative rules that it does not exclude, such as:

$$Dc_i\neg\delta \Rightarrow \delta$$
$$Dc_i\delta \;\&\; Dc_j\neg\delta \Rightarrow \delta$$

In both these cases, a symbolic action counts as being in favour of $\neg\delta$, but it nevertheless has the effect of supporting δ. Such counterproductive or contravening effects of symbolic action should not be allowed in a well-organized legal system.[6]

To express this more precisely, let us transform the antecedent into its disjunctive normal form. If pertinence is satisfied, then the antecedent α of a potestative rule $\alpha \Rightarrow \delta$ is a truth-functional combination of sentences of the form $Dc_i[\neg]\delta$. According to the disjunctive normal form theorem, α can be written in the form $\varepsilon_1 \vee \ldots \vee \varepsilon_m$, where each ε_k is a conjunction of sentences of the form $[\neg]Dc_i[\neg]\delta$. Intuitively, each disjunct ε_k is a sufficient condition for δ.

Each conjunct in ε_k has one of the four forms $Dc_i\delta$, $Dc_i\neg\delta$, $\neg Dc_i\delta$, or $\neg Dc_i\neg\delta$. That $Dc_i\delta$ and $\neg Dc_i\neg\delta$ appear in this position, in a potestative rule with δ as its consequent, is appropriate. The fact that some individual i declares that δ, or refrains from declaring that $\neg\delta$, can be a good reason for the legal system to endorse δ. However, the same cannot be said of $Dc_i\neg\delta$. In a well-ordered normative system, the fact that someone declares against δ should not contribute to a verdict in favour of δ. Such rules would encourage nontruthful symbolic actions and thereby distort the whole system of symbolic actions. For a simple example, consider the rule $Dc_i\neg\delta \Rightarrow \delta$, where δ denotes that the police are required to investigate a certain crime in which i is a victim. This rule makes much less sense than $Dc_i\delta \Rightarrow \delta$. If we wish to let i influence the legal status of δ, then it is a rule of the latter type that should be adopted.

[6] The requirement that they be excluded is akin to the decision-theoretic postulate of nonnegative response that was introduced in Section 7.3.

Conjuncts of the form $\neg Dc_i\delta$ are equally implausible, and for the same reason. (Consider the rule $\neg Dc_i\delta \Rightarrow \delta$, with δ interpreted as above.) Hence, we can reject two of the four types of conjuncts and arrive at the following:

Definition 13.4. *A potestative rule* $\alpha \Rightarrow \delta$ *is* canonical *if and only if* α *is logically equivalent with a disjunction* $\varepsilon_1 \vee \ldots \vee \varepsilon_n$, *such that each disjunct* ε_i *has the form:*

$$Dc_{s_1}\delta \;\&\ldots\&\; Dc_{s_k}\delta \;\&\; \neg Dc_{t_1}\neg\delta \;\&\ldots\&\; \neg Dc_{t_m}\neg\delta,$$

where $s_1 \ldots s_k$ *and* $t_1 \ldots t_m$ *are sequences of elements of* $I_C \cup I_V$, *and at least one of them is nonempty.*[7]

Canonical potestative rules are proposed as a general framework for expressing how norms can depend on symbolic actions.

13.2 TYPES OF LEGAL RELATIONS

In order to apply canonical potestative rules to the classification of legal relations, some further simplifications are useful.

Each disjunct ε_k in the antecedent is a sufficient condition for the consequent. Therefore, it is a reasonable approximation to replace each disjunctive potestative rule $\varepsilon_1 \vee \ldots \vee \varepsilon_n \Rightarrow \delta$ by the n nondisjunctive rules $\varepsilon_1 \Rightarrow \delta, \ldots \varepsilon_n \Rightarrow \delta$.

The consequent δ of a canonical potestative rule $\alpha \Rightarrow \delta$ may be any normative sentence, including molecular sentences such as $Oq \& Pr$. For most purposes, however, it will be sufficient to consider consequents of the two simple forms Oq and $\neg Oq$ or, equivalently, of the two forms Oq and Pq.

Definition 13.5. *A potestative rule is a* permissive rule *if it has the form* $\alpha \Rightarrow Pq$. *It is an* obligative rule *if it has the form* $\alpha \Rightarrow Oq$.

Nondisjunctive canonical permissive rules have the following form:

$$Dc_{s_1}Pq \;\&\ldots\&\; Dc_{s_k}Pq \;\&\; \neg Dc_{t_1}\neg Pq \;\&\ldots\&\; \neg Dc_{t_m}\neg Pq \Rightarrow Pq$$

and nondisjunctive canonical obligative rules have the form:

[7] A representation theorem for canonical potestative rules is reported in Hansson 1996c.

$$Dc_{s_1}Oq \mathbin{\&} \ldots \mathbin{\&} Dc_{s_k}Oq \mathbin{\&} \neg Dc_{t_1}\neg Oq \mathbin{\&} \ldots \mathbin{\&} \neg Dc_{t_m}\neg Oq \Rightarrow Oq,$$

where in both cases $s_1 \ldots s_k$ and $t_1 \ldots t_m$ are sequences of elements of $I_C \cup I_V$.

A further simplification can be achieved in the case when only one agent's symbolic actions need to be taken into account. In that case, there are three types of canonical permissive rules, namely:

$$Dc_iPq \Rightarrow Pq,$$
$$\neg Dc_i\neg Pq \Rightarrow Pq, \text{ and}$$
$$Dc_iPq \mathbin{\&} \neg Dc_i\neg Pq \Rightarrow Pq.$$

Up to now, no logical principles have been introduced for the *Dc* operator. A minimal such principle is *declarative consistency*, $\neg(Dc_i\delta \mathbin{\&} Dc_i\neg\delta)$. This is a plausible property, given the interpretation of Dc_i in terms of what *counts as* symbolic action. Of course an individual may make, in the relevant court, both a declaration that she wants a divorce and one that she does not want a divorce. However, the court, and the legal system in general, should not count both of these statements as (serious) declarations. At least one of them is cancelled by the other.

It follows from declarative consistency that the first and third of the three listed types of permissive rules have logically equivalent antecedents. Therefore, the third, more complex expression may be deleted from the list.

Similarly, after the above simplifications there are three types of canonical obligative rules, namely:

$$Dc_iOq \Rightarrow Oq,$$
$$\neg Dc_i\neg Oq \Rightarrow Oq, \text{ and}$$
$$Dc_iOq \mathbin{\&} \neg Dc_i\neg Oq \Rightarrow Oq.$$

The third of these can be omitted due to declarative consistency. Hence, in the case with only one declarant (index of the *Dc* operator), we have four canonical types of permissive or obligative potestative rules. To this can be added the two simple types of nonconditional permissive and obligative rules: $\top \Rightarrow Oq$ and $\top \Rightarrow Pq$. In this way, we arrive at the following six types of legal relations. They form a basic typology for potentially declaration-dependent legal relations.

(1) $\top \Rightarrow Oq$ (*categorical obligation*)
(2) $\top \Rightarrow Pq$ (*categorical permission*)

(3) $Dc_iOq \Rightarrow Oq$ (*claimable obligation*)
(4) $Dc_iPq \Rightarrow Pq$ (*grantable permission*)
(5) $\neg Dc_i\neg Oq \Rightarrow Oq$ (*revocable obligation*)
(6) $\neg Dc_i\neg Pq \Rightarrow Pq$ (*revocable permission*)

The sentence q in these expressions represents the *descriptive contents* of the respective rules. It contains some description of human action or behaviour. As was explained above, the distinctions with respect to agency available in the Kanger–Lindahl typologies have been repressed here. More elaborate typologies, which include these distinctions, can be obtained by introducing Kanger's agency operator into the present framework. Then q should be replaced by each of the four variants $[\neg]E_i[\neg]r$. Assuming the usual interdefinability between O and P, the original atomic Kanger types are covered by these variants of the first two of our types, categorical obligation and permission. Our types 3, 4, 5, and 6 have no counterparts in Kanger's theory.

All the six basic types represent forms of legal relations that have practical relevance, and it is not difficult to find examples of each of them:

(1′) Alan is forbidden to exert physical violence on Betty.
(2′) Betty is allowed to squander her own money.
(3′) If Alan demands that Betty shall repay this loan, then she has to do so.
(4′) If Alan allows Betty to enter this park, then she may do so.
(5′) Alan is forbidden to enter Betty's house unless she lets him do so.
(6′) Alan is allowed to walk along King's Street unless the police forbid him to walk there.

13.3 RIGHTS

Previous studies of legal relations have to a large extent aimed at explicating the concept of a (legal) right. It should therefore be of interest to investigate how the basic typology of potestative rules relates to that concept.

Legal philosophy has two major traditions in its treatment of rights, namely the benefit theory and the claims theory. According to the benefit theory, someone has a right if she is the beneficiary of another's duty or obligation. According to the claims theory, someone has a right if she may, at her option, demand the execution of someone else's duty. Such a person "has the choice of asserting or not asserting that right."[8]

The common language phrase to "have a right" (technically, to be a right-holder) is ambiguous between being a beneficiary to a right and being able to activate it by making a claim. A baby is normally said to have rights, although it cannot claim them. On the other hand, someone who can activate through her claims a benefit for someone else may also be said to have a right. The classic example of this is the son whose friend has promised him to care for his aging mother.

Rights are involved legal complexes that typically include "a host of legal liberties, powers, and duties of various officials playing diverse roles in the legal system that creates, protects, and enforces the legal right."[9] Strictly speaking, what I would call my "right" to get back the car that I loaned a friend is not a single, simple legal relation. Instead, it is a complex or cluster of legal relations of various types, including my friend's duty to return my car and various claims that I have on legal authorities to act on my behalf in the case of his noncompliance. In general, such a complex has one component that is the mainspring of the whole complex. The other components would have no motive in the absence of this one component. It will be called the *central legal relation* of the complex.[10] As an example, a property right involves a number of obligations for those other than the owner, but these are all subsidiary to the central legal relation, which is in this case the permission for the owner to use and dispose of the property as she wishes.[11]

The nature of a complex of legal relations depends to a large degree on its central legal relation. As a working hypothesis, we can therefore assume that an examination of the central legal relation of a complex is in most cases sufficient to determine if the complex is a right, in the common sense of the word. In what follows, each of the six types of legal relations that were introduced in Section 13.2 will be examined from this point of view.

[8] Williams 1978, p. 187. Cf. also Lyons 1969; Feinberg 1970.

[9] Wellman 1975, pp. 52–55.

[10] This terminology was also used in Hansson 1990b.

[11] Strictly speaking, this is not one single legal relation but a set of legal relations, one for each of the actions that the owner is entitled to perform. In general, what is called here "the central legal relation" is often a set of closely related legal relations that are all of the same type and that all have the same declarant. The singular form will be used here, for linguistic convenience.

It will be understood that a right is always something that someone has, so that if there is a right, then there is a right-holder. However, it is not immediately clear who this right-holder is. There are, generally speaking, three candidates: (1) The declarant, denoted by the index of the *Dc* operator; (2) The agent(s). The descriptive content (p in $\mathsf{T} \Rightarrow Op$, p in $Dc_i Pp \Rightarrow Pp$, etc.) represents some action, and therefore refers to some agent; (3) The beneficiary. Typical legal relations are intended to further the interest of some person. If there is both a declarant and a beneficiary, then they are in typical cases, but not in all cases, identical.

Many legal complexes that are centred around a *categorical obligation* have the purpose to protect a position-holder, who is then a beneficiary. It is customary to describe such a legal complex as a right, with the beneficiary of the central legal relation as a right-holder.

Although the central legal relation of such a complex does not have a declarant, the beneficiary is in general also a declarant in subsidiary legal relations, such as those of enforcement. However, the role as a declarant of subsidiary legal relations does not seem to be necessary for her to be called a right-holder. One can easily imagine a legal system in which violence is effectively proceeded against irrespective of the claims (or even permissions) of victims or prospective victims. It does not seem natural to say that this would deprive the individuals of their right not to be physically attacked.

Furthermore, there are complexes that are centred around a categorical obligation that has a beneficiary, such as a baby, who is unable to make declarations that activate subsidiary legal relations. Such complexes are still called rights. There is no ground, either, in the structure of legal relations to deny "that animals have rights, for instance the right not to be hurt without some good reason."[12]

Some complexes are centred around a categorical obligation that does not have a (specific) beneficiary. One example is the obligation to stop at a stop sign in a situation when no other road-user benefits from your doing so. No right seems to be involved here.

A complex centred around a *categorical permission* will normally be called a right, with the agent of the central legal relation as a right-holder. If there is a beneficiary, then she will not be regarded as a right-holder unless she is identical with the agent. For instance, I am allowed

[12] Brandt 1959, p. 440.

to give you a sum of money. This is a right that I have, not a right that you have.

A complex centred around a *claimable obligation* is always regarded as a right, with the declarant (claimant) of the central legal relation as the right-holder. She will be called a right-holder even if she is not a beneficiary. Suppose, for instance, that the central legal relation is that you shall receive a large sum of money if a certain person (who does not act on your behalf) commands that you shall be given the money. She, as the claimant, would be considered to be a right-holder, whereas you as a beneficiary would not.

In most practical cases, the claimant and the beneficiary of a (central) claimable obligation are identical. When they are different persons, the claimant is commonly someone who is supposed to act on behalf of the beneficiary, such as a parent for a child. In these cases, the person on whose behalf the claim can be made (the child) may be called a right-holder. However, it is not as a beneficiary but as a "claimant by proxy" that she has that position.

Complexes centred around a *grantable permission* are also rights. Although they are not frequently referred to, such legal complexes are very common. As an example, Alan can make it permissible for Betty to use his car. This is a right with the declarant (permitter) of the central legal relation (Alan), not its beneficiary (Betty) or its agent (also Betty) as a right-holder.

A complex centred around a *revocable obligation* is also a right, with the declarant as a right-holder. Suppose that you must pay me a sum of money unless I absolve you from paying it. Then I am clearly a right-holder.

In most practical examples of revocable obligations, the declarant is also a beneficiary. It is, however, her role as a declarant and not her role as a beneficiary that makes her a right-holder. This can be seen from the following example. Suppose that as a result of betting you have to pay €20 to the Queen of England unless I absolve you from doing this. This is clearly a right, with me and not the Queen of England as a right-holder. (Although neither I nor she has the right to receive the money, I have a right to have you perform a certain action.)

A complex centred around a *revocable permission* is a somewhat less clear case. Although a legal complex of this type would normally be called a right, it is not quite clear who is the right-holder. Suppose that a widow is allowed to give away property inherited from her husband to a friend unless her child explicitly disallows her to do so.

Table 4. *Which legal relations are rights?*

Type of central legal relation of the complex	Is it a right?	Who is the right-holder?
Categorical obligation without a beneficiary	No	–
Categorical obligation with a beneficiary	Yes	The beneficiary
Categorical permission	Yes	The agent
Claimable obligation	Yes	The declarant
Grantable permission	Yes	The declarant
Revocable obligation	Yes	The declarant
Revocable permission	Yes	The declarant, the agent?

The beneficiary (the friend) is clearly not a right-holder. The declarant (the child) would probably be regarded as a right-holder, but it is not clear whether the agent (the widow) is also a right-holder. Since revocable permissions are seldom explicitly referred to, common usage of the word "right" gives no clear indication for this case.

The above conclusions for the six main types of legal complexes are summarized in Table 4. They can also be summarized as follows:

(1) A legal relation of one of these types is a right unless it is a categorical obligation without a beneficiary.
(2) A declarant is always a right-holder.
(3) The agent of a categorical permission is a right-holder.
(4) The same applies to the beneficiary of a categorical obligation, if there is one.

One of the most contested issues in the theory of rights is whether or not all rights have a correlative obligation. Our exposition of various types of legal relations indicates that this is not the case. The table contains several types of legal complexes that are rights although their central legal relation does not imply an obligation. It also contains one type of legal complex (categorical obligation without a beneficiary) that is an obligation without a correlative right.

13.4 A COMPARISON WITH HOHFELD'S TYPOLOGY

The most influential classification of legal relations is the typology proposed by Wesley N. Hohfeld. It is a standard against which any new classification has to be compared.

All legal relations in Hohfeld's typology are relations between two persons. He distinguished between eight types of legal relations, divided into two groups. The first group contains four types: right, duty, no-right, and privilege. They are interdefinable in purely logical terms. "X has a right against Y that he shall stay off the former's land" is equivalent with "Y is under a duty toward X to stay off the place." Furthermore, to have a no-right means not to have the corresponding right, and "[t]he privilege of entering is the negation of a duty to stay off."[13]

The second group of four contains power, liability, disability, and immunity. They are interdefinable in the same way as the first group. Power, which is often treated as the primitive notion among the four, is a legal capacity to change someone else's rights or other legal relations. Neither power nor duty is definable in terms of the other.[14]

Deontic logic and Hohfeldian analysis are not as easily compatible as they may seem at first glance. There are two major reasons for this. First, whereas deontic logic contains an expression for duty (namely the O operator) that can be used for Hohfeldian duty (and for the three interdefinable types), it does not contain any expression for the capacity to change a legal relation. Therefore, power and the other members of Hohfeld's second group cannot be expressed solely with deontic logic. Deontic typologies such as Kanger's correspond only to the first half of Hohfeld's typology.[15] This problem can only be solved by an addition to the deontic language. Potestative rules can be used for this purpose. $Dc_i\delta \Rightarrow \delta$ corresponds, roughly, to the statement that i has the

[13] Hohfeld [1919] 1964, pp. 38–39.

[14] Kamba 1974, p. 258.

[15] Fitch (1967) proposed that Hohfeldian power can be represented by the expression $Cando_ip$, which means that i has the ability to cause p by striving for something that includes p. This is not an adequate definition. Anybody can make his neighbour a widow by killing her husband, but this does not give him the legal power to make her a widow. Hohfeld himself was careful to point this out: "If, for example, we consider the ordinary property owner's power of alienation, it is necessary to distinguish between the *legal* power, the *physical* power to do the things necessary for the 'exercise' of the legal power, and finally the *privilege* of doing these things" (Hohfeld [1919] 1964, p. 58). Fitch's proposal mistakes physical power for legal power.

Stig and Helle Kanger (1966) considered $\neg O \neg E_i p$ to represent a (legal) power held by i, thus making the other mistake that Hohfeld warned against, namely that of confusing a legal power with a privilege (permission) to make use of it.

power to activate the norm δ.[16] (Hence, the first two of our six basic types – categorical obligation and categorical permission – should be used to translate Hohfeld's first group, whereas the last four types are related to his second group.)

The second problem in the relationship between deontic logic and Hohfeldian analysis concerns the O operator itself. To introduce this problem, it is useful to enrich the language with an operator E_i denoting agency. An expression in deontic logic such as OE_iq tells us about a duty that i has, namely in this case a duty to see to it that q. However, contrary to a Hohfeldian duty, this expression does not tell us who is the "counterparty" toward whom this duty is directed (i.e., it does not tell us who has the right that correlates with this duty). Whereas a Hohfeldian duty refers to a relation between two persons, only one of these persons is mentioned in standard formalizations of duties in deontic logic. Therefore, information about the identity of the counterparty is lost when Hohfeldian types are translated into deontic logic. This loss of information has been called the "counterparty problem."[17]

The counterparty problem is connected with a more basic difference between the two frameworks, namely their different approaches to the compositional nature of actual legal relations. As was observed in Section 13.3, the central legal relation of a legal complex determines much of the nature of the complex. (Therefore, the whole complex is named after it. This terminological practice makes the distinction between a legal complex and its central legal relation linguistically cumbersome. It is seldom made, but for the purposes of logical analysis it is indispensable.)

Hohfeld's types refer to legal complexes *in toto*. Simple deontic sentences, on the other hand, refer to much smaller units that are typically the components of legal complexes. The counterparty of a Hohfeldian duty does not appear in the (deontic) duty that is its central legal relation. To identify it, we must have access to the subsidiary legal relations of the legal complex.[18] The counterparty problem is therefore a direct

[16] More precisely, it means that i has the power to do this by performing a symbolic action that is recognized by the legal system as a declaration by i to the effect that δ.

[17] Makinson 1986.

[18] Cf. Makinson 1986.

consequence of the fact that the representation in deontic logic of a legal complex refers only to its central legal relation. This can be generalized to a *reconstruction problem*: To what extent can we reconstruct the whole complex from our knowledge of the central legal relation?

It is not difficult to find examples of how one and the same legal relation can be the central legal relation of two different complexes. For instance, I am not allowed to litter the street in front of your house. This prohibition (or duty) may either lack a counterparty or have you as a counterparty (so that you have special legal capacities that relate to it). It cannot be seen from the central legal relation alone which of these is the case. Therefore, the reconstruction problem (and as this example shows, also its special case, the counterparty problem) is unresolvable unless the formal structure is enriched in one way or the other.[19]

The Hohfeldian types do not either fully specify the whole legal complex. There are many ways in which I may have a legal right that you return the car that you borrowed from me: There may be various legal sanctions, various legal means of recovery or compensation may be available to me, and so on. Therefore, Hohfeldian analysis also has a reconstruction problem, although it is of a somewhat different nature. It should be granted, though, that Hohfeldian analysis solves one part of the reconstruction problem, namely the identification of counterparties, better than our typology based on potestative rules.[20]

On the other hand, Hohfeld's typology has a serious limitation of another nature. As was noted above, Hohfeld represented all legal relations as binary: They have two parties, one of which is the holder and the other the counterparty of the relation in question.[21] This is clearly a limitation. Actual legal complexes may (at least in principle) involve

[19] The most common way to do this is to provide the deontic operator O with an index that identifies the counterparty. According to Bengt Hansson (1970b) and David Makinson (1986, p. 420), a deontic logic that is adequate for the representation of Hohfeld's categories should have as its primitive notion an operator O_{xy}, "x has an obligation to y to do . . . ." In a similar vein, Lars Lindahl (1994) replaced the O operator with the composite expression $N(\neg p \rightarrow W(i,j))$, where N denotes necessity and $W(i,j)$ denotes that i is wronged by j.

[20] This refers primarily to Hohfeld's first group. In the second group, the counterparty can in typical cases be identified as the claimant (i in Dc_iOp) or permitter (i in Dc_iPp).

[21] Cf. Raz 1970, pp. 179–181, and MacCormick 1982, p. 348.

only one person. They may also involve more than two persons, as in the following examples:

(1) According to Swedish law, the passport authority is required to issue a passport for a minor if one of the parents requires this to be done and the other parent permits it (in writing).
(2) According to Swedish law, a widow(er) may retain undivided possession of the estate provided that none of the deceased's children objects to this.

Neither Hohfeld's typology nor the sixfold classification introduced in Section 13.2 contains categories that correspond to these examples. However, the present typology differs from Hohfeld's in being derived within a general framework that has the expressive power necessary for these examples. The passport example can be expressed as follows:

$$Dc_iOp \,\&\, Dc_jPp \Rightarrow Op$$

The example of the undivided estate can be expressed as a generalized form of a revocable permission:

$$\neg Dc_{i_1}\neg Pp \,\&\, \ldots \,\&\, \neg Dc_{i_n}\neg Pp \Rightarrow Pp$$

These examples confirm the versatility of the general framework of potestative rules. As compared to the various typologies, this framework provides us with greater expressive power, but – as is so often the case – only at the price of a loss in terms of simplicity.

Epilogue

14

Afterthought

In the Preface, I pointed out that formal representations of values and norms can be useful both for the clarification of basic philosophical issues and for applications in subjects such as economics, jurisprudence, decision theory, and social choice theory. In this final chapter, some indications will be given of how the results reported in the previous chapters can contribute to these two purposes.

14.1 PHILOSOPHICAL RELEVANCE

The discussions that a formalized treatment of philosophical subject matter gives rise to can be divided into three categories:

(1) New aspects on issues already discussed in informal philosophy.
(2) Issues not previously discussed in informal philosophy, but with a clear philosophical interest.
(3) Issues that are peculiar to the chosen formalism and have no bearing on philosophical issues that can be expressed without the formalism.

To mention just one example from each of these categories, logical approaches to the classification of legal relations belong to category (1), preference transitivity to category (2), and the necessitation paradoxes in deontic logic to category (3). When assessing the philosophical relevance of a formalized approach, it may be useful to compare how much discussion of types (1) and (2) on the one hand, and (3) on the other, it gives rise to.

It is not difficult to find examples of how results from the previous chapters can be applied to philosophical problems, giving rise to discussions that belong to categories (1) and (2) in the above classification. To begin with, we have investigated the logical relationships between norms and values, between monadic and dyadic value predicates, and between obligations and rights.[1] Results from these investigations can be used for exploring the connections between value-based, norm-based, and rights-based moral theories. One of the issues that can be treated in this way is the classic question of whether consequentialist and deontological moral theories are of an essentially different nature or whether one of them can be reduced to the other (at least in terms of its recommendations).

We have also studied the relationships between normative rules and the norms that actually apply in a particular situation.[2] These relationships have bearing on the connections between moral theories that differ in their use of rules, for instance between act and rule utilitarianism.

A major purpose of developing a less paradoxical deontic logic (Chapter 10) is to provide moral and legal philosophy with useful tools for precise analysis. One area of application, namely moral dilemmas, was indicated in Chapter 11. Other topics that can be attacked with a more realistic deontic logic include prima facie norms, defeasible norms, "ought implies can," and residual obligations.[3]

14.2 APPLICATIONS IN OTHER DISCIPLINES

The formal work undertaken in Chapters 5–7 was focused on the relationship between preferences that refer to wholes (complete alternatives) and preferences that refer to smaller units (states of affairs). The distinction between these two types of preferences is often missing in treatments of preferences in the social sciences. In some cases, its introduction can contribute to increased realism.

Richard Wollheim's so-called paradox of the outvoted democrat is a simple example of this.[4] Wollheim's democrat considers a certain social state p to be better than its negation $\neg p$. When a democratic decision has been made in favour of $\neg p$, she will prefer $\neg p$ to p, since she wants the majority's will to be respected. But she still basically prefers p to $\neg p$. How can she both prefer p to $\neg p$ and $\neg p$ to p?

[1] Section 10.2, Chapter 8, and Section 13.3.
[2] Chapter 12.
[3] Residual obligations, such as duties of compensation, arise when other obligations are not complied with. Williams 1965.
[4] Wollheim 1962.

The paradox can be resolved if we realize that both these preferences are ceteris paribus, but differ in the aspects of the world that they keep constant.[5] Therefore, the paradox vanishes if we specify in an exact manner both (1) the nature of the complete alternatives and (2) the relationships between preferences over these complete alternatives and preferences that refer to more restricted states of affairs.[6]

Such specifications are also useful in social choice theory. This is a discipline that develops models in which collective decisions are derived as aggregations of individual preferences. In the standard (Arrovian) framework, the alternatives to which these preferences refer are assumed to coincide with the options that the procedure has been set up to decide between. Hence, if a committee has to decide between the three alternatives A, B, and C, the decision is treated as an aggregation of individual preferences over the set $\{A,B,C\}$.

In real life, individuals who take part in collective decision procedures often have other types of preferences that cannot be reduced to preferences over the options. A committee member may prefer to be part of the winning coalition, prefer that the decision be taken by as large a majority as possible (preference for consensus), and so on. Such "procedural" preferences can be introduced into the formal representation if more inclusive alternatives are chosen. Consider a three-person committee that is required to make a majority decision between the three alternatives A, B, and C. In the extended model, the preferences of each participant should be represented by a preference relation, not over the set $\{A,B,C\}$ but over the set $\{\langle A,A,A\rangle, \langle A,A,B\rangle, \ldots \langle C,C,C\rangle\}$ of voting patterns. (The voting pattern $\langle A,B,C\rangle$ consists in that the first participant votes for A, the second for B, and the third for C.) A participant with a preference for consensus will prefer $\langle A,A,A\rangle$ to $\langle A,A,B\rangle$, although the decision outcome is the same in the two cases. Typically, preference for consensus will be a ceteris paribus preference; in our example, the participant may very well also prefer $\langle A,A,B\rangle$ to $\langle B,B,B\rangle$ since she (ceteris paribus) prefers the outcome A to the outcome B.[7]

Our preferences change all the time, in response to new opportunities, to habituation, to the boredom of repetition, and not least to influences from others. Indeed, a large part of all economic activities aim

[5] Formally, they may be constructed with different similarity relations.
[6] Hansson 1993e. [7] Hansson 1992a, 1996e.

at inducing changes in other people's preferences. In spite of this, formalized economic theory standardly assumes that preferences are constant. There is an urgent need for formalized accounts of how preferences and norms are continuously transformed in real economies. Dynamic models of preferences and norms, such as those introduced in Chapters 4 and 12, can be used as building blocks in such models.

14.3 A DIFFICULT TRADE-OFF

In Chapter 1, I remarked that formalization involves a trade-off between simplicity and faithfulness to the original. Generally speaking, it is impossible to combine as much simplicity as we want with as much realism as we want. To formalize is therefore to sail between the Scylla of undue complexity and the Charybdis of unrealism.

In my view, there has often been a tendency to give too high priority to simplicity and consequently too low priority to realism. It is a sure sign of oversimplification that the simplicity gained when constructing a model is lost in an involved process of interpreting it. In order to make the formal treatment of values and norms more useful, both for philosophy and for the social sciences, it is often necessary to introduce into the formalism some of the complications of real life that have previously been abstracted from.

Proofs

PROOFS FOR CHAPTER 2

PROOF OF OBSERVATION 2.2. *Part 1*: *Left-to-right*: From $A>B$, it follows by the definition of $\geq$ that $A\geq B$. Furthermore, it follows from the asymmetry of preference that $\neg(B>A)$ and from the incompatibility of preference and indifference that $\neg(A\equiv B)$, i.e., by the symmetry of indifference, $\neg(B\equiv A)$. Thus, $\neg(B>A \vee B\equiv A)$, i.e., by the definition of $\geq$, $\neg(B\geq A)$. *Right-to-left*: It follows from $A\geq B$, according to the definition of $\geq$, that either $A>B$ or $A\equiv B$. By the same definition, it follows from $\neg(B\geq A)$ that $\neg(B\equiv A)$. By the symmetry of indifference, $\neg(A\equiv B)$, so that $A>B$ may be concluded.

Part 2: *Left-to-right*: It follows from $A\equiv B$, by the definition of $\geq$, that $A\geq B$. By the symmetry of indifference, $A\equiv B$ yields $B\equiv A$ so that, by the definition of $\geq$, $B\geq A$. *Right-to-left*: It follows from the definition of $\geq$ and $A\geq B$ & $B\geq A$ that $(A>B \vee A\equiv B)$ & $(B>A \vee B\equiv A)$. By the symmetry of indifference, $(A>B \vee A\equiv B)$ & $(B>A \vee A\equiv B)$. By the asymmetry of preference, $A>B$ is incompatible with $B>A$. We may conclude that $A\equiv B$.

PROOF OF THEOREM 2.4. *Part 1*: For one direction, suppose that acyclicity does not hold. Then there are $A_1, \ldots A_n \in \mathcal{A}$ such that $A_1>A_2, \ldots A_{n-1}>A_n$ and $A_n>A_1$. Weak eligibility is not satisfied for the subset $\{A_1, \ldots A_n\}$ of $\mathcal{A}$.

For the other direction, suppose for *reductio* that acyclicity but not restrictable weak eligibility is satisfied. We are going to show that $\mathcal{A}$ is infinite, contrary to the assumptions. Since restrictable weak eligibility is violated, there is some subset $\mathcal{B}$ of $\mathcal{A}$ for which weak eligibility does not hold. Let $A_1 \in \mathcal{B}$. Since weak eligibility is not satisfied, there is some $A_2 \in \mathcal{B}$ such that $A_2>A_1$. Similarly, there is some A_3 such that $A_3>A_2$, etc. If any two elements on the list $A_1, A_2, A_3 \ldots$ are identical, then acyclicity is violated. Thus, $\mathcal{B}$ is infinite, and consequently so is $\mathcal{A}$, contrary to the conditions.

Part 2: For one direction, suppose that $\geq$ satisfies restrictable strong eligibility. To see that it satisfies completeness, let A, $B \in \mathcal{A}$. Since strong eligibility holds restrictably for $\mathcal{A}$, strong eligibility holds for the subset $\{A,B\}$ of $\mathcal{A}$, so that either $A \geq B$ or $B \geq A$. Since restrictable strong eligibility implies restrictable weak eligibility, acyclicity follows from Part 1.

For the other direction, suppose for *reductio* that $\geq$ is complete and acyclic but violates restrictable strong eligibility. There must be some subset $\mathcal{B}$ of $\mathcal{A}$ for which strong eligibility does not hold. Let $A_1 \in \mathcal{B}$. There is then some $A_2 \in \mathcal{B}$ such that $\neg(A_1 \geq A_2)$. By completeness, $A_2 > A_1$. Similarly, there is some $A_3 \in \mathcal{B}$ such that $\neg(A_2 \geq A_3)$ and consequently $A_3 > A_2$, etc. Suppose that any two elements of the list A_1, A_2, A_3, . . . are identical. Then acyclicity is violated. Thus, $\mathcal{B}$ is infinite, and so is $\mathcal{A}$, contrary to the conditions.

Part 3: First suppose that $\geq$ satisfies restrictable top-transitive weak eligibility. Acyclicity follows from Part 1. For PI-transitivity, let A, B, and C be three elements of $\mathcal{A}$ such that $A > B$ and $B \equiv C$. Suppose that $\neg(A > C)$. Then $\neg(X > C)$ for all $X \in \{A, B, C\}$, and by top-transitive weak eligibility for that set $B \equiv C$ yields $\neg(X > B)$ for all $X \in \{A, B, C\}$, contrary to $A > B$. We may conclude that $A > C$.

For the other direction, suppose that acyclicity and PI-transitivity are satisfied. It follows from Part 1 that $\geq$ satisfies restrictable weak eligibility. For top-transitivity, let $\mathcal{B}$ be a subset of $\mathcal{A}$ and A and B two elements of $\mathcal{B}$ such that $A \equiv B$ and that for all $X \in \mathcal{B}$, $\neg(X > B)$. For *reductio*, suppose that for some $C \in \mathcal{B}$, $C > A$. Then it follows from $A \equiv B$ and PI-transitivity that $C > B$, contrary to the conditions. We may conclude that $\neg(X > A)$ holds for all $X \in \mathcal{B}$, so that top-transitivity of weak eligibility holds for $\geq$ in $\mathcal{B}$. Since this applies to all subsets $\mathcal{B}$ of $\mathcal{A}$, top-transitivity of weak eligibility holds restrictably in $\mathcal{A}$.

Part 4: First suppose that restrictable top-transitive strong eligibility is satisfied. Completeness follows from Part 2. For transitivity, let $A \geq B$ and $B \geq C$. Since top-transitivity of strong eligibility holds restrictably for $\mathcal{A}$, top-transitive strong eligibility holds for $\{A,B,C\}$. There are three cases:

Case i, $A \equiv B$: By completeness, $B \geq B$, so that $B \geq X$ for all $X \in \{A,B,C\}$. It follows from top-transitivity of strong eligibility, as applied to $\{A,B,C\}$, that $A \geq X$ for all $X \in \{A,B,C\}$, so that $A \geq C$.

Case ii, $A>B$ and $B>C$: Suppose that $C>A$. Then strong eligibility does not hold for $\{A,B,C\}$, contrary to the conditions. It follows that $\neg(C>A)$ and by completeness that $A\geq C$.

Case iii, $A>B$ and $B\equiv C$: Suppose that $C\geq A$. By completeness $C\geq C$, so that $C\geq X$ for all $X\in\{A,B,C\}$. By top-transitivity and $B\equiv C$, $B\geq X$ for all $X\in\{A,B,C\}$, so that $B\geq A$, contrary to $A>B$. By this contradiction, $\neg(C\geq A)$. By completeness, $A\geq C$.

For the other direction, suppose that completeness and transitivity are satisfied. Transitivity implies acyclicity, so that restrictable strong eligibility follows from Part 2. For top-transitivity, let $\mathcal{B}$ be a subset of $\mathcal{A}$ with $A, B\in\mathcal{B}$ and such that $A\equiv B$ and that $A\geq C$ for all $C\in\mathcal{B}$. Then for all C, $B\geq A$ and $A\geq C$ yield $B\geq C$, so that top-transitivity of strong eligibility holds in $\mathcal{B}$. Since this applies to all subsets $\mathcal{B}$ of $\mathcal{A}$, top-transitivity of strong eligibility holds restrictably in $\mathcal{A}$.

Part 5: For one direction, suppose that $\geq$ satisfies restrictable exclusive weak eligibility. It follows from Part 1 that acyclicity is satisfied.

To see that antisymmetry holds, suppose to the contrary that this is not the case. Then there are two distinct elements A and B of $\mathcal{A}$ such that $A\geq B$ & $B\geq A$. Then $\neg(A>B)$ & $\neg(B>A)$, contrary to the exclusive weak eligibility of the subset $\{A,B\}$ of $\mathcal{A}$.

To see that completeness holds, suppose to the contrary that this is not the case. Then there are elements A and B of $\mathcal{A}$ such that $\neg(A\geq B)$ & $\neg(B\geq A)$. This too implies $\neg(A>B)$ & $\neg(B>A)$, again contrary to the exclusive weak eligibility of $\{A,B\}$.

For the other direction of the proof, suppose that completeness, acyclicity, and antisymmetry are satisfied. It follows from Part 1 that restrictable weak eligibility holds. Suppose that restrictable exclusive weak eligibility does not hold. Then there is a set $\mathcal{B}\subseteq\mathcal{A}$ with two distinct elements $A_1, A_2\in\mathcal{B}$, such that $\neg(X>A_1)$ and $\neg(X>A_2)$ both hold for all $X\in\mathcal{B}$. It follows that $\neg(A_2>A_1)$ and $\neg(A_1>A_2)$. By completeness, $\neg(A_2>A_1)$ yields $A_1\geq A_2$ and $\neg(A_1>A_2)$ yields $A_2\geq A_1$. By antisymmetry, $A_1\geq A_2\rightarrow\neg(A_2\geq A_1)$. It follows from this contradiction that restrictable exclusive weak eligibility holds.

Part 6: For one direction, suppose that $\geq$ satisfies restrictable exclusive strong eligibility. It follows from Part 2 that completeness and acyclicity are satisfied. For antisymmetry, suppose to the contrary that it does not hold. Then there are two distinct A and B in $\mathcal{A}$ such that $A\geq B$ and $B\geq A$. By completeness, $A\geq A$ and $B\geq B$. This contradicts the exclusive strong eligibility of $\{A,B\}$. It can be concluded that antisymmetry holds.

For the other direction of the proof, suppose that completeness, acyclicity, and antisymmetry are satisfied. It follows from Part 2 that restrictable strong eligibility holds. Suppose that restrictable exclusive strong eligibility does not hold. Then there is a set $\mathcal{B} \subseteq \mathcal{A}$ with two distinct elements $A_1, A_2 \in \mathcal{B}$, such that for all $X \in \mathcal{B}$, $A_1 \geq X$ and $A_2 \geq X$. It follows that $A_1 \geq A_2$ and $A_2 \geq A_1$, contrary to antisymmetry. This contradiction concludes the proof.

PROOFS FOR CHAPTER 3

PROOF OF OBSERVATION 3.4. *Part 1, inclusion*: By the inclusion and monotony properties of Cn_0, $A \subseteq \text{Cn}_0(A) \subseteq \text{Cn}_0(s(T) \cup A) = \text{Cn}_T(A)$.

Part 2, monotony: Let $A \subseteq B$. Then $s(T) \cup A \subseteq s(T) \cup B$ and, by the monotony of Cn_0, $\text{Cn}_0(s(T) \cup A) \subseteq \text{Cn}_0(s(T) \cup B)$, i.e., $\text{Cn}_T(A) \subseteq \text{Cn}_T(B)$.

Part 3, iteration: One direction of iteration follows directly from inclusion. For the other direction of this property, suppose that $\chi \in \text{Cn}_T(\text{Cn}_T(A))$. Then $\chi \in \text{Cn}_0(s(T) \cup \text{Cn}_0(s(T) \cup A))$. By inclusion and monotonicity for Cn_0, $s(T) \subseteq \text{Cn}_0(s(T) \cup A)$, so that $s(T) \cup \text{Cn}_0(s(T) \cup A) = \text{Cn}_0(s(T) \cup A)$. We thus have $\chi \in \text{Cn}_0(\text{Cn}_0(s(T) \cup A))$ and, by the iteration property for Cn_0, $\chi \in \text{Cn}_0(s(T) \cup A)$, i.e., $\chi \in \text{Cn}_T(A)$.

Part 4, supraclassicality: Let $\chi \in \text{Cn}_0(A)$. Then by monotony for Cn_0, $\chi \in \text{Cn}_0(s(T) \cup A)$, i.e., $\chi \in \text{Cn}_T(A)$.

Part 5, deduction: $\psi \in \text{Cn}_T(A \cup \{\chi\})$ holds if and only if $\psi \in \text{Cn}_0(s(T) \cup A \cup \{\chi\})$, (by deduction for Cn_0) if and only if $(\chi \rightarrow \psi) \in \text{Cn}_0(s(T) \cup A)$, if and only if $(\chi \rightarrow \psi) \in \text{Cn}_T(A)$.

Part 6, compactness: Let $\chi \in \text{Cn}_T(A)$, i.e., $\chi \in \text{Cn}_0(s(T) \cup A)$. By the compactness of Cn_0, there are finite subsets S of $s(T)$ and A' of A such that $\chi \in \text{Cn}_0(S \cup A')$. By monotony for Cn_0, $\chi \in \text{Cn}_0(s(T) \cup A')$, i.e., $\chi \in \text{Cn}_T(A')$.

PROOF OF OBSERVATION 3.10.
R is *T-obeying* iff $s(T)\uparrow|\mathbf{R}| \subseteq [\mathbf{R}]$ (Definition 3.9)
iff $s(T)\uparrow|\mathbf{R}| \subseteq \cap\{[R] \mid R \in \mathbf{R}\}$ (Definition 3.9)
iff $s(T)\uparrow|R| \subseteq [R]$ for all $R \in \mathbf{R}$
iff each element of **R** is *T*-obeying (Definition 3.8)

Definition 3.16. *A subset S of $\mathcal{L}_\mathcal{U}$ is a* Henkin set *if and only if for all $\alpha \in \mathcal{L}_\mathcal{U}\uparrow|S|$:*

either $\alpha \in S$ or $\neg\alpha \in S$, but not both.

S' is a Henkin extension *of S if and only if:*

(1) $S \subseteq S'$,
(2) $|S| = |S'|$, and
(3) S' is a Henkin set.

Lemma 3.17. *A preference set S is a Henkin set if and only if for all X, $Y \in |S|$: either $X \geq Y \in S$ or $\neg(X \geq Y) \in S$.*

PROOF OF LEMMA 3.17. For the nontrivial direction, suppose that for all $X, Y \in |S|$: either $X \geq Y \in S$ or $\neg(X \geq Y) \in S$. By induction on the number of sentential connectives, we can show that for all $\alpha \in \mathcal{L}_{\mathcal{U}}\uparrow|S|$: either $\alpha \in S$ or $\neg\alpha \in S$.

Lemma 3.18. *If S is a T-consistent preference set, then it has a T-consistent Henkin extension.*

PROOF OF LEMMA 3.18. Let S be a T-consistent preference set. According to Lemma 3.17, it is sufficient to show that there is some T-consistent preference set S' such that $|S| = |S'|$, $S \subseteq S'$, and for every X, $Y \in |S|$: either $X \geq Y \in S'$ or $\neg(X \geq Y) \in S'$. Let $\alpha_1, \ldots \alpha_k \ldots$ be a list of all the formulas of the form $X \geq Y$ with $X, Y \in |S|$, such that $X \geq Y \notin S$ and $\neg(X \geq Y) \notin S$. Let $S_0, \ldots S_k \ldots$ be a series of sets that satisfies the following two conditions:

(1) $S_0 = S$
(2) for each k such that $0<k$, S_{k+1} is a T-consistent set such that either $S_{k+1} = S_k \cup \{\alpha_{k+1}\}$ or $S_{k+1} = S_k \cup \{\neg\alpha_{k+1}\}$.

To see that the construction is possible, suppose to the contrary that for some T-consistent S_k, both $S_k \cup \{\alpha_{k+1}\}$ and $S_k \cup \{\neg\alpha_{k+1}\}$ are T-inconsistent. Then, equivalently, both $T \cup S_k \cup \{\alpha_{k+1}\}$ and $T \cup S_k \cup \{\neg\alpha_{k+1}\}$ are Cn_0-inconsistent; hence, $T \cup S_k$ is Cn_0-inconsistent, so that S_k is T-inconsistent, contrary to the conditions.

Let $S' = \mathrm{Cn}_T(S_0) \cup \mathrm{Cn}_T(S_1) \cup \ldots \cup \mathrm{Cn}_T(S_k) \cup \ldots$. It follows directly that S' has the desired properties.

Lemma 3.19. *Let S be a T-consistent set. Then S is the intersection of its T-consistent Henkin extensions.*

PROOF OF LEMMA 3.19. Let $\mathcal{H}$ be the set of T-consistent Henkin extensions of S. It follows from Lemma 3.18 that $\mathcal{H}$ is nonempty, so that $S \subseteq \cap\mathcal{H}$. In order to prove $\cap\mathcal{H} \subseteq S$, let $\alpha \notin S$. We are going to prove that $\alpha \notin \cap\mathcal{H}$. Excluding a trivial case, we assume that $\alpha \in \mathcal{L}_{\mathcal{U}}\uparrow|S|$.

It follows from $\alpha \notin S$ and the Cn_T-closure of S that $S \cup \{\neg\alpha\}$ is T-consistent. According to Lemma 3.18, there is some Henkin extension S' of $S \cup \{\neg\alpha\}$. Since $S' \in \mathcal{H}$, and $\alpha \notin S'$, we may conclude that $\alpha \notin \cap\mathcal{H}$.

Lemma 3.20. *Let R be a reflexive subset of $\mathcal{U}\times\mathcal{U}$. If R is T-obeying, then* $[R] = \text{Cn}_T([R])\uparrow|R|$.

PROOF OF LEMMA 3.20. It follows from Definition 3.8 that $[R]$ is a Henkin set. Thus, $[R] = (\text{Cn}_0([R]))\uparrow|R|$. Since R is T-obeying, we also have $(s(T)\uparrow|R|) \subseteq [R]$. Thus, $[R] = (\text{Cn}_0((s(T)\uparrow|R|) \cup [R]))\uparrow|R| = (\text{Cn}_T([R]))\uparrow|R|$.

PROOF OF THEOREM 3.11. *Part I*: For one direction, let $\mathbf{R}$ be a T-obeying preference model. We need to show that:

(1) $[\mathbf{R}] = (\text{Cn}_T([\mathbf{R}]))\uparrow|\mathbf{R}|$
(2) For all α, if $\alpha \in [\mathbf{R}]$, then $\neg\alpha \notin [\mathbf{R}]$.

Ad I:1: It follows directly from Observation 3.4 that $[\mathbf{R}] \subseteq \text{Cn}_T([\mathbf{R}])$, thus $[\mathbf{R}] \subseteq (\text{Cn}_T([\mathbf{R}]))\uparrow|\mathbf{R}|$. It remains to be shown that $(\text{Cn}_T([\mathbf{R}]))\uparrow|\mathbf{R}| \subseteq [\mathbf{R}]$. Let $\alpha \in (\text{Cn}_T([\mathbf{R}]))\uparrow|\mathbf{R}|$. For every $R \in \mathbf{R}$, we have $[\mathbf{R}] \subseteq [R]$ and thus $\text{Cn}_T([\mathbf{R}]) \subseteq \text{Cn}_T([R])$, so that $\alpha \in \text{Cn}_T([R])$. According to Lemma 3.20, $\alpha \in [R]$ for every $R \in \mathbf{R}$, thus by Definition 3.9, $\alpha \in [\mathbf{R}]$.

Ad I:2: Suppose to the contrary that $\alpha \in [\mathbf{R}]$ and $\neg\alpha \in [\mathbf{R}]$. Let $R \in \mathbf{R}$. Then by Definition 3.9, $\alpha \in [R]$ and $\neg\alpha \in [R]$, contrary to Definition 3.8.

Part II: For the other direction, let S be a T-consistent preference set. We need to construct a T-obeying preference model $\mathbf{R}$ such that $S = [\mathbf{R}]$. The following construction will be used:

$$\mathbf{R} = \{R \mid (|R| = |S|) \,\&\, (S \subseteq [R])\}$$

We need to verify that:

(1) $\mathbf{R}$ is a preference model,
(2) $\mathbf{R}$ is T-obeying, and
(3) $S = [\mathbf{R}]$.

Ad II:1: For every $R \in \mathbf{R}$, it follows from $X \geq X \in T$, the T-consistency of S, and $S \subseteq [R]$, that R is reflexive. In order to show that $\mathbf{R}$ is a preference model, it remains to show that $\mathbf{R}$ is nonempty. It follows from Lemma 3.18 that there is some S' such that $S \subseteq S'$, $|S| = |S'|$, and S' is a Henkin set. Let R be such that $|R| = |S|$ and that for all $X, Y \in |R|$, $\langle X,Y \rangle \in R$ if and only if $X \geq Y \in S'$. Then $S' = [R]$, so that $S \subseteq [R]$.

Ad II:2: Since S is a T-consistent preference set, we have $s(T){\uparrow}|S| \subseteq (\mathrm{Cn}_0(s(T) \cup S)){\uparrow}|S| = (\mathrm{Cn}_T(S)){\uparrow}|S| = S$.

If $R \in \mathbf{R}$, then $|S| = |R|$ and we have $s(T){\uparrow}|R| = s(T){\uparrow}|S| \subseteq S \subseteq [R]$. Thus, R is T-obeying. Since this holds for all $R \in \mathbf{R}$, we may conclude (using Observation 3.10) that $\mathbf{R}$ is T-obeying.

Ad II:3: It follows directly from the definition and from the nonemptiness of $\mathbf{R}$ (Part II:1) that $S \subseteq [\mathbf{R}]$. In order to prove that $[\mathbf{R}] \subseteq S$, let $\alpha \notin S$. Then $\mathrm{Cn}_T(S \cup \{\neg\alpha\})$ is T-consistent, and it follows from Lemma 3.18 that there is some T-consistent Henkin extension S' of S such that $\neg\alpha \in S'$. In the same way as in Part II:1, it follows that there is some R such that $S' = [R]$. Since $S \subseteq S' = [R]$, $R \in \mathbf{R}$, and we have $\alpha \notin [\mathbf{R}]$.

PROOF OF OBSERVATION 3.13. *Part 1*: Let $R \in \mathbf{R}$ satisfy antisymmetry, and let $X \geq Y, Y \geq X \in [\mathbf{R}]$. Then $X \geq Y, Y \geq X \in [R]$, from which it follows that $X = Y$.

Part 2: Let $X_1 > X_2, X_2 > X_3, \ldots X_{n-1} > X_n, X_n > X_1 \in [\mathbf{R}]$ and $R \in \mathbf{R}$. It follows directly from Definition 3.9 that $X_1 > X_2, X_2 > X_3, \ldots X_{n-1} > X_n, X_n > X_1 \in [R]$.

Part 3: Let $R \in \mathbf{R}$, and let X be such that for all Y, $Y > X \notin [R]$. It follows directly from Definition 3.9 that for all Y, $Y > X \notin [\mathbf{R}]$.

PROOF OF OBSERVATION 3.14. *Part 1*: Let $X \in |\mathbf{R}|$. Then for all $R \in \mathbf{R}$, we have $X \in |R|$. Since the given property is satisfied for R, it follows that $X \geq X \in [R]$. This holds for all $R \in \mathbf{R}$, and we may conclude that $X \geq X \in [\mathbf{R}]$.

Parts 2, 3, 5, 7, and 9: In the same way as Part 1.

Part 4: Suppose that *TRANSITIVITY* holds for each $R \in \mathbf{R}$ and that $X \geq Y, Y \geq Z \in [\mathbf{R}]$. It follows that for each $R \in \mathbf{R}$, $X \geq Y, Y \geq Z \in [R]$. Due to *TRANSITIVITY*, we then have $X \geq Z \in [R]$. Since this holds for each $R \in \mathbf{R}$, we can conclude that $X \geq Z \in [\mathbf{R}]$.

Part 6, 8, 10, 11, and 12: In the same way as Part 4.

PROOF OF OBSERVATION 3.15.
Parts 1, 2, and 3: Let $\mathbf{R} = \{\{A{\geq}A, B{\geq}B, A{\geq}B\}, \{A{\geq}A, B{\geq}B, B{\geq}A\}\}$.

Part 4: Let $\mathbf{R} = \{R_1, R_2\}$, and

$$R_1 = \{A{\geq}A,\ B{\geq}B,\ C{\geq}C,\ D{\geq}D,\ A{\geq}B,\ B{\geq}A,\ C{\geq}A,\ C{\geq}B\}$$

$$R_2 = \{A{\geq}A,\ B{\geq}B,\ C{\geq}C,\ D{\geq}D,\ A{\geq}B,\ B{\geq}A,\ D{\geq}A,\ C{\geq}B\}$$

It is easy to verify that both R_1 and R_2 satisfy the property. To see that $\mathbf{R}$ does not, note: (1) $A{\equiv}B \in [\mathbf{R}]$. (2) Since neither $A{>}A$, $B{>}A$, nor $C{>}A$ is an element of $[R_2]$, neither of them is in $[\mathbf{R}]$. Since $D{>}A \notin [R_1]$, $D{>}A \notin [\mathbf{R}]$. Hence, $X{>}A \notin [\mathbf{R}]$ for all X. (3) Since $C{>}B \in [R_1]$ and $C{>}B \in [R_2]$, we have $C{>}B \in [\mathbf{R}]$.

PROOFS FOR CHAPTER 4

PROOF OF OBSERVATION 4.4. *Parts 1 and 2*: Directly from Definition 4.3.

Part 3: Let $\alpha \in [\mathbf{R}]$. For one direction, let $R' \in \mathbf{R}*_{\mathcal{B}}\alpha$, and suppose for contradiction that $R' \notin \mathbf{R}$. Let R'' be any element of $\mathbf{R}$. We then have $\langle R'',\mathbf{R}\rangle \sqsubset \langle R',\mathbf{R}\rangle$, contrary to Definition 4.3. Hence, $R' \in \mathbf{R}$.

For the other direction, let $R \in \mathbf{R}$. Then R can be shown to satisfy the three conditions of Definition 4.3. Clauses (1) and (2) are trivial. For (3), suppose to the contrary that there is some T-obeying preference ordering R' such that $\alpha \in [R']$ and $\langle R',\mathbf{R}\rangle \sqsubset \langle R,\mathbf{R}\rangle$. Then, according to Definition 4.2, there is some $R'' \in \mathbf{R}$ such that $\langle R',R''\rangle \sqsubset \langle R,R\rangle$ which is impossible since $R\Delta R = (R\Delta R){\downarrow}\mathcal{B} = \emptyset$.

Part 4: From Definition 4.3.

Part 5: Let $\neg\alpha \notin [\mathbf{R}]$. Then there is some $R \in \mathbf{R}$ such that $\alpha \in R$. It follows from Definition 4.3 that $R \in \mathbf{R}*_{\mathcal{B}}\alpha$. Furthermore, if $R' \notin \mathbf{R}$, then $\langle R,\mathbf{R}\rangle \sqsubset \langle R',\mathbf{R}\rangle$. The rest is obvious.

Part 6: Let $\neg\alpha \notin [\mathbf{R}]$. It follows from Part 2 of the present observation that $\alpha \in [\mathbf{R}*_{\mathcal{B}}\alpha]$ and from Part 5 that $\mathbf{R}*_{\mathcal{B}}\alpha \subseteq \mathbf{R}$ so that $[\mathbf{R}] \subseteq [\mathbf{R}*_{\mathcal{B}}\alpha]$. Thus, $[\mathbf{R}] \cup \{\alpha\} \subseteq [\mathbf{R}*_{\mathcal{B}}\alpha]$. It follows from Theorem 3.11 and Part 1 of the present observation that $[\mathbf{R}*_{\mathcal{B}}\alpha]$ is a T-consistent preference set, so that $[\mathbf{R}*_{\mathcal{B}}\alpha] = \mathrm{Cn}_T([\mathbf{R}*_{\mathcal{B}}\alpha])\!\uparrow\!|\mathbf{R}*_{\mathcal{B}}\alpha| = \mathrm{Cn}_T([\mathbf{R}*_{\mathcal{B}}\alpha])\!\uparrow\!|\mathbf{R}|$. It follows from this and $[\mathbf{R}] \cup \{\alpha\} \subseteq [\mathbf{R}*_{\mathcal{B}}\alpha]$ that $\mathrm{Cn}_T([\mathbf{R}] \cup \{\alpha\})\!\uparrow\!|\mathbf{R}| \subseteq [\mathbf{R}*_{\mathcal{B}}\alpha]$.

For the other direction of the proof, let $\beta \notin \mathrm{Cn}_T([\mathbf{R}] \cup \{\alpha\})\!\uparrow\!|\mathbf{R}|$. We are going to show that $\beta \notin [\mathbf{R}*_{\mathcal{B}}\alpha]$. By exclusion of a trivial case, we may assume that $\beta \in \mathcal{L}_{\mathcal{U}}\!\uparrow\!|\mathbf{R}|$.

It follows from $\beta \notin Cn_T([\mathbf{R}] \cup \{\alpha\})$ that $\alpha \rightarrow \beta \notin Cn_T([\mathbf{R}])$, hence $\alpha \rightarrow \beta \notin [\mathbf{R}]$, hence there is some $R \in \mathbf{R}$ such that $\alpha \rightarrow \beta \notin [R]$. It follows that $\neg\alpha \notin [R]$, so that $\alpha \in [R]$. It follows from $\alpha \in [R]$, by Part 5 of the present observation, that $R \in \mathbf{R}*_{\mathcal{B}}\alpha$. Furthermore, it follows from $\alpha \rightarrow \beta \notin [R]$ that $\beta \notin [R]$. From $R \in \mathbf{R}*_{\mathcal{B}}\alpha$ and $\beta \notin [R]$, we may conclude that $\beta \notin [\mathbf{R}*_{\mathcal{B}}\alpha]$. This concludes the proof.

Part 7: Let $\neg\beta \notin [\mathbf{R}*_{\mathcal{B}}\alpha]$.

In order to show that $(\mathbf{R}*_{\mathcal{B}}\alpha)*_{\mathcal{B}}\beta \subseteq \mathbf{R}*_{\mathcal{B}}(\alpha\&\beta)$, suppose that this is not the case. Then there is some $R \in (\mathbf{R}*_{\mathcal{B}}\alpha)*_{\mathcal{B}}\beta$ such that $R \notin \mathbf{R}*_{\mathcal{B}}(\alpha\&\beta)$. Due to Part 5 of the present observation, it follows from $\neg\beta \notin [\mathbf{R}*_{\mathcal{B}}\alpha]$ and $R \in (\mathbf{R}*_{\mathcal{B}}\alpha)*_{\mathcal{B}}\beta$ that $R \in \mathbf{R}*_{\mathcal{B}}\alpha$ and $\beta \in [R]$. Thus, $\alpha\&\beta \in [R]$. It follows from $R \notin \mathbf{R}*_{\mathcal{B}}(\alpha\&\beta)$ that there is some R' such that $\alpha\&\beta \in [R']$ and $\langle R',\mathbf{R}\rangle \sqsubset \langle R,\mathbf{R}\rangle$. This, however, is impossible since $\alpha\&\beta \in [R]$, $R \in \mathbf{R}$ and $\mathbf{R}$ is T-obeying. We may conclude from this contradiction that $(\mathbf{R}*_{\mathcal{B}}\alpha)*_{\mathcal{B}}\beta \subseteq \mathbf{R}*_{\mathcal{B}}(\alpha\&\beta)$.

For the other direction of the proof, we are first going to show that $\mathbf{R}*_{\mathcal{B}}(\alpha\&\beta) \subseteq \mathbf{R}*_{\mathcal{B}}\alpha$. Suppose not. Then there is some $R \in \mathbf{R}*_{\mathcal{B}}(\alpha\&\beta)$ such that $R \notin \mathbf{R}*_{\mathcal{B}}\alpha$. Since $\alpha \in [R]$ we may conclude that there is some $R' \in \mathbf{R}*_{\mathcal{B}}\alpha$ such that $\langle R',\mathbf{R}\rangle \sqsubset \langle R,\mathbf{R}\rangle$. Since $\neg\beta \notin [\mathbf{R}*_{\mathcal{B}}\alpha]$, there is some $R'' \in \mathbf{R}*_{\mathcal{B}}\alpha$ such that $\beta \in [R'']$. It follows from $R' \in \mathbf{R}*_{\mathcal{B}}\alpha$ and $R'' \in \mathbf{R}*_{\mathcal{B}}\alpha$ that $\langle R'',\mathbf{R}\rangle \sqsubseteq \langle R',\mathbf{R}\rangle$. Since $\sqsubseteq$ is transitive, it follows that $\langle R'',\mathbf{R}\rangle \sqsubset \langle R,\mathbf{R}\rangle$. Since $\alpha\&\beta \in [R'']$, this contradicts $R \in \mathbf{R}*_{\mathcal{B}}(\alpha\&\beta)$. We may conclude that $\mathbf{R}*_{\mathcal{B}}(\alpha\&\beta) \subseteq \mathbf{R}*_{\mathcal{B}}\alpha$. It follows that $\mathbf{R}*_{\mathcal{B}}(\alpha\&\beta) \subseteq \{R \in \mathbf{R}*_{\mathcal{B}}\alpha \mid \beta \in [R]\} = (\mathbf{R}*_{\mathcal{B}}\alpha)*_{\mathcal{B}}\beta$. (Part 5 of the present observation is needed for the last step.)

Part 8: *I*: We are first going to show that $\mathbf{R}*_{\mathcal{B}}(\alpha\vee\beta) \subseteq \mathbf{R}*_{\mathcal{B}}\alpha \cup \mathbf{R}*_{\mathcal{B}}\beta$. Let $R \in \mathbf{R}*_{\mathcal{B}}(\alpha\vee\beta)$. It follows from $\alpha\vee\beta \in [R]$, since $[R]$ is a Henkin set, that either $\alpha \in [R]$ or $\beta \in [R]$.

In the case when $\alpha \in [R]$, suppose for *reductio* that $R \notin \mathbf{R}*_{\mathcal{B}}\alpha$. Then there is some R' such that $\alpha \in [R']$ and $\langle R',\mathbf{R}\rangle \sqsubset \langle R,\mathbf{R}\rangle$. But since $\alpha\vee\beta \in [R']$, this contradicts $R \in \mathbf{R}*_{\mathcal{B}}(\alpha\vee\beta)$. Hence, $R \in \mathbf{R}*_{\mathcal{B}}\alpha$.

In the other case, when $\beta \in [R]$, it follows in the same way that $R \in \mathbf{R}*_{\mathcal{B}}\beta$.

II: The next step is to show that if $(\mathbf{R}*_{\mathcal{B}}\alpha) \cap (\mathbf{R}*_{\mathcal{B}}(\alpha\vee\beta)) \neq \emptyset$, then $\mathbf{R}*_{\mathcal{B}}\alpha \subseteq \mathbf{R}*_{\mathcal{B}}(\alpha\vee\beta)$.

Let $(\mathbf{R}*_{\mathcal{B}}\alpha) \cap (\mathbf{R}*_{\mathcal{B}}(\alpha\vee\beta)) \neq \emptyset$. Then there is some R' such that $R' \in \mathbf{R}*_{\mathcal{B}}\alpha$ and $R' \in \mathbf{R}*_{\mathcal{B}}(\alpha\vee\beta)$. Let R be any element of $\mathbf{R}*_{\mathcal{B}}\alpha$. Then $\langle R,\mathbf{R}\rangle \sqsubseteq \langle R',\mathbf{R}\rangle \sqsubseteq \langle R,\mathbf{R}\rangle$. Since $\alpha\vee\beta \in [R]$ we may conclude that $R \in \mathbf{R}*_{\mathcal{B}}(\alpha\vee\beta)$.

III: It follows from Part I that either $\mathbf{R}*_{\mathcal{B}}\alpha \cap \mathbf{R}*_{\mathcal{B}}(\alpha\vee\beta) \neq \emptyset$ or $\mathbf{R}*_{\mathcal{B}}\beta \cap \mathbf{R}*_{\mathcal{B}}(\alpha\vee\beta) \neq \emptyset$. There are three cases:

Case i, $\mathbf{R}*_{\mathcal{B}}\alpha \cap \mathbf{R}*_{\mathcal{B}}(\alpha\vee\beta) = \emptyset$: Then $\mathbf{R}*_{\mathcal{B}}\beta \cap \mathbf{R}*_{\mathcal{B}}(\alpha\vee\beta) \neq \emptyset$, and we can conclude from Part II that $\mathbf{R}*_{\mathcal{B}}\beta \subseteq \mathbf{R}*_{\mathcal{B}}(\alpha\vee\beta)$. Part I yields $\mathbf{R}*_{\mathcal{B}}(\alpha\vee\beta) \subseteq \mathbf{R}*_{\mathcal{B}}\beta$.

Case ii, $\mathbf{R}*_{\mathcal{B}}\beta \cap \mathbf{R}*_{\mathcal{B}}(\alpha\vee\beta) = \emptyset$: Then, by a symmetrical proof, $\mathbf{R}*_{\mathcal{B}}(\alpha\vee\beta) = \mathbf{R}*_{\mathcal{B}}\alpha$.

Case iii, $\mathbf{R}*_{\mathcal{B}}\alpha \cap \mathbf{R}*_{\mathcal{B}}(\alpha\vee\beta) \neq \emptyset$ and $\mathbf{R}*_{\mathcal{B}}\alpha \cap \mathbf{R}*_{\mathcal{B}}(\alpha\vee\beta) \neq \emptyset$: We have $\mathbf{R}*_{\mathcal{B}}(\alpha\vee\beta) \subseteq \mathbf{R}*_{\mathcal{B}}\alpha \cup \mathbf{R}*_{\mathcal{B}}\beta$ according to I and $\mathbf{R}*_{\mathcal{B}}\alpha \cup \mathbf{R}*_{\mathcal{B}}\beta \subseteq \mathbf{R}*_{\mathcal{B}}(\alpha\vee\beta)$ according to II. This finishes the proof.

Part 9: $(\mathbf{R}_1\cup\mathbf{R}_2)*_{\mathcal{B}}\alpha$ is the set of T-consistent preference relations validating α that are as close as possible to some element of $\mathbf{R}_1\cup\mathbf{R}_2$. There are three cases:

Case 1, it holds for all R in $(\mathbf{R}_1\cup\mathbf{R}_2)*_{\mathcal{B}}\alpha$ that some element of $\mathbf{R}_1$ is closer to R than is any element of $\mathbf{R}_2$: (Intuitively speaking, the distance from α to $\mathbf{R}_1$ is smaller than that from α to $\mathbf{R}_2$.) Then $(\mathbf{R}_1\cup\mathbf{R}_2)*_{\mathcal{B}}\alpha$ coincides with the set of T-consistent preference relations validating α that are as close as possible to some element of $\mathbf{R}_1$, i.e., it coincides with $\mathbf{R}_1*_{\mathcal{B}}\alpha$.

Case 2, it holds for all R in $(\mathbf{R}_1\cup\mathbf{R}_2)*_{\mathcal{B}}\alpha$ that some element of $\mathbf{R}_2$ is closer to R than is any element of $\mathbf{R}_1$: Then for reasons analogous to Case 1, $(\mathbf{R}_1\cup\mathbf{R}_2)*_{\mathcal{B}}\alpha$ coincides with $\mathbf{R}_2*_{\mathcal{B}}\alpha$.

Case 3: In the remaining case, $(\mathbf{R}_1\cup\mathbf{R}_2)*_{\mathcal{B}}\alpha$ coincides with the set of T-consistent preference relations validating α that are as close as possible to some element of either $\mathbf{R}_1$ or $\mathbf{R}_2$, i.e., it coincides with $\mathbf{R}_1*_{\mathcal{B}}\alpha \cup \mathbf{R}_2*_{\mathcal{B}}\alpha$.

PROOF OF OBSERVATION 4.6. *Part 1*: From Part 1 of Observation 4.4.

Part 2: Directly from the definition.

Part 3: Let $\alpha \notin [\mathbf{R}]$. Then it follows from Part 5 of Observation 4.4 that $\mathbf{R}*_{\mathcal{B}}\neg\alpha \subseteq \mathbf{R}$.

Part 4: From Part 2 of Observation 4.4.

Part 5: From Part 4 of Observation 4.4.

Part 6: From Part 8 of Observation 4.4

Part 7: According to Definition 4.5, $(\mathbf{R}_1\cup\mathbf{R}_2)\div_{\mathcal{B}}\alpha = (\mathbf{R}_1\cup\mathbf{R}_2) \cup ((\mathbf{R}_1\cup\mathbf{R}_2)*_{\mathcal{B}}\neg\alpha)$, hence according to Part 9 of Observation 4.4, it is equal to one of the following:

(1) $(\mathbf{R}_1\cup\mathbf{R}_2) \cup \mathbf{R}_1*_{\mathcal{B}}\neg\alpha = (\mathbf{R}_1 \cup \mathbf{R}_1*_{\mathcal{B}}\neg\alpha) \cup \mathbf{R}_2 = \mathbf{R}_1\div_{\mathcal{B}}\alpha \cup \mathbf{R}_2$.

(2) $(\mathbf{R}_1\cup\mathbf{R}_2) \cup \mathbf{R}_2*_{\mathcal{B}}\neg\alpha = \mathbf{R}_1 \cup (\mathbf{R}_2 \cup \mathbf{R}_2*_{\mathcal{B}}\neg\alpha) = \mathbf{R}_1 \cup \mathbf{R}_2\div_{\mathcal{B}}\alpha$.

(3) $(\mathbf{R}_1\cup\mathbf{R}_2) \cup \mathbf{R}_1*_{\mathcal{B}}\neg\alpha \cup \mathbf{R}_2*_{\mathcal{B}}\neg\alpha = (\mathbf{R}_1 \cup \mathbf{R}_1*_{\mathcal{B}}\neg\alpha) \cup (\mathbf{R}_2 \cup \mathbf{R}_2*_{\mathcal{B}}\neg\alpha)$
$= \mathbf{R}_1\div_{\mathcal{B}}\alpha \cup \mathbf{R}_2\div_{\mathcal{B}}\alpha$.

PROOF OF OBSERVATION 4.7. *Part 1*: Let $\alpha \in [\mathbf{R}]$. Then $\neg\alpha \notin [\mathbf{R}]$, hence according to Part 2 of Observation 4.6, $\neg\alpha \notin [\mathbf{R}\div_{\mathcal{B}}\alpha]$. It follows from Part 5 of Observation 4.4 that $(\mathbf{R}\div_{\mathcal{B}}\alpha)*_{\mathcal{B}}\alpha = \{R \in \mathbf{R}\div_{\mathcal{B}}\alpha \mid \alpha \in [R]\} = \{R \in \mathbf{R}\cup\mathbf{R}*_{\mathcal{B}}(\neg\alpha) \mid \alpha \in [R]\} = \mathbf{R}$.

Part 2: There are two cases.

Case a, $\neg\alpha \notin [\mathbf{R}]$: It follows from Observation 4.6, Part 3, that $\mathbf{R}\div_{\mathcal{B}}(\neg\alpha) = \mathbf{R}$.

Case b, $\neg\alpha \in [\mathbf{R}]$: We have $\mathbf{R}\div_{\mathcal{B}}(\neg\alpha) = \mathbf{R} \cup \mathbf{R}*_{\mathcal{B}}\alpha$. Since $\alpha \notin [R]$ for all $R \in \mathbf{R}$ and $\alpha \in [R]$ for all $R \in \mathbf{R}*_{\mathcal{B}}\alpha$, it follows from Observation 4.4, Part 5, that $(\mathbf{R} \cup \mathbf{R}*_{\mathcal{B}}\alpha)*_{\mathcal{B}}\alpha = \mathbf{R}*_{\mathcal{B}}\alpha$.

Part 3: According to Observation 4.6, Part 4 (success), we have $\neg\alpha \notin [\mathbf{R}\div_{\mathcal{B}}(\neg\alpha)]$. Part 2 of the present observation yields $\mathbf{R}*_{\mathcal{B}}\alpha = (\mathbf{R}\div_{\mathcal{B}}(\neg\alpha))*_{\mathcal{B}}\alpha$. It follows from Part 6 of Observation 4.4 that $[(\mathbf{R}\div_{\mathcal{B}}(\neg\alpha))*_{\mathcal{B}}\alpha] = \mathrm{Cn}_T([\mathbf{R}\div_{\mathcal{B}}(\neg\alpha)] \cup \{\alpha\})\!\uparrow\!|\mathbf{R}\div_{\mathcal{B}}(\neg\alpha)|$. Since $|\mathbf{R}\div_{\mathcal{B}}(\neg\alpha)| = |\mathbf{R}|$, we can conclude that $[\mathbf{R}*_{\mathcal{B}}\alpha] = (\mathrm{Cn}_T([\mathbf{R}\div_{\mathcal{B}}(\neg\alpha)] \cup \{\alpha\}))\!\uparrow\!|\mathbf{R}|$.

Parts 4–5: Directly from Definition 4.5.

PROOF OF OBSERVATION 4.9. The proofs of Parts 1–3 are straightforward.

Part 4: Since $\mathbf{R}$ is T-obeying, for every $R \in \mathbf{R}$, we have $s(T)\!\uparrow\!|\mathbf{R}\ominus A| \subseteq s(T)\!\uparrow\!|\mathbf{R}| \subseteq [R]$, so that $s(T)\!\uparrow\!|\mathbf{R}\ominus A| \subseteq [R\!\uparrow\!|\mathbf{R}\ominus A|] = [R\!\downarrow\!\{A\}]$. It follows that $s(T)\!\uparrow\!|\mathbf{R}\ominus A| \subseteq [\mathbf{R}\ominus A]$.

PROOF OF OBSERVATION 4.11. Parts 1, 2, and 4 follow directly from Definition 4.10.

Part 3: In the proof, we are going to employ the multiple operation $\mathbf{R}\oplus\{A,B\}$, such that $R \in \mathbf{R}\oplus\{A,B\}$ iff:

(1) $|R| = |\mathbf{R}|\cup\{A,B\}$,
(2) There is some $R' \in \mathbf{R}$ such that $= R' = R\!\uparrow\!|\mathbf{R}|$, and
(3) $s(T)\!\uparrow\!(|\mathbf{R}|\cup\{A,B\}) \subseteq [R]$.

We are going to show that $\mathbf{R}\oplus A\oplus B = \mathbf{R}\oplus\{A,B\}$.

First let $R \in \mathbf{R}\oplus A\oplus B$. It follows directly from the definition that $R \in \mathbf{R}\oplus\{A,B\}$. Thus, $\mathbf{R}\oplus A\oplus B \subseteq \mathbf{R}\oplus\{A,B\}$.

Next, let $R \in \mathbf{R}\oplus\{A,B\}$. Let $R' = R\!\uparrow\!(|\mathbf{R}|\cup\{A\})$. It follows from the definitions that $R' \in \mathbf{R}\oplus A$ and (using R' in clause 2) that $R \in (\mathbf{R}\oplus A)\oplus B$. We may conclude that $\mathbf{R}\oplus\{A,B\} \subseteq (\mathbf{R}\oplus A)\oplus B$.

We have proved that $\mathbf{R}\oplus A\oplus B = \mathbf{R}\oplus\{A,B\}$. By a symmetrical argument, $\mathbf{R}\oplus B\oplus A = \mathbf{R}\oplus\{A,B\}$. We may conclude that $\mathbf{R}\oplus A\oplus B = \mathbf{R}\oplus B\oplus A$.

Part 5: For one direction, let $\alpha \in (\text{Cn}_T([\mathbf{R}]))\uparrow(|\mathbf{R}| \cup \{A\})$. Let $R \in \mathbf{R}\oplus A$. According to Definition 4.10, there is some $R' \in \mathbf{R}$ such that $R' = R\uparrow|\mathbf{R}|$. It follows from $R' \in \mathbf{R}$ and $\alpha \in (\text{Cn}_T([\mathbf{R}]))\uparrow(|\mathbf{R}|\cup\{A\})$ that $\alpha \in \text{Cn}_T([R'])$, i.e., $\alpha \in \text{Cn}_0(s(T)\cup[R'])$. Since $\alpha \in \mathcal{L}_U\uparrow(|\mathbf{R}|\cup\{A\})$, $\alpha \in \text{Cn}_0((s(T))\uparrow(|\mathbf{R}|\cup\{A\})\cup[R'])$. Since R is T-obeying, $(s(T))\uparrow(|\mathbf{R}|\cup\{A\}) \subseteq [R]$. Since we also have $[R'] \subseteq [R]$, we may conclude that $\alpha \in \text{Cn}_0([R])$. Since $[R]$ is a Henkin set, $[R] = \text{Cn}_0([R])$, so that $\alpha \in [R]$. Since this holds for all $R \in \mathbf{R}\oplus A$, we may conclude that $\alpha \in [\mathbf{R}\oplus A]$.

For the other direction, let $\alpha \in [\mathbf{R}\oplus A]$. In order to show that $\alpha \in \text{Cn}_T([\mathbf{R}])$, suppose that this is not the case. Then there is some $R' \in \mathbf{R}$ such that $\alpha \notin \text{Cn}_T([R'])$. It follows that there is some T-consistent Henkin set S with $|S| = |\mathbf{R}|\cup\{A\}$, such that $[R']\cup\{\neg\alpha\} \subseteq S$. Let R be the relation such that for all X and Y, $\langle X,Y\rangle \in R$ iff $X\geq Y \in S$. It follows from Definition 4.10 that $R \in \mathbf{R}\oplus A$. Since $\neg\alpha \in [R]$, this contradicts $\alpha \in [\mathbf{R}\oplus A]$. We may conclude that $\alpha \in \text{Cn}_T([\mathbf{R}])$. Since $\alpha \in [\mathbf{R}\oplus A]$, $\alpha \in (\text{Cn}_T([\mathbf{R}]))\uparrow(|\mathbf{R}|\cup\{A\})$ follows directly.

PROOF OF OBSERVATION 4.12. Directly from Observations 4.8 and 4.10.

PROOF FOR CHAPTER 5

PROOF OF OBSERVATION 5.6. Conditions (1) and (2) of Definition 5.4 coincide with the equally numbered conditions of Definition 5.5. To see that condition (3) of Definition 5.4 is satisfied, let $A, A' \in \mathcal{A}$ and $A \neq A'$. Without loss of generality, we can assume that there is some $\alpha \in A\backslash A'$. It follows from condition (3) of Definition 5.5 that $\neg\alpha \in A'$. Hence, $\{\alpha, \neg\alpha\} \subseteq A\cup A'$, so that condition (3) of Definition 5.4 is satisfied.

PROOFS FOR CHAPTER 6

PROOF OF OBSERVATION 6.8. Since A is the only representation of a in $\mathcal{A}$, it is also the only representation of $a/_{\mathcal{A}}b$ in $\mathcal{A}$. Similarly, B is the only representation of $b/_{\mathcal{A}}a$ in $\mathcal{A}$. Definition 6.4 yields $f(\langle a/_{\mathcal{A}}b, b/_{\mathcal{A}}a\rangle) = \langle A,B\rangle$. According to Definition 6.5, $a\geq_f b$ iff $A\geq B$.

PROOF OF THE COROLLARY TO OBSERVATION 6.8. If $A = \text{Cn}(\{a\})$, then A is a representation of a, and according to Definition 5.4 it is the only representation of a in $\mathcal{A}$.

PROOF OF OBSERVATION 6.11.

$p \equiv_f q$	
iff $p \geq_f q$ & $q \geq_f p$	Definition 6.10
iff $A \geq B$ for all $\langle A,B \rangle \in f(\langle p/_{\mathcal{A}}\, q, q/_{\mathcal{A}} p \rangle)$	Definition 6.5
and $B \geq A$ for all $\langle B,A \rangle \in f(\langle q/_{\mathcal{A}} p, p/_{\mathcal{A}}\, q \rangle)$	
iff $A \geq B$ and $B \geq A$ for all $\langle A,B \rangle \in f(\langle p/_{\mathcal{A}} q, q/_{\mathcal{A}} p \rangle)$	Symmetry of f
iff $p \cong_f q$	Definition 6.10

PROOF OF OBSERVATION 6.14. *Centring*: Suppose to the contrary that centring does not hold for f. Then, according to Definition 6.7, there are $A_1, A_2 \in \mathcal{A}$ and $x, y \in \cup\mathcal{A}$ such that $\vDash_{\mathcal{A}} x \leftrightarrow y$, $\langle A_1,A_2 \rangle \in f(\langle x,y \rangle)$, and $A_1 \neq A_2$. It follows from $A_1 \neq A_2$, using (T4), that $\neg T(A_1,A_2,A_2,A_2)$. On the other hand, according to Definition 6.13, it follows from $x, y \in A_2$ that $T(A_1,A_2,A_2,A_2)$. This contradiction concludes the proof.

Symmetry: Suppose to the contrary that symmetry is not satisfied. Then there are $x, y \in \cup\mathcal{A}$ and $A, B \in \mathcal{A}$ such that $\langle A,B \rangle \in f(\langle x,y \rangle)$ and $\langle B,A \rangle \notin f(\langle y,x \rangle)$. It follows by Definition 6.13 from $\langle B,A \rangle \notin f(\langle y,x \rangle)$ that there are $A', B' \in \mathcal{A}$ such that $x \in A'$, $y \in B'$, and $\neg T(B,A,B',A')$.

On the other hand, it follows according to Definition 6.13, from $\langle A,B \rangle \in f(\langle x,y \rangle)$, $x \in A'$, and $y \in B'$, that $T(A,B,A',B')$. We can use (T5) to obtain $T(B,A,A,B)$ and $T(A',B',B',A')$. Two applications of (T2) to $T(B,A,A,B)$, $T(A,B,A',B')$, and $T(A',B',B',A')$ provide us with $T(B,A,B',A')$, contrary to what was just shown. This contradiction concludes the proof.

PROOF OF OBSERVATION 6.16. *Part 1*: This can be shown with any similarity relation T in a model such that $f(\langle x,y \rangle) = \{\langle A,B \rangle \mid x \in A \in \mathcal{A}$ and $y \in B \in \mathcal{A}\}$ for all $x, y \in \cup\mathcal{A}$. (In this model, any degree of similarity is sufficient.)

Part 2: This can be shown with a model such that, for two particular elements x and y of $\cup\mathcal{A}$:

(1) $f(\langle x,y \rangle)$ is the set of representations $\langle A,B \rangle$ in $\mathcal{A}$ of $\langle x,y \rangle$ such that $T(A,B,A',B')$ holds for all $A', B' \in \mathcal{A}$ such that $x \in A'$ and $y \in B'$;
(2) $f(\langle y,x \rangle) = \{\langle B,A \rangle \mid y \in B \in \mathcal{A}$ and $x \in A \in \mathcal{A}\}$

Hence, the construction coincides with the similarity-maximizing construction for $\langle x,y \rangle$, whereas for $\langle y,x \rangle$ it employs the same construction that was used for Part 1 of this proof.

PROOF OF OBSERVATION 6.17. *Part 1*: Let f be a similarity-maximizing representation function, based on T. We need to show that the two conditions given in Definition 6.15 are satisfied. It is obvious that the first of them is satisfied. For the second, let its antecedent hold, i.e., let $\langle A,B\rangle \in f(\langle x,y\rangle)$, $x \in A'$, $y \in B'$, and $T(A',B',A,B)$. We need to show that $\langle A',B'\rangle \in f(\langle x,y\rangle)$.

This can be done, according to Definition 6.13, by showing that it holds for all A'' and B'' with $x \in A''$ and $y \in B''$ that $T(A',B',A'',B'')$. Let $x \in A''$ and $y \in B''$. Then it follows from $\langle A,B\rangle \in f(\langle x,y\rangle)$, again according to Definition 6.13, that $T(A,B,A'',B'')$. It follows from $T(A',B',A,B)$ and $T(A,B,A'',B'')$, according to (T2), that $T(A',B',A'',B'')$. With this, the proof is completed.

Part 2: For this purpose, we can, just as in the proof of Part 1 of Observation 6.16, use a model such that $f(\langle x,y\rangle) = \{\langle A,B\rangle \mid x \in A \in \mathcal{A}$ and $y \in B \in \mathcal{A}\}$ for all $x, y \in \cup\mathcal{A}$. This representation function is similarity-satisficing with respect to all similarity relations. However, it is not similarity-maximizing since it does not satisfy centring (cf. Observation 6.14).

PROOF OF OBSERVATION 6.19. *Part 1*: Let T be the relation such that $T(V,X,Y,Z)$ if and only if $V{=}X$ or $Y{\neq}Z$. It can straightforwardly be shown that T satisfies (T1)–(T5) of Definition 6.12, hence it is a similarity relation. It also follows from Definitions 6.13 and 6.18 that $\geq_f$ is similarity-maximizing with respect to T.

Part 2: Let $\mathcal{A} = \{A,B,C\}$, with $p,\neg q \in A$, $\neg p,q \in B$, and $p, \neg q \in C$. Let $A{>}B{>}C$, and let T be a similarity relation over $\mathcal{A}$ such that similarity coincides with closeness in the following representation:

$$A \qquad\qquad B \qquad\qquad\qquad\qquad C$$

Let f be the similarity-maximizing relation that is based on T. Then $p{>}_f q$. However, for the maximal centred relation based on $\mathcal{A}$ and $\geq$, $\neg(p{>}_f q)$.

PROOF OF OBSERVATION 6.20. Directly from Definition 6.5.

PROOF OF THE COROLLARY TO OBSERVATION 6.20. *Part 1* follows directly from the observation and Definition 6.7. For *Part 2*, let $\mathcal{A} = \{A,B\}$ with $a \in A$ and $a \in B$. Let $\geq$ be an exclusionary preference relation on $\mathcal{A}$, such that $A{>}B$. Let f be as in the proof of Observation 6.16, Part 1.

PROOF OF OBSERVATION 6.21. Let $\geq$ be complete and such that $A_2>B_2$ and $B_1>A_1$. Then it follows from $\langle B_2,A_2\rangle \in f(\langle q/_{\mathcal{A}}p,p/_{\mathcal{A}}q\rangle)$ and $\neg(B_2\geq A_2)$ that $\neg(q\geq_f p)$. Similarly, it follows from $\langle A_1,B_1\rangle \in f(\langle p/_{\mathcal{A}}q,q/_{\mathcal{A}}p\rangle)$ and $\neg(A_1\geq B_1)$ that $\neg(p\geq_f q)$.

PROOF OF OBSERVATION 6.22. Let $\mathcal{A} = \{A, B, C, D\}$, with $A = \text{Cn}(\{p, q, \neg r\})$, $B = \text{Cn}(\{\neg p, \neg q, r\})$, $C = \text{Cn}(\{p, \neg q, \neg r\})$, and $D = \text{Cn}(\{\neg p, q, r\})$. Furthermore, let $\geq$ be a weak ordering (complete and transitive) such that $A>B>C>D$. We obtain $p\geq_f q$ and $q\geq_f r$, but $\neg(p\geq_f r)$.

PROOF OF OBSERVATION 6.23. Since p, q, and r are pairwise $\mathcal{A}$-incompatible, notation in this proof can be simplified by writing p instead of $p/_{\mathcal{A}}q$, etc.

Part 1: Let $\geq$ be transitive and let $p\geq_f q$ and $q\geq_f r$. Let $\langle A,C\rangle \in f(\langle p,r\rangle)$. Since $q \in \cup\mathcal{A}$, there is some $B \in \mathcal{A}$ such that $q \in B$. Clearly, $\langle A,B\rangle \in f(\langle p,q\rangle)$. It follows from $p\geq_f q$ that $A\geq B$. Similarly, it follows from $\langle B,C\rangle \in f(\langle q,r\rangle)$ and $q\geq_f r$ that $B\geq C$. By the transitivity of $\geq$, $A\geq C$. Since this holds for all $\langle A,C\rangle \in f(\langle p,r\rangle)$, we may conclude that $p\geq_f r$.

Part 2: Let $\equiv$ be transitive and let $p\equiv_f q$ and $q\equiv_f r$. Let $\langle A,C\rangle \in f(\langle p,r\rangle)$. Let B be some element of $\mathcal{A}$ such that $q \in B$. It follows from $p\equiv_f q$ that $p\geq_f q$ and $q\geq_f p$. We may conclude from $p\geq_f q$ and $\langle A,B\rangle \in f(\langle p,q\rangle)$ that $A\geq B$, and similarly from $q\geq_f p$ and $\langle B,A\rangle \in f(\langle q,p\rangle)$ that $B\geq A$. Thus, $A\equiv B$. In the same way, it follows that $B\geq C$ and $C\geq B$, so that $B\equiv C$. By the transitivity of $\equiv$, $A\equiv C$, i.e., $A\geq C$ and $C\geq A$.

Since $A\geq C$ holds for all $\langle A,C\rangle \in f(\langle p,r\rangle)$, we have $p\geq_f r$. Since $C\geq A$ holds for all $\langle A,C\rangle \in f(\langle p,r\rangle)$, i.e., for all $\langle C,A\rangle \in f(\langle r,p\rangle)$, $r\geq_f p$. Thus, $p\equiv_f r$.

Part 3: Let $>$ be transitive and let $p\gg_f q$ and $q\gg_f r$. Let $\langle A,C\rangle \in f(\langle p,r\rangle)$. Let B be some element of $\mathcal{A}$ such that $q \in B$. It follows from $p\gg_f q$ and $\langle A,B\rangle \in f(\langle p,q\rangle)$ that $A>B$. Similarly, it follows from $q\gg_f r$ and $\langle B,C\rangle \in f(\langle q,r\rangle)$ that $B>C$. Since $>$ is transitive, we have $A>C$. This holds for all $\langle A,C\rangle \in f(\langle p,r\rangle)$, and we may conclude that $p\gg_f r$.

Part 4: Let IP-transitivity hold, and let $p\equiv_f q$ and $q\gg_f r$. Let $\langle A,C\rangle \in f(\langle p,r\rangle)$. Let B be some element of $\mathcal{A}$ such that $q \in B$. We may conclude in the same way as in Part 2 that $A\equiv B$. Furthermore, it follows from $q\gg_f r$ and $\langle B,C\rangle \in f(\langle q,r\rangle)$ that $B>C$. By the IP-transitivity of $\geq$, $A>C$. Since this holds for all $\langle A,C\rangle \in f(\langle p,r\rangle)$, we may conclude that $p\gg_f r$.

Part 5: Follows in the same way as Part 4.

Part 6: Let $\geq$ be complete and IP-transitive, and let $p\equiv_f q$ and $q>_f r$. We are first going to show that $p\geq_f r$. For that purpose, let $\langle A,C\rangle \in f(\langle p,r\rangle)$, and let B be some element of $\mathcal{A}$ such that $q \in B$. It can be concluded from $p \equiv_f q$, in the same way as in Part 2, that $A \equiv B$.

It follows from $q >_f r$ that $q \geq_f r$. Since $\langle B,C \rangle \in f(\langle q,r \rangle)$, we have $B \geq C$, i.e., either $B>C$ or $B \equiv C$. If $B>C$, then it follows directly from $A \equiv B$, due to the IP-transitivity of $\geq$, that $A>C$, so that $A \geq C$. In the other case, when $B \equiv C$, suppose that $A \geq C$ does not hold. By the completeness of $\geq$, we then have $C>A$. However, $B \equiv C$ and $C>A$ yield, due to the IP-transitivity of $\geq$, $B>A$, contrary to $A \equiv B$. It follows from this contradiction that $A \geq C$. Thus, if $\langle A,C \rangle \in f(\langle p,r \rangle)$, then $A \geq C$. We may conclude that $p \geq_f r$.

Next, we are going to show that $\neg(r \geq_f p)$. It follows from $q >_f r$ that $\neg(r \geq_f q)$. Thus, there are elements Z and Y of $\mathcal{A}$ such that $r \in Z$, $q \in Y$, and $\neg(Z \geq Y)$. Since $\geq$ is complete, $Y>Z$. Let X be any element of $\mathcal{A}$ such that $p \in X$. It follows from $p \equiv_f q$ that $X \equiv Y$. We can now use the IP-property of $\geq$ to obtain $X>Z$. Thus, $\langle Z,X \rangle \in f(\langle r,p \rangle)$ and $\neg(Z \geq X)$, so that $\neg(r \geq_f p)$. It follows from $p \geq_f r$ and $\neg(r \geq_f p)$ that $p >_f r$.

Part 7: Follows in the same way as Part 6.

PROOF OF OBSERVATION 6.24. Let $\mathcal{A} = \{A,B,C,D\}$ be contextually complete, with $r \in A$, $p \in B$, $q \in C$, and $r \in D$. Let T be a similarity relation over $\mathcal{A}$ such that similarity coincides with closeness in the following representation:

$A \qquad\qquad B \qquad\qquad C \qquad\qquad D$

(The distances A–B, B–C, and C–D are the same.) Furthermore, let $\geq$ be a weak ordering (complete and transitive) over $\mathcal{A}$.

Parts 1 and 3: Let $A>B>C>D$.
Part 2: Let $A>B \equiv C \equiv D$.
Parts 4 and 6: Let $A>B \equiv C>D$.
Parts 5 and 7: Let $A>B>C \equiv D$.

PROOF OF OBSERVATION 6.25. Let $\mathcal{A} = \{A,B_1,B_2,C\}$, with $p\&s \in A$, $q\&s \in B_1$, $q\&\neg s \in B_2$, and $r\&s \in C$. Furthermore, let $\geq$ be such that $A>B_1 \equiv B_2 \equiv A \equiv C \equiv B_1$ and $B_2>C$. Then $>$ is transitive ($\geq$ is quasitransitive), and we have $p >_f q$ and $q >_f r$ but $p \equiv_f r$.

PROOF OF OBSERVATION 6.26. Since $q_1, \ldots q_n$ are pairwise $\mathcal{A}$-incompatible, notation in this proof can be simplified by writing q_1 instead of $q_1 /_{\mathcal{A}} q_2$, etc.

Part 1: The observation will be proved in its converse form.

Let $\{q_1, \ldots q_m\}$ be a subset of $\{q_1, \ldots q_n\}$ such that $q_1 \gg_f q_2 \gg_f \ldots q_m \gg_f q_1$. Let $\{Q_1, \ldots Q_m\}$ be a subset of $\mathcal{A}$, such that for each $1 \leq k \leq m$, $q_k \in Q_k$. It then follows that $Q_1 > Q_2 > \ldots > Q_m > Q_1$.

Part 2: Let $\mathcal{A} = \{A, B_1, B_2, C_1, C_2\}$, with $q_1 \in A$, $q_2 \in B_1$, $q_2 \in B_2$, $q_3 \in C_1$, and $q_3 \in C_2$. Let $\geq$ be such that $C_1 > A$, $A > B_1$ and $B_2 > C_2$, and that indifference holds for all other combinations (i.e., $X \geq Y$ if $\langle X,Y \rangle \in (\mathcal{A} \times \mathcal{A}) \setminus \{\langle A,C_1 \rangle, \langle B_1,A \rangle, \langle C_2,B_2 \rangle\}$.) Then $>$ is acyclic, but $>_f$ is not, since $q_1 > q_2$, $q_2 > q_3$ and $q_3 > q_1$.

PROOF OF OBSERVATION 6.27. For both parts of the proof, the same counterexample can be used as in the part of the proof of Observation 6.24 that refers to property (1).

PROOF OF OBSERVATION 6.28. *Part 1*: Let $p \geq_f q$, and let $\langle A,B \rangle \in f(\langle \neg q /_{\mathcal{A}} \neg p, \neg p /_{\mathcal{A}} \neg q \rangle)$. It follows from $p \nvDash_{\mathcal{A}} q$ that $\neg q /_{\mathcal{A}} \neg p$ is equivalent to $p /_{\mathcal{A}} q$, and from $q \nvDash_{\mathcal{A}} p$ that $\neg p /_{\mathcal{A}} \neg q$ is equivalent to $q /_{\mathcal{A}} p$. Thus, $\langle A,B \rangle \in f(\langle p /_{\mathcal{A}} q, q /_{\mathcal{A}} p \rangle)$. It follows from $p \geq_f q$ that $A \geq B$. Since this holds for all $\langle A,B \rangle \in f(\langle \neg q /_{\mathcal{A}} \neg p, \neg p /_{\mathcal{A}} \neg q \rangle)$, we may conclude that $\neg q \geq_f \neg p$.

Part 2: From Part 1.

Part 3: Let $p >_f q$, i.e., $p \geq_f q$ and $\neg(q \geq_f p)$. It follows from $p \geq_f q$, according to Part 1, that $\neg q \geq_f \neg p$.

It follows from $\neg(q \geq_f p)$ that there is some A and some B such that $\langle B,A \rangle \in f(\langle q /_{\mathcal{A}} p, p /_{\mathcal{A}} q \rangle)$ and $\neg(B \geq A)$. It follows from $p \nvDash_{\mathcal{A}} q$ that $\neg q /_{\mathcal{A}} \neg p$ is equivalent to $p /_{\mathcal{A}} q$, and from $q \nvDash_{\mathcal{A}} p$ that $\neg p /_{\mathcal{A}} \neg q$ is equivalent to $q /_{\mathcal{A}} p$. Thus, $\langle B,A \rangle \in f(\langle \neg p /_{\mathcal{A}} \neg q, \neg q /_{\mathcal{A}} \neg p \rangle)$. From this and $\neg(B \geq A)$ follows $\neg(\neg p \geq_f \neg q)$.

From $\neg q \geq_f \neg p$ and $\neg(\neg p \geq_f \neg q)$ it follows that $\neg q >_f \neg p$.

Part 4: Suppose that $p \gg_f q$, and let $\langle A,B \rangle \in f(\langle \neg q /_{\mathcal{A}} \neg p, \neg p /_{\mathcal{A}} \neg q \rangle)$. It follows in the same way as in Part 1 that $\langle A,B \rangle \in f(\langle p /_{\mathcal{A}} q, q /_{\mathcal{A}} p \rangle)$. It follows from $p \gg_f q$ that $A > B$. Since this holds for all $\langle A,B \rangle \in f(\langle \neg q /_{\mathcal{A}} \neg p, \neg p /_{\mathcal{A}} \neg q \rangle)$, we may conclude that $\neg q \gg_f \neg p$.

PROOF OF OBSERVATION 6.29. For all $\langle A,B \rangle \in \mathcal{A} \times \mathcal{A}$, $\langle A,B \rangle \in f(p /_{\mathcal{A}} q, q /_{\mathcal{A}} p \rangle)$ iff $\langle A,B \rangle \in f(\langle p \& \neg q /_{\mathcal{A}} q \& \neg p, q \& \neg p /_{\mathcal{A}} p \& \neg q \rangle)$. The proof proceeds as that of Observation 6.28.

PROOF OF OBSERVATION 6.30. For *Part 1*, we have the following equalities: $f(\langle p /_{\mathcal{A}} (p \vee q), (p \vee q) /_{\mathcal{A}} p \rangle) = f(\langle p,q \rangle) = f(\langle p /_{\mathcal{A}} q, q /_{\mathcal{A}} p \rangle)$. *Part 2* is proved in the same way. For *Parts 3 and 4*, let $\mathcal{A} = \{A,B\}$, with $p, \neg q \in A, \neg p, q \in B$, and $A > B$.

PROOF OF OBSERVATION 6.31. *Part 1*: Let $\mathcal{A} = \{A,B,C\}$, with $p,\neg q \in A$, $\neg p,q \in B$, and $p,q \in C$. Let $A>B>C$. Then $f(\langle p/_{\mathcal{A}}q,q/_{\mathcal{A}}p\rangle) = \{\langle A,B\rangle\}$ and $A\geq B$, so that $p\geq_f q$. However, it follows from $\langle C,B\rangle \in f(\langle p/_{\mathcal{A}}p\vee q,p\vee q/_{\mathcal{A}}p\rangle)$ and $B>C$ that $\neg(p\geq_f(p\vee q))$.

Part 2: Let $\mathcal{A} = \{A,B,C\}$, with $p,q \in A$, $p,\neg q \in B$, and $\neg p,q \in C$. Let $A>B>C$. Then $f(\langle p/_{\mathcal{A}}q,q/_{\mathcal{A}}p\rangle) = \{\langle B,C\rangle\}$ and $B\geq C$, so that $p\geq_f q$. However, it follows from $\langle B,A\rangle \in f(\langle p\vee q/_{\mathcal{A}}q,q/_{\mathcal{A}}p\vee q\rangle)$ and $A>B$ that $\neg((p\vee q)\geq_f q)$.

Part 3: Let $\mathcal{A}$ be as in Part 2, but let $A>B\equiv C$.

PROOF OF OBSERVATION 6.32. *Part 1*: Let $\mathcal{A} = \{A,B\}$, with $\neg p,q \in A$ and $p,q \in B$, and let $A>B$.

Part 2: Let $\mathcal{A}$ be as in Part 1, but let $B>A$.

PROOF OF OBSERVATION 6.33. *Part 1, left to right*: Let $(p\vee q)\geq_f r$. For symmetry reasons, it is sufficient to prove that $p\geq_f r$. Let $\langle A,B\rangle \in f(\langle p/_{\mathcal{A}}r,r/_{\mathcal{A}}p\rangle)$. Then $(p\vee q)\&\neg r \in A$, hence $p\vee q/_{\mathcal{A}}r \in A$. Furthermore, since p, q, and r are mutually exclusive in $\mathcal{A}$, $p \notin B$ and $q \notin B$. Since $\mathcal{A}$ is contextually complete, $\neg(p\vee q) \in B$. Thus, $r\&\neg(p\vee q) \in B$, hence $r/_{\mathcal{A}}p\vee q \in B$. We therefore have $\langle A,B\rangle \in f(\langle p\vee q/_{\mathcal{A}}r,r/_{\mathcal{A}}p\vee q\rangle)$. It can be concluded from this and $(p\vee q)\geq_f r$ that $A\geq B$. Since this holds for all $\langle A,B\rangle \in f(\langle p/_{\mathcal{A}}r,r/_{\mathcal{A}}p\rangle)$, we may conclude that $p\geq_f r$.

Part 1, right to left: Let $p\geq_f r$ and $q\geq_f r$. We are going to show that $(p\vee q)\geq_f r$. Let $\langle A,B\rangle \in f(\langle p\vee q/_{\mathcal{A}}r,r/_{\mathcal{A}}p\vee q\rangle)$. Since $\mathcal{A}$ is contextually complete, it follows from $p \in \cup\mathcal{A}$, $q \in \cup\mathcal{A}$, and $p\vee q \in A$ that either $p \in A$ or $q \in A$. Since p, q, and r are pairwise mutually exclusive elements of $\cup\mathcal{A}$, either $p\&\neg r \in A$ or $q\&\neg r \in A$. It follows from $r \in B$, since p, q, and r are pairwise mutually exclusive elements of $\cup\mathcal{A}$, that $r\&\neg p \in B$ and $r\&\neg q \in B$. Thus, either $\langle A,B\rangle \in f(\langle p/_{\mathcal{A}}r,r/_{\mathcal{A}}p\rangle)$ or $\langle A,B\rangle \in f(\langle q/_{\mathcal{A}}r,r/_{\mathcal{A}}q\rangle)$. In the first case, $A\geq B$ follows from $p\geq_f r$ and in the second case from $q\geq_f r$. Since $A\geq B$ holds for all $\langle A,B\rangle \in f(\langle p\vee q/_{\mathcal{A}}r,r/_{\mathcal{A}}p\vee q\rangle)$, we may conclude that $(p\vee q)\geq_f r$.

Part 2 follows in the same way as Part 1.

Part 3:

$(p\vee q)\equiv_f r$	
iff $(p\vee q)\geq_f r$ & $r\geq_f(p\vee q)$	Definition 6.10
iff $p\geq_f r$ & $q\geq_f r$ & $r\geq_f p$ & $r\geq_f q$	Parts 1 and 2 of present proof
iff $p\equiv_f r$ & $q\equiv_f r$	Definition 6.10

Part 4: Directly from Part 3 by substitution and the symmetry of $\equiv_f$.

Part 5, left to right: Let $(p \vee q) \gg_f r$. For symmetry reasons, it is sufficient to prove that $p \gg_f r$. Let $\langle A,B \rangle \in f(\langle p/_{\mathcal{A}} r, r/_{\mathcal{A}} p \rangle)$. It follows in the same way as in the proof of Part 1, left to right, that $\langle A,B \rangle \in f(\langle p \vee q/_{\mathcal{A}} r, r/_{\mathcal{A}} p \vee q \rangle)$. It follows from this and $(p \vee q) \gg_f r$ that $A>B$. Since this holds for all $\langle A,B \rangle \in f(\langle p/_{\mathcal{A}} r, r/_{\mathcal{A}} p \rangle)$, we may conclude that $p \gg_f r$.

Part 5, right to left: Let $p \gg_f r$ and $q \gg_f r$. We are going to show that $(p \vee q) \gg_f r$. Let $\langle A,B \rangle \in f(\langle p \vee q/_{\mathcal{A}} r, r/_{\mathcal{A}} p \vee q \rangle)$. It follows in the same way as in the proof of Part 1, right to left, that either $\langle A,B \rangle \in f(\langle p/_{\mathcal{A}} r, r/_{\mathcal{A}} p \rangle)$ or $\langle A,B \rangle \in f(\langle q/_{\mathcal{A}} r, r/_{\mathcal{A}} q \rangle)$. In the first case, $A>B$ follows from $p \gg_f r$ and in the second case from $q \gg_f r$. Since $A>B$ holds for all $\langle A,B \rangle \in f(\langle p \vee q/_{\mathcal{A}} r, r/_{\mathcal{A}} p \vee q \rangle)$, we may conclude that $(p \vee q) \gg_f r$.

Part 6 follows in the same way as Part 5.

Part 7, left to right: Let $(p \vee q) >_f r$. It follows from $(p \vee q) >_f r$ that $(p \vee q) \geq_f r$, so that according to Part 1, $p \geq_f r$ and $q \geq_f r$. We are going to show that $p >_f r \vee q >_f r$.

It follows from $(p \vee q) >_f r$ that $\neg(r \geq_f (p \vee q))$, so that there are $A \in \mathcal{A}$ and $B \in \mathcal{A}$ such that $\langle A,B \rangle \in f(\langle r/_{\mathcal{A}}(p \vee q), (p \vee q)/_{\mathcal{A}} r \rangle)$ and $\neg(A \geq B)$. In the same way as in Part 2, right to left, it follows that either $\langle A,B \rangle \in f(\langle r/_{\mathcal{A}} p, p/_{\mathcal{A}} r \rangle)$ or $\langle A,B \rangle \in f(\langle r/_{\mathcal{A}} q, q/_{\mathcal{A}} r \rangle)$. In the first case, we obtain $\neg(r \geq_f p)$, which with $p \geq_f r$ yields $p >_f r$. In the second case, we obtain $\neg(r \geq_f q)$, which with $q \geq_f r$ yields $q >_f r$.

Part 7, right to left: Let $p \geq_f r$ & $q \geq_f r$ & $(p >_f r \vee q >_f r)$. It follows from $p \geq_f r$ & $q \geq_f r$, according to Part 1, that $(p \vee q) \geq_f r$. There are two (symmetric) cases.

If $p >_f r$, then $\neg(r \geq_f p)$, and there are $A \in \mathcal{A}$ and $B \in \mathcal{A}$ such that $\langle A,B \rangle \in f(\langle r/_{\mathcal{A}} p, p/_{\mathcal{A}} r \rangle)$ and $\neg(A \geq B)$. Since $\mathcal{A}$ is contextually complete, it follows from the mutual exclusivity of p, q, and r that $r \& \neg(p \vee q) \in A$, hence $r/_{\mathcal{A}} (p \vee q) \in A$. Clearly, $(p \vee q) \& \neg r \in B$, hence $(p \vee q)/_{\mathcal{A}} r \in B$. We now have $\langle A,B \rangle \in f(\langle r/_{\mathcal{A}} (p \vee q), (p \vee q)/_{\mathcal{A}} r \rangle)$ and $\neg(A \geq B)$. It follows from this that $\neg(r \geq_f (p \vee q))$.

If $q >_f r$, then $\neg(r \geq_f (p \vee q))$ follows in the same way. We may conclude that $(p \vee q) \geq_f r$ and $\neg(r \geq_f (p \vee q))$, so that $(p \vee q) >_f r$.

Part 8 can be proved in the same way as Part 7.

Part 9: Let $\mathcal{A} = \{A,B,C\}$, with $p \in A$, $q \in B$, and $r \in C$. Furthermore, let $\geq$ be a weak ordering (complete and transitive) such that $A>B \equiv C$. Then $(p \vee q) >_f r$ but $q \equiv_f r$.

Part 10: Let $\mathcal{A}$ be as in the proof of Part 9, and let $\geq$ be a weak ordering such that $A \equiv B > C$. Then $p >_f (q \vee r)$ but $p \equiv_f q$.

PROOF OF OBSERVATION 6.34. Counterexamples can be constructed as follows:

Parts 1, 2, 5, 6, 7, and 8: Let $\mathcal{A} = \{A,B,C,D\}$, with $p \in A$, $q \in B$, $r \in C$, and $p \in D$. Let $A>B>C>D$, and let T be a similarity relation on $\mathcal{A}$ such that similarity coincides with closeness in the following representation:

$A \qquad B \qquad C \qquad\quad D$

(The distances A–B and B–C are of equal length, and the distance C–D is 1.5 times longer than B–C.)

Part 3: Let $\mathcal{A} = \{A,B,C\}$, with $p \in A$, $q \in B$, and $r \in C$. Let $A>B\equiv C$ and let T be a similarity relation on $\mathcal{A}$ such that similarity coincides with closeness in the following representation:

$A \qquad B \qquad C$

(The distances A–B and B–C are of equal length.)

Part 4: Let $\mathcal{A}$ and T be as in Part 3, but let $A\equiv B>C$.

PROOF OF OBSERVATION 6.35. *Part 1*: Let $(p\vee q)\geq_f r$. Let $\langle X,Y\rangle \in f(\langle p\vee q/_{\mathcal{A}}r, r/_{\mathcal{A}}p\vee q\rangle)$. Then either $p \in X$ or $q \in X$.

If $p \in X$, then let $\langle X'',Y''\rangle \in f(\langle p/_{\mathcal{A}}r, r/_{\mathcal{A}}p\rangle)$. It follows from the construction of f via T that $\langle X'',Y''\rangle \in f(\langle p\vee q/_{\mathcal{A}}r, r/_{\mathcal{A}}p\vee q\rangle)$, hence it follows from $(p\vee q)\geq_f r$ that $X''\geq Y''$. Since this holds for all $\langle X'',Y''\rangle \in f(\langle p/_{\mathcal{A}}r, r/_{\mathcal{A}}p\rangle)$, we have $p\geq_f r$.

If $q \in X$, then $q\geq_f r$ follows in the same way.

Part 2: Let $\mathcal{A} = \{A,B,C,D,E\}$ with $p,\neg q,r \in A$, $\neg p,q,r \in B$, $p,\neg q,\neg r \in C$, $\neg p,q,\neg r \in D$, and $\neg p,\neg q,r \in E$. Furthermore, let $A\equiv B>C\equiv D>E$. We then have $f(\langle p\vee q/_{\mathcal{A}}r, r/_{\mathcal{A}}p\vee q\rangle) = \{\langle C,E\rangle,\langle D,E\rangle\}$, so that $p\vee q\geq_f r$. It follows from $\langle C,B\rangle \in f(\langle p/_{\mathcal{A}}r, r/_{\mathcal{A}}p\rangle)$ and $B>C$ that $\neg(p\geq_f r)$. It follows from $\langle D,A\rangle \in f(\langle q/_{\mathcal{A}}r, r/_{\mathcal{A}}q\rangle)$ and $D>A$ that $\neg(q\geq_f r)$.

Part 3: Let $\mathcal{A} = \{A,B,C\}$, with $p \in A$, $r \in B$, and $q \in C$. Let $A>B>C$. Then $f(\langle p/_{\mathcal{A}}r, r/_{\mathcal{A}}p\rangle) = \{\langle A,B\rangle\}$ and $A\geq B$ so that $p\geq_f r$. However, it follows from $\langle C,B\rangle \in f(\langle p\vee q/_{\mathcal{A}}r, r/_{\mathcal{A}}p\vee q\rangle)$ and $B>C$ that $\neg(p\vee q\geq_f r)$.

Part 4 follows in the same way as Part 1.

Part 5: Let $\mathcal{A} = \{A,B,C,D,E\}$ with $p,\neg q,\neg r \in A$, $\neg p,q,\neg r \in B$, $\neg p,\neg q,r \in C$, $p,\neg q,r \in D$, and $p,q,\neg r \in E$. Let $A>B\equiv C>D\equiv E$. Then $f(\langle p/_{\mathcal{A}}q\vee r, q\vee r/_{\mathcal{A}}p\rangle) = \{\langle A,B\rangle,\langle A,C\rangle\}$, so that $p\geq_f(q\vee r)$. Furthermore, $\langle D,B\rangle \in f(\langle p/_{\mathcal{A}}q, q/_{\mathcal{A}}p\rangle)$, from which follows $\neg(p\geq_f q)$, and $\langle E,C\rangle \in f(\langle p/_{\mathcal{A}}r, r/_{\mathcal{A}}p\rangle)$, from which follows $\neg(p\geq_f r)$.

Part 6: Let $\mathcal{A} = \{A,B,C\}$, with $r \in A$, $p \in B$, and $q \in C$. Let $A>B>C$. Then $f(\langle p/_{\mathcal{A}}q, q/_{\mathcal{A}}p\rangle) = \{\langle B,C\rangle\}$ and $B \geq C$ so that $p \geq_f q$. However, it follows from $\langle B,A\rangle \in f(\langle p/_{\mathcal{A}}q\vee r, q\vee r/_{\mathcal{A}}p\rangle)$ and $A>B$ that $\neg(p \geq_f q\vee r)$.

PROOF OF OBSERVATION 6.36. *Part 1*: Let $\mathcal{A} = \{A,B,C,D,E\}$ with $p \in A$, $q \in B$, $r \in C$, $p \in D$, and $q \in E$. Let $A>B>C>D>E$, and let f be based on a similarity relation that assigns similarity in inverse proportionality to distances in the following representation:

$A \quad B \quad C \qquad\qquad\qquad\qquad D \quad E$

We may then have $f(\langle p\vee q, r\rangle) = \{\langle A,C\rangle,\langle B,C\rangle\}$, $f(\langle p,r\rangle) = \{\langle A,C\rangle,\langle D,C\rangle\}$, and $f(\langle q,r\rangle) = \{\langle B,C\rangle,\langle E,C\rangle\}$.

Part 2: Let $\mathcal{A} = \{A,B,C,D,E\}$ with $q \in A$, $r \in B$, $p \in C$, $q \in D$, and $r \in E$. Let $A>B>C>D>E$, and let f be based on a similarity relation that assigns similarity in inverse proportionality to distances in the following representation:

$A \quad B \qquad\qquad\qquad\qquad C \quad D \quad E$

We may then have $f(\langle p, q\vee r\rangle) = \{\langle C,D\rangle,\langle C,E\rangle\}$, $f(\langle p,q\rangle) = \{\langle C,A\rangle,\langle C,D\rangle\}$, and $f(\langle p,r\rangle) = \{\langle C,B\rangle,\langle C,E\rangle\}$.

PROOFS FOR CHAPTER 7

PROOF OF OBSERVATIONS 7.5 AND 7.6. Directly from Definition 7.4.

PROOF OF OBSERVATION 7.7. *Parts 1 and 2*: Let $\mathcal{A} = \{A,B,C\}$, $p,\neg q \in A$, $\neg p,q \in B$, and $\neg p,\neg q \in C$. Furthermore, let $v(A) = 5$, $v(B) = 4$, $v(C) = 0$, $w(A) = 0.2$, and $w(B) = w(C) = 0.4$.

Part 3: Let $\mathcal{A} = \{A,B,C\}$, p, $\neg q \in A$, $\neg p,q \in B$, and $\neg p,\neg q \in C$. Furthermore, let $v(A) = v(B) = 6$, $v(C) = 0$, $w(A) = w(C) = 0.25$, and $w(B) = 0.5$.

Parts 4–6: If p and q are $\mathcal{A}$-incompatible, then $repr(p) = repr(p\&\neg q)$ and $repr(q) = repr(q\&\neg p)$.

Part 7: Let $\mathcal{A} = \{A,B,C\}$, $p,q \in A$, $p,\neg q \in B$, and $\neg p,q \in C$. Furthermore, let $v(A) = 5$, $v(B) = 0$, $v(C) = 1$, $w(A) = w(B) = 0.1$, and $w(C) = 0.8$.

Part 8: Let $\mathcal{A} = \{A,B,C\}$, $p,q \in A$, $p,\neg q \in B$, and $\neg p,q \in C$. Furthermore, let $v(A) = 3$, $v(B) = 15$, $v(C) = 11$, $w(A) = w(B) = 0.2$, and $w(C) = 0.6$.

Definition 7.30. *Let $\mathcal{A}$ be a finite sentential alternative set. Let v be a value assignment and w a weight assignment on $\mathcal{A}$. Let $\mathcal{B} \subseteq \mathcal{A}$. Then:*

$$W(\mathcal{B}) = \sum_{X \in \mathcal{B}} w(X)$$

$$EU(\mathcal{B}) = \frac{\sum_{X \in \mathcal{B}} w(X) \times v(X)}{\sum_{X \in \mathcal{B}} w(X)}$$

Lemma 7.31. *Let $\mathcal{A}$ be a finite sentential alternative set. Let v be a value assignment and w a weight assignment on $\mathcal{A}$. Let $\mathcal{B}_1$ and $\mathcal{B}_2$ be subsets of $\mathcal{A}$ such that $\mathcal{B}_1 \cap \mathcal{B}_2 = \emptyset$. Then:*

$$EU(\mathcal{B}_1 \cup \mathcal{B}_1) = \frac{W(\mathcal{B}_1) \times EU(\mathcal{B}_1) + W(\mathcal{B}_2) \times EU(\mathcal{B}_2)}{W(\mathcal{B}_1) + W(\mathcal{B}_2)}$$

PROOF OF LEMMA 7.31. Left to the reader.

PROOF OF OBSERVATION 7.8. *Part 1*:

$p \geq_w q$
iff $EU(p) \geq EU(q)$
iff $EU(p) \times W(q) \geq W(q) \times EU(q)$
iff $EU(p) \times (W(p)+W(q)) \geq W(p) \times EU(p) + W(q) \times EU(q)$
iff $EU(p) \geq \dfrac{W(p) \times EU(p) + W(q) \times EU(q)}{W(p) + W(q)}$
iff $EU(p) \geq EU(p \vee q)$ (Lemma 7.31)
iff $p \geq_w (p \vee q)$

Parts 2 and 3: Let $\mathcal{A} = \{A,B,C\}$, $p,q \in A$, $p,\neg q \in B$, and $\neg p,q \in C$. Furthermore, let $v(A) = 0$, $v(B) = v(C) = 1$, and $w(A) = w(B) = w(C) = 1/3$.

Part 4 can be proved in the same way as Part 1.

Parts 5 and 6: Let $\mathcal{A} = \{A,B,C\}$, $p,q \in A$, $p,\neg q \in B$, and $\neg p,q \in C$. Furthermore, let $v(A) = 8$, $v(B) = v(C) = 2$, and $w(A) = w(B) = w(C) = 1/3$.

Parts 7 and 8: Let $\mathcal{A} = \{A,B\}$, $p,\neg q \in A$, and $\neg p,q \in B$. Furthermore, let $v(A) = 2$, $v(B) = 0$, and $w(A) = w(B) = 0.5$.

PROOF OF OBSERVATION 7.9. *Part 1*: Let $(p \vee q) \geq_w r$. According to Observation 7.6, either $p \geq_w q$ or $q \geq_w p$. If $p \geq_w q$, then Part 1 of Observation 7.8

yields $p{\geq_w}(p{\vee}q)$. Due to transitivity (Observation 7.6), we can combine $p{\geq_w}(p{\vee}q)$ and $(p{\vee}q){\geq_w}r$ to obtain $p{\geq_w}r$. If $q{\geq_w}p$, then $q{\geq_w}r$ can be obtained in the same way.

Part 2: Let $\mathcal{A} = \{A,B,C,D\}$, $p,q,\neg r \in A$, $p,\neg q,\neg r \in B$, $\neg p,q,\neg r \in C$, and $\neg p,\neg q,r \in D$. Furthermore, let $v(A) = 8$, $v(B) = v(C) = 20$, $v(D) = 15$, and $w(A) = w(B) = w(C) = w(D) = 0.25$.

Part 3: Let $\mathcal{A} = \{A,B,C,D\}$, $p,q,\neg r \in A$, $p, \neg q, \neg r \in B$, $\neg p,q, \neg r \in C$, and $\neg p, \neg q,r \in D$. Furthermore, let $v(A) = 8$, $v(B) = v(C) = 20$, $v(D) = 16$, and $w(A) = w(B) = w(C) = w(D) = 0.25$.

Part 4: Let $\mathcal{A} = \{A,B,C\}$, $p,\neg q,\neg r \in A$, $\neg p,q,\neg r \in B$, and $\neg p,\neg q,r \in C$. Furthermore, let $v(A) = 0$, $v(B) = 4$, $v(C) = 1$, and $w(A) = w(B) = w(C) = 1/3$.

Part 5: Let $\mathcal{A} = \{A,B,C\}$, $p, \neg q, \neg r \in A$, $\neg p,q,\neg r \in B$, and $\neg p,\neg q,r \in C$. Furthermore, let $v(A) = 0$, $v(B) = 4$, $v(C) = 2$, and $w(A) = w(B) = w(C) = 1/3$.

Part 6: Can be proved in the same way as Part 1.

Part 7: Let $\mathcal{A} = \{A,B,C,D\}$, $p,\neg q,\neg r \in A$, $\neg p,q,r \in B$, $\neg p,q,\neg r \in C$, and $\neg p,\neg q,r \in D$. Furthermore, let $v(A) = 9$, $v(B) = 16$, $v(C) = v(D) = 4$, and $w(A) = w(B) = w(C) = w(D) = 0.25$.

Part 8: Let $\mathcal{A} = \{A,B,C,D\}$, $p,\neg q,\neg r \in A$, $\neg p,q,r \in B$, $\neg p,q,\neg r \in C$, and $\neg p,\neg q,r \in D$. Furthermore, let $v(A) = 8$, $v(B) = 16$, $v(C) = v(D) = 4$, and $w(A) = w(B) = w(C) = w(D) = 0.25$.

Part 9: Let $\mathcal{A} = \{A,B,C\}$, $p,\neg q, \neg r \in A$, $\neg p,q,\neg r \in B$, and $\neg p, \neg q,r \in C$. Furthermore, let $v(A) = 3$, $v(B) = 4$, $v(C) = 0$, and $w(A) = w(B) = w(C) = 1/3$.

Part 10: Let $\mathcal{A} = \{A,B,C\}$, $p,\neg q,\neg r \in A$, $\neg p,q,\neg r \in B$, and $\neg p,\neg q,r \in C$. Furthermore, let $v(A) = 2$, $v(B) = 4$, $v(C) = 0$, and $w(A) = w(B) = w(C) = 1/3$.

Part 11: Let $\mathcal{A} = \{A,B,C\}$, $p,\neg q,\neg r \in A$, $\neg p,q,\neg r \in B$, and $\neg p,\neg q,r \in C$. Furthermore, let $v(A) = 4$, $v(B) = 0$, $v(C) = 3$, and $w(A) = w(B) = w(C) = 1/3$.

Part 12: Let $\mathcal{A} = \{A,B,C\}$, $p,\neg q,\neg r \in A$, $\neg p,q,\neg r \in B$, and $\neg p,\neg q,r \in C$. Furthermore, let $v(A) = 1$, $v(B) = 0$, $v(C) = 4$, and $w(A) = w(B) = w(C) = 1/3$.

PROOF OF OBSERVATION 7.10. *Part 1*: Let $\mathcal{A} = \{A,B\}$, $p,q \in A$, and $\neg p,q \in B$. Furthermore, let $v(A) = 0$, $v(B) = 2$, and $w(A) = w(B) = 0.5$.

Part 2: Let $\mathcal{A} = \{A,B\}$, $p,q \in A$, and $\neg p,q \in B$. Furthermore, let $v(A) = 2, = v(B) = 0$, and $w(A) = w(B) = 0.5$.

PROOF OF THEOREM 7.14. Let $\mathcal{B}_1$ and $\mathcal{B}_2$ be two subsets of $\mathcal{A}$ such that $\mathfrak{min}(\mathcal{B}_1)\equiv\mathfrak{min}(\mathcal{B}_2)$ and $\mathfrak{max}(\mathcal{B}_1)\equiv\mathfrak{max}(\mathcal{B}_2)$. Without loss of generality, we may let

$$\mathcal{B}_1 = \{X_1,X_2, \dots X_n\} \text{ with } X_1\geq X_2\geq \dots \geq X_n, \text{ and}$$
$$\mathcal{B}_2 = \{Y_1,Y_2, \dots Y_m\} \text{ with } Y_1\geq Y_2\geq \dots \geq Y_m.$$

The limiting case when $X_1 \equiv X_n$ is trivial, and we will assume that $X_1 \equiv X_n$ is not the case.

We will call a subset $\mathcal{D}$ of $\mathcal{A}$ *multilevelled* if and only if there are X, Y, $Z \in \mathcal{D}$ such that $X>Y>Z$. The series $\Phi_0 \dots \Phi_n$ is formed recursively as follows:

(1) $\Phi_0 = \mathcal{B}_1$
(2) If Φ_m is not multilevelled, then $\Phi_{m+1} = \Phi_m$
(3) If Φ_m is multilevelled, then $\Phi_{m+1} = \Phi_m \backslash X$ for some $X \in \Phi_m \backslash (\mathfrak{min}(\Phi_m) \cup \mathfrak{max}(\Phi_m))$

It follows from Part (1) of nonnegative response that $\Phi_{m+1} \geq' \Phi_m$. (Treat any element of $\mathfrak{max}(\mathcal{B}_1)$ as replacing X.) Furthermore, it follows from Part (2) of nonnegative response that $\Phi_m \geq' \Phi_{m+1}$. (Treat any element of $\mathfrak{min}(\mathcal{B}_1)$ as replacing Y.) Since $\geq'$ is transitive, $\Phi_0 \equiv' \Phi_n$.

Since $\Phi_n = \mathfrak{min}(\mathcal{B}_1)\cup\mathfrak{max}(\mathcal{B}_1)$, it follows that $\mathcal{B}_1 \equiv' (\mathfrak{min}(\mathcal{B}_1)\cup\mathfrak{max}(\mathcal{B}_1))$. In the same way, it follows that $\mathcal{B}_2 \equiv' (\mathfrak{min}(\mathcal{B}_2)\cup\mathfrak{max}(\mathcal{B}_2))$. It follows from positionality that $(\mathfrak{min}(\mathcal{B}_1)\cup\mathfrak{max}(\mathcal{B}_1)) \equiv' (\mathfrak{min}(\mathcal{B}_2)\cup\mathfrak{max}(\mathcal{B}_2))$, and we can conclude that $\mathcal{B}_1 \equiv' \mathcal{B}_2$.

PROOF OF OBSERVATION 7.16. Directly from Definitions 7.11, 7.13, and 7.15.

PROOF OF THEOREM 7.19. That an interval maximin relation satisfies the two conditions follows straightforwardly. For the other direction, let $\geq'$ be an extremal relation that satisfies the two conditions. To prove that it is an interval maximin relation, we need to show that:

(1) If $X_1>X_2$, then $\langle X_1,Y_1\rangle >' \langle X_2,Y_2\rangle$
(2) If $X_1\equiv X_2$, and $Y_1\equiv Y_2$, then $\langle X_1,Y_1\rangle \equiv' \langle X_2,Y_2\rangle$
(3) If $X_1\equiv X_2$, and $Y_1>Y_2$, then $\langle X_1,Y_1\rangle >' \langle X_2,Y_2\rangle$

(1) follows from cautiousness. (2) follows from Definition 7.13 ($\geq'$ is an extremal relation), and (3) from sensitivity.

PROOF OF OBSERVATION 7.23. For completeness of $\geq_i$, it is sufficient to note that since $\geq$ is complete, either $\mathfrak{min}(p) \geq \mathfrak{min}(q)$ or $\mathfrak{min}(q) \geq \mathfrak{min}(p)$. The completeness of the other three relations also follows from the completeness of $\geq$.

For transitivity of $\geq_i$, let $p \geq_i q \geq_i r$. Then $\mathfrak{min}(p) \geq \mathfrak{min}(q) \geq \mathfrak{min}(r)$, hence $\mathfrak{min}(p) \geq \mathfrak{min}(r)$, hence $p \geq_i r$. The transitivity of $\geq_x$ follows in the same way.

For the transitivity of $\geq_{ix}$, let $p \geq_{ix} q \geq_{ix} r$. Then clearly $\mathfrak{min}(p) \geq \mathfrak{min}(q)$ and $\mathfrak{min}(q) \geq \mathfrak{min}(r)$. If either $\mathfrak{min}(p) > \mathfrak{min}(q)$ or $\mathfrak{min}(q) > \mathfrak{min}(r)$, then $\mathfrak{min}(p) > \mathfrak{min}(r)$, hence $p \geq_{ix} r$. If both $\mathfrak{min}(p) \equiv \mathfrak{min}(q)$ and $\mathfrak{min}(q) \equiv \mathfrak{min}(r)$, then $\mathfrak{max}(p) \geq \mathfrak{max}(q)$ and $\mathfrak{max}(q) \geq \mathfrak{max}(r)$, hence again $p \geq_{ix} r$. The transitivity of $\geq_{xi}$ is proved in the same way.

PROOF OF OBSERVATION 7.24. *Parts 1a, 1b, 2a, 2b, 3a, 3b, 4a, and 4b*: Let $\mathcal{A} = \{A, B, C, D\}$, $p, q \in A$, $\neg p, q \in B$, $p, \neg q \in C$, and $p, q \in D$. Let $A \equiv B > C \equiv D$.

Part 1c, 2c, 3c, and 4c: Let $\mathcal{A} = \{A, B, C, D\}$, $\neg p, \neg q \in A$, $p, \neg q \in B$, $\neg p, q \in C$, and $\neg p, \neg q \in D$. Let $A \equiv B > C \equiv D$.

Parts 5a and 5b: Let $\mathcal{A} = \{A, B, C, D\}$, with A, B, C, and D as in Parts 1a, etc., and let $\delta = 0.5$ and $v(A) = v(B) = 10$ and $v(C) = v(D) = 0$.

Part 5c: Let $\mathcal{A} = \{A, B, C\}$, $p, q \in A$, $\neg p, \neg q \in B$, and $\neg p, q \in C$. Furthermore, let $\delta = 0.5$, $v(A) = 10$, and $v(B) = v(C) = 0$.

Definition 7.32. *For any sentence p,* $\text{MIN}(p)$ *is an arbitrary element of* $\mathfrak{min}(p)$*, and* $\text{MAX}(p)$ *an arbitrary element of* $\mathfrak{max}(p)$.

Lemma 7.33. *Let $\mathcal{A}$ be a sentential alternative set and $\geq$ a transitive and complete relation on $\mathcal{A}$. Let $p, q \in \mathcal{L}_{\mathcal{A}}$. Then:*

(1) $\text{MAX}(p) \geq \text{MAX}(p\&q)$
(2) $\text{MAX}(p \vee q) \geq \text{MAX}(p)$
(3) $\text{MIN}(p\&q) \geq \text{MIN}(p)$
(4) $\text{MIN}(p) \geq \text{MIN}(p \vee q)$

Furthermore, if $\mathcal{A}$ is contextually complete, then:

(5) If $\text{MAX}(p) \geq \text{MAX}(q)$*, then* $\text{MAX}(p) \equiv \text{MAX}(p \vee q)$.
(6) If $\text{MIN}(p) \geq \text{MIN}(q)$*, then* $\text{MIN}(q) \equiv \text{MIN}(p \vee q)$.

(The lemma will be used in what follows without explicit reference.)

PROOF OF LEMMA 7.33. *Part 1*: If $p\&q \notin \cup\mathcal{A}$, then this follows from Definition 7.12, Part 1. Otherwise, it follows from Definition 7.11 since $p \in \text{MAX}(p\&q)$.

Parts 2, 3, and 4 follow in the same way as Part 1.

Part 5: Let $\text{MAX}(p) \geq \text{MAX}(q)$. If $q \notin \cup\mathcal{A}$, then $p \dashv\vDash_{\mathcal{A}} p \lor q$, and the desired result follows directly. In the principal case, $q \in \cup\mathcal{A}$. It follows from Definition 7.12 that $p \in \cup\mathcal{A}$. Since $\mathcal{A}$ is contextually complete, it follows from $p \lor q \in \text{MAX}(p \lor q)$ that either $p \in \text{MAX}(p \lor q)$ or $q \in \text{MAX}(p \lor q)$. In the former case, it follows from Definition 7.11 that $\text{MAX}(p) \geq \text{MAX}(p \lor q)$. In the latter case, we obtain $\text{MAX}(q) \geq \text{MAX}(p \lor q)$ in the same way, and since $\text{MAX}(p) \geq \text{MAX}(q)$, transitivity yields $\text{MAX}(p) \geq \text{MAX}(p \lor q)$ in this case as well. It follows from Part 2 that $\text{MAX}(p \lor q) \geq \text{MAX}(p)$.

Part 6 follows in the same way as Part 5.

PROOF OF OBSERVATION 7.25. *Parts 1a, 1b, 2a, 2b, 3a, 3b, 4a, 4b*: Let $\mathcal{A} = \{A, B, C, D\}$, $p, q \in A$, $\neg p, q \in B$, $p, \neg q \in C$, and $p, q \in D$. Furthermore, let $A \equiv B > C \equiv D$.

Part 1c: Let $p >_i q$. Then $\text{MIN}(p) > \text{MIN}(q)$. Clearly, $\text{MIN}(p\&\neg q) \geq \text{MIN}(p)$. Furthermore, it follows from $\text{MIN}(p) > \text{MIN}(q)$ that $\neg p \in \text{MIN}(q)$, hence $q\&\neg p \in \text{MIN}(q)$, hence $\text{MIN}(q) \equiv \text{MIN}(q\&\neg p)$. We can apply transitivity to $\text{MIN}(p\&\neg q) \geq \text{MIN}(p)$, $\text{MIN}(p) > \text{MIN}(q)$, and $\text{MIN}(q) \equiv \text{MIN}(q\&\neg p)$, and obtain $\text{MIN}(p\&\neg q) > \text{MIN}(q\&\neg p)$, so that $(p\&\neg q) >_i (q\&\neg p)$.

Part 2c: Let $p >_x q$. Then $\text{MAX}(p) > \text{MAX}(q)$. Clearly, $\text{MAX}(q) \geq \text{MAX}(q\&\neg p)$. Furthermore, it follows from $\text{MAX}(p) > \text{MAX}(q)$ that $\neg q \in \text{MAX}(p)$, hence $p\&\neg q \in \text{MAX}(p)$, hence $\text{MAX}(p\&\neg q) \equiv \text{MAX}(p)$. We can apply transitivity to $\text{MAX}(p\&\neg q) \equiv \text{MAX}(p)$, $\text{MAX}(p) > \text{MAX}(q)$, and $\text{MAX}(q) \geq \text{MAX}(q\&\neg p)$, and obtain $\text{MAX}(p\&\neg q) > \text{MAX}(q\&\neg p)$, so that $(p\&\neg q) >_x (q\&\neg p)$.

Part 3c: Let $\mathcal{A} = \{A,B,C,D\}$, $p,\neg q \in A$, $\neg p,q \in B$, $p,\neg q \in C$, and $p,q \in D$. Let $A>B>C>D$.

Part 4c: Let $\mathcal{A} = \{A,B,C,D\}$, $p,q \in A$, $\neg p,q \in B$, $p,\neg q \in C$, and $\neg p,q \in D$. Let $A>B>C>D$.

Parts 5a and 5b: Let $\mathcal{A} = \{A,B,C,D\}$, $p,q \in A$, $\neg p,q \in B$, $p,\neg q \in C$, and $p,q \in D$. Let $v(A) = v(B) = 10$, $v(C) = v(D) = 0$, and $\delta = 0.5$.

Part 5c: Let $\mathcal{A} = \{A,B,C,D\}$, $p\&\neg q \in A \cap C$, $\neg p\&q \in B$, $p\&q \in D$, $v(A) = 10$, $v(B) = 8$, $v(C) = v(D) = 0$, and $\delta = 0.5$.

PROOF OF OBSERVATION 7.26. *Part 1a*: Left to right: Let $(p \vee q) \geq_i r$. Then $\text{MIN}(p \vee q) \geq \text{MIN}(r)$. Since $\text{MIN}(p) \geq \text{MIN}(p \vee q)$, transitivity yields $\text{MIN}(p) \geq \text{MIN}(r)$, hence $p \geq_i r$. We can prove $q \geq_i r$ in the same way.

Right to left: Let $p \geq_i r$ and $q \geq_i r$. Then $\text{MIN}(p) \geq \text{MIN}(r)$ and $\text{MIN}(q) \geq \text{MIN}(r)$. Since either $\text{MIN}(p \vee q) \equiv \text{MIN}(p)$ or $\text{MIN}(p \vee q) \equiv \text{MIN}(q)$, we can use transitivity to obtain $\text{MIN}(p \vee q) \geq \text{MIN}(r)$, hence $(p \vee q) \geq_i r$.

Part 1b: Let $\mathcal{A} = \{A,B,C\}$, $p,\neg q,\neg r \in A$, $\neg p,\neg q,r \in B$, $\neg p,q,\neg r \in C$ and $A > B > C$.

Part 2a: Left to right: Let $(p \vee q) \geq_x r$, i.e., $\text{MAX}(p \vee q) \geq \text{MAX}(r)$. Since either $\text{MAX}(p) \equiv \text{MAX}(p \vee q)$ or $\text{MAX}(q) \equiv \text{MAX}(p \vee q)$, we can use transitivity to obtain either $\text{MAX}(p) \geq \text{MAX}(r)$ or $\text{MAX}(q) \geq \text{MAX}(r)$, hence either $p \geq_x r$ or $q \geq_x r$.

Right to left: Let $p \geq_x r$, i.e., $\text{MAX}(p) \geq \text{MAX}(r)$. We have $\text{MAX}(p \vee q) \geq \text{MAX}(p)$, and transitivity yields $\text{MAX}(p \vee q) \geq \text{MAX}(r)$, i.e., $(p \vee q) \geq_x r$. If $q \geq_x r$, then $(p \vee q) \geq_x r$ follows in the same way.

Part 2b: Let $\mathcal{A} = \{A,B,C\}$, $\neg p,q,\neg r \in A$, $\neg p,\neg q,r \in B$, $p,\neg q,\neg r \in C$, and $A > B > C$.

Part 3a: Let $(p \vee q) \geq_{ix} r$. There are two cases.

Case *i*, $\text{MIN}(p \vee q) > \text{MIN}(r)$: Then, since $\text{MIN}(p) \geq \text{MIN}(p \vee q)$, transitivity yields $\text{MIN}(p) > \text{MIN}(r)$, hence $p \geq_{ix} r$.

Case *ii*, $\text{MIN}(p \vee q) \equiv \text{MIN}(r)$ and $\text{MAX}(p \vee q) \geq \text{MAX}(r)$: Since either $\text{MAX}(p) \equiv \text{MAX}(p \vee q)$ or $\text{MAX}(q) \equiv \text{MAX}(p \vee q)$, it follows from transitivity that either $\text{MAX}(p) \geq \text{MAX}(r)$ or $\text{MAX}(q) \geq \text{MAX}(r)$.

In the case when $\text{MAX}(p) \geq \text{MAX}(r)$, we make use of $\text{MIN}(p) \geq \text{MIN}(p \vee q)$ and $\text{MIN}(p \vee q) \equiv \text{MIN}(r)$ to obtain $\text{MIN}(p) \geq \text{MIN}(r)$. It follows from $\text{MAX}(p) \geq \text{MAX}(r)$ and $\text{MIN}(p) \geq \text{MIN}(r)$ that $p \geq_{ix} r$.

In the case when $\text{MAX}(q) \geq \text{MAX}(r)$, we obtain $q \geq_{ix} r$ in the same way.

Part 3b: Let $\mathcal{A} = \{A,B,C,D\}$, $\neg p,q,\neg r \in A$, $\neg p,\neg q,r \in B$, $p,\neg q,\neg r \in C$, $\neg p,\neg q,r \in D$, and $A > B > C \equiv D$.

Part 3c: Let $\mathcal{A} = \{A,B,C\}$, $p,\neg q,\neg r \in A$, $\neg p,\neg q,r \in B$, $\neg p,q,\neg r \in C$ and $A > B > C$.

Part 4a: Let $(p \vee q) \geq_{xi} r$. There are two cases.

Case *i*: $\text{MAX}(p \vee q) > \text{MAX}(r)$. Since either $\text{MAX}(p) \equiv \text{MAX}(p \vee q)$ or $\text{MAX}(q) \equiv \text{MAX}(p \vee q)$, transitivity yields either $\text{MAX}(p) > \text{MAX}(r)$ or $\text{MAX}(q) > \text{MAX}(r)$, hence either $p \geq_{xi} r$ or $q \geq_{xi} r$.

Case *ii*: $\text{MAX}(p\vee q)\equiv\text{MAX}(r)$ and $\text{MIN}(p\vee q)\geq\text{MIN}(r)$. Since $\text{MIN}(p)\geq\text{MIN}(p\vee q)$, it follows from transitivity that $\text{MIN}(p)\geq\text{MIN}(r)$. In the same way, we obtain $\text{MIN}(q)\geq\text{MIN}(r)$.

It follows from $\text{MAX}(p\vee q)\equiv\text{MAX}(r)$, since either $\text{MAX}(p)\equiv\text{MAX}(p\vee q)$ or $\text{MAX}(q)\equiv\text{MAX}(p\vee q)$, that either $\text{MAX}(p)\equiv\text{MAX}(r)$ or $\text{MAX}(q)\equiv\text{MAX}(r)$.

If $\text{MAX}(p)\equiv\text{MAX}(r)$, then we combine this with $\text{MIN}(p)\geq\text{MIN}(r)$ to obtain $p\geq_{xi}r$. In the other case, we combine $\text{MAX}(q)\equiv\text{MAX}(r)$ and $\text{MIN}(q)\geq\text{MIN}(r)$ to obtain $q\geq_{xi}r$.

Part 4b: Let $\mathcal{A} = \{A,B,C\}$, $\neg p,q,\neg r \in A$, $\neg p,\neg q,r \in B$, $p,\neg q,\neg r \in C$, and $A>B>C$.

Part 4c: Let $\mathcal{A} = \{A,B,C,D\}$, $p,\neg q,\neg r \in A$, $\neg p,\neg q,r \in B$, $\neg p,\neg q,r \in C$, $\neg p,q,\neg r \in D$, and $A\equiv B>C>D$.

Part 5a: Let $(p\vee q)\geq_{E}r$. Either $\text{MAX}(p\vee q)\equiv\text{MAX}(p)$ or $\text{MAX}(p\vee q)\equiv\text{MAX}(q)$. In the first case, since $\text{MIN}(p)\geq\text{MIN}(p\vee q)$, we have $p\geq_{E}(p\vee q)$, hence by transitivity $p\geq_{E}r$. In the second case, $q\geq_{E}r$ follows in the same way.

Part 5b: Let $\mathcal{A}$ be as in Part 4b, and let $v(A) = 10$, $v(B) = 4$, $v(C) = 0$ and $\delta = 0.5$.

Part 5c: Let $\mathcal{A} = \{A,B,C\}$, $p, \neg q,\neg r \in A$, $\neg p,\neg q,r \in B$, $\neg p,q,\neg r \in C$, $v(A) = 10$, $v(B) = 9$, $v(C) = 0$, and $\delta = 0.5$.

PROOF OF OBSERVATION 7.27. *Part 1a*: Left to right: Let $p\geq_{i}(q\vee r)$. Then $\text{MIN}(p)\geq\text{MIN}(q\vee r)$. Since either $\text{MIN}(q\vee r)\equiv\text{MIN}(q)$ or $\text{MIN}(q\vee r)\equiv\text{MIN}(r)$, transitivity yields either $\text{MIN}(p)\geq\text{MIN}(q)$ or $\text{MIN}(p)\geq\text{MIN}(r)$, hence either $p\geq_{i}q$ or $p\geq_{i}r$.

Right to left: For symmetry reasons, we may assume that $p\geq_{i}q$. Then $\text{MIN}(p)\geq\text{MIN}(q)$. Since $\text{MIN}(q)\geq\text{MIN}(q\vee r)$, transitivity yields $\text{MIN}(p)\geq\text{MIN}(q\vee r)$, hence $p\geq_{i}(q\vee r)$.

Part 1b: Let $\mathcal{A} = \{A,B,C\}$, $\neg p,q,\neg r \in A$, $p,\neg q,\neg r \in B$, $\neg p,\neg q,r \in C$, and $A>B>C$.

Part 2a: Left to right: Let $p\geq_{x}(q\vee r)$. Then $\text{MAX}(p)\geq\text{MAX}(q\vee r)$. Since $\text{MAX}(q\vee r)\geq\text{MAX}(q)$, transitivity yields $\text{MAX}(p)\geq\text{MAX}(q)$, so that $p\geq_{x}q$. We can obtain $p\geq_{x}r$ in the same way.

Right to left: Let $p\geq_{x}q$ and $p\geq_{x}r$. Then $\text{MAX}(p)\geq\text{MAX}(q)$ and $\text{MAX}(p)\geq\text{MAX}(r)$. Furthermore, either $\text{MAX}(q)\equiv\text{MAX}(q\vee r)$ or $\text{MAX}(r)\equiv\text{MAX}(q\vee r)$. In both cases, it follows by transitivity that $\text{MAX}(p)\geq\text{MAX}(q\vee r)$, hence $p\geq_{x}(q\vee r)$.

Part 2b: Let $\mathcal{A} = \{A,B,C\}$, $\neg p,\neg q,r \in A$, $p,\neg q,\neg r \in B$, $\neg p,q,\neg r \in C$ and $A>B>C$.

Part 3a: Let $p \geq_{ix} (q \vee r)$. There are two cases.

Case *i*, $\text{MIN}(p) > \text{MIN}(q \vee r)$: Since either $\text{MIN}(q \vee r) \equiv \text{MIN}(q)$ or $\text{MIN}(q \vee r) \equiv \text{MIN}(r)$, transitivity yields either $\text{MIN}(p) > \text{MIN}(q)$ or $\text{MIN}(p) > \text{MIN}(r)$, hence either $p \geq_{ix} q$ or $p \geq_{ix} r$.

Case *ii*, $\text{MIN}(p) \equiv \text{MIN}(q \vee r)$ and $\text{MAX}(p) \geq \text{MAX}(q \vee r)$: We have either $\text{MIN}(q \vee r) \equiv \text{MIN}(q)$ or $\text{MIN}(q \vee r) \equiv \text{MIN}(r)$.

If $\text{MIN}(q \vee r) \equiv \text{MIN}(q)$, then transitivity yields $\text{MIN}(p) \equiv \text{MIN}(q)$. Since $\text{MAX}(q \vee r) \geq \text{MAX}(q)$, transitivity also yields $\text{MAX}(p) \geq \text{MAX}(q)$. We can conclude from this that $p \geq_{ix} q$.

If $\text{MIN}(q \vee r) \equiv \text{MIN}(q)$, then $p \geq_{ix} r$ follows in the same way.

Part 3b: Let $\mathcal{A} = \{A,B,C\}$, $\neg p, q, \neg r \in A$, $p, \neg q, \neg r \in B$, $\neg p, \neg q, r \in C$, and $A > B > C$.

Part 3c: Let $\mathcal{A} = \{A,B,C,D\}$, $\neg p, \neg q, r \in A$, $p, \neg q, \neg r \in B$, $p, \neg q, \neg r \in C$, $\neg p, q, \neg r \in D$, and $A > B > C \equiv D$.

Part 4a: Let $p \geq_{xi} (q \vee r)$. There are two cases.

Case *i*, $\text{MAX}(p) > \text{MAX}(q \vee r)$: Since $\text{MAX}(q \vee r) \geq \text{MAX}(q)$, it follows by transitivity that $\text{MAX}(p) > \text{MAX}(q)$, hence $p \geq_{xi} q$.

Case *ii*, $\text{MAX}(p) \equiv \text{MAX}(q \vee r)$ and $\text{MIN}(p) \geq \text{MIN}(q \vee r)$: Since either $\text{MIN}(q \vee r) \equiv \text{MIN}(q)$ or $\text{MIN}(q \vee r) \equiv \text{MIN}(r)$, it follows by transitivity that either $\text{MIN}(p) \geq \text{MIN}(q)$ or $\text{MIN}(p) \geq \text{MIN}(r)$.

If $\text{MIN}(p) \geq \text{MIN}(q)$, then we use $\text{MAX}(p) \equiv \text{MAX}(q \vee r)$ and $\text{MAX}(q \vee r) \geq \text{MAX}(q)$ to obtain $\text{MAX}(p) \geq \text{MAX}(q)$. It follows from $\text{MIN}(p) \geq \text{MIN}(q)$ and $\text{MAX}(p) \geq \text{MAX}(q)$ that $p \geq_{xi} q$.

If $\text{MIN}(p) \geq \text{MIN}(r)$, then $p \geq_{xi} r$ can be obtained in the same way.

Part 4b: Let $\mathcal{A} = \{A,B,C,D\}$, $p, \neg q, \neg r \in A$, $\neg p, q, \neg r \in B$, $p, \neg q, \neg r \in C$, $\neg p, \neg q, r \in D$, and $A \equiv B > C > D$.

Part 4c: Let $\mathcal{A} = \{A,B,C\}$, $\neg p, \neg q, r \in A$, $p, \neg q, \neg r \in B$, $\neg p, q, \neg r \in C$ and $A > B > C$.

Part 5a: Let $p \geq_E (q \vee r)$. Either $\text{MIN}(q \vee r) \equiv \text{MIN}(q)$ or $\text{MIN}(q \vee r) \equiv \text{MIN}(r)$. In the first case, since $\text{MAX}(q \vee r) \geq \text{MAX}(q)$, we have $(q \vee r) \geq_E r$, hence by transitivity $p \geq_E r$. In the second case, $q \geq_E r$ follows in the same way.

Part 5b: Let $\mathcal{A} = \{A,B,C\}$, $\neg p, q, \neg r \in A$, $p, \neg q, \neg r \in B$, $\neg p, \neg q, r \in C$, $v(A) = 10$, $v(B) = 8$, $v(C) = 0$, and $\delta = 0.5$.

Part 5c: Let $\mathcal{A} = \{A,B,C\}$, $\neg p, \neg q, r \in A$, $p, \neg q, \neg r \in B$, $\neg p, q, \neg r \in C$, $v(A) = 10$, $v(B) = 2$, $v(C) = 0$, and $\delta = 0.5$.

PROOF OF OBSERVATION 7.28. *Part 1*: It follows from $\vDash_{\mathcal{A}} p \rightarrow q$ that $\text{MIN}(p) \geq \text{MIN}(q)$, hence $p \geq_i q$.

Part 2: It follows from $\vDash_{\mathcal{A}} p \rightarrow q$ that $\text{MAX}(q) \geq \text{MAX}(p)$, hence $q \geq_x p$.

Parts 3a and 4a: Let $\mathcal{A} = \{A,B\}$, $\neg p,q \in A$, $p,q \in B$, and $A>B$.

Part 3b and 4b: Let $\mathcal{A} = \{A,B\}$, $p,q \in A$, $\neg p,q \in B$, and $A>B$.

Part 5a: Let $\mathcal{A}$ be as in Parts 3a and 4a, and let $v(A) = 10$, $v(B) = 0$, and $\delta = 0.5$.

Part 5b: Let $\mathcal{A}$ be as in Parts 3b and 4b, and let $v(A) = 10$, $v(B) = 0$, and $\delta = 0.5$.

PROOF OF OBSERVATION 7.29. *Parts 1a and 1b*: Let $p \geq_i q$. Then $\text{MIN}(p) \geq \text{MIN}(p \vee q)$ and $\text{MIN}(p \vee q) \equiv \text{MIN}(q)$.

Parts 2a and 2b: Let $p \geq_x q$. Then $\text{MAX}(p) \equiv \text{MAX}(p \vee q)$ and $\text{MAX}(p \vee q) \geq \text{MAX}(q)$.

Part 3a: Let $p \geq_{ix} q$. There are two cases.

Case *i*, $\text{MIN}(p) > \text{MIN}(q)$: Then $\text{MIN}(p \vee q) \equiv \text{MIN}(q)$. Transitivity yields $\text{MIN}(p) > \text{MIN}(p \vee q)$, so that $p \geq_{ix} (p \vee q)$. We can then combine $\text{MIN}(p \vee q) \equiv \text{MIN}(q)$ with $\text{MAX}(p \vee q) \geq \text{MAX}(q)$ to obtain $(p \vee q) \geq_{ix} q$.

Case *ii*, $\text{MIN}(p) \equiv \text{MIN}(q)$ and $\text{MAX}(p) \geq \text{MAX}(q)$: It follows from $\text{MIN}(p) \equiv \text{MIN}(q)$ that $\text{MIN}(p) \equiv \text{MIN}(p \vee q)$. It also follows from $\text{MAX}(p) \geq \text{MAX}(q)$ that $\text{MAX}(p) \equiv \text{MAX}(p \vee q)$. We can combine $\text{MIN}(p) \equiv \text{MIN}(p \vee q)$ and $\text{MAX}(p) \equiv \text{MAX}(p \vee q)$ to obtain $p \geq_{ix} (p \vee q)$.

We also have $\text{MIN}(p \vee q) \equiv \text{MIN}(q)$ and $\text{MAX}(p \vee q) \geq \text{MAX}(q)$. This is sufficient to prove $(p \vee q) \geq_{ix} q$.

Parts 3b and 4b: Let $\mathcal{A} = \{A,B,C\}$, $p,\neg q \in A$, $\neg p,q \in B$, $\neg p,q \in C$, and $A \equiv B > C$.

Parts 3c and 4c: Let $\mathcal{A} = \{A,B,C\}$, $p,\neg q \in A$, $p,\neg q \in B$, $\neg p,q \in C$, and $A > B \equiv C$.

Parts 3d and 4d: Let $p \equiv_{ix} q$ or $p \equiv_{xi} q$. Then $\text{MIN}(p) \equiv \text{MIN}(q)$ and $\text{MAX}(p) \equiv \text{MAX}(q)$. From this follows that $\text{MIN}(p) \equiv \text{MIN}(p \vee q)$ and $\text{MAX}(p) \equiv \text{MAX}(p \vee q)$, hence $p \equiv_{ix} (p \vee q)$ and $p \equiv_{xi} (p \vee q)$.

Part 4a: Let $p \geq_{xi} q$. There are two cases.

Case *i*, $\text{MAX}(p) > \text{MAX}(q)$: Then $\text{MAX}(p) \equiv \text{MAX}(p \vee q)$. We can combine this with $\text{MIN}(p) \geq \text{MIN}(p \vee q)$ to obtain $p \geq_{xi} (p \vee q)$. It follows from $\text{MAX}(p \vee q) > \text{MAX}(q)$ that $(p \vee q) \geq_{xi} q$.

Case *ii*, $\text{MAX}(p) \equiv \text{MAX}(q)$ and $\text{MIN}(p) \geq \text{MIN}(q)$: It follows from $\text{MAX}(p) \equiv \text{MAX}(q)$ that $\text{MAX}(p) \equiv \text{MAX}(p \vee q)$. We can combine this with $\text{MIN}(p) \geq \text{MIN}(p \vee q)$ to obtain $p \geq_{xi} (p \vee q)$.

It follows from $\text{MIN}(p) \geq \text{MIN}(q)$ that $\text{MIN}(p \vee q) \equiv \text{MIN}(q)$. We can combine this with $\text{MAX}(p \vee q) \geq \text{MAX}(q)$ to obtain $(p \vee q) \geq_{xi} q$.

Part 5a: Let $p \geq_E q$. Then:

$$\delta \cdot v_{MAX}(p) + (1-\delta) \cdot v_{MIN}(p) \geq \delta \cdot v_{MAX}(q) + (1-\delta) \cdot v_{MIN}(q)$$

or equivalently:

$$\delta \cdot (v_{MAX}(p) - v_{MAX}(q)) + (1-\delta) \cdot (v_{MIN}(p) - v_{MIN}(q)) \geq 0$$

There are four cases:

(1) $v_{MAX}(p \vee q) = v_{MAX}(p)$ and $v_{MIN}(p \vee q) = v_{MIN}(p)$. Then $(p \vee q) \equiv_E p$, and the rest is obvious.

(2) $v_{MAX}(p \vee q) = v_{MAX}(p)$ and $v_{MIN}(p \vee q) = v_{MIN}(q)$. Then:
$v_\delta(p) = \delta \cdot v_{MAX}(p) + (1-\delta) \cdot v_{MIN}(p)$
$\geq \delta \cdot v_{MAX}(p) + (1-\delta) \cdot v_{MIN}(q) = v_\delta(p \vee q)$
$\geq \delta \cdot v_{MAX}(q) + (1-\delta) \cdot v_{MIN}(q)$
$= v_\delta(q)$

(3) $v_{MAX}(p \vee q) = v_{MAX}(q)$ and $v_{MIN}(p \vee q) = v_{MIN}(p)$. Then $v_{MAX}(q) \geq v_{MAX}(p)$ and $v_{MIN}(q) \geq v_{MIN}(p)$. Since $p \geq_E q$, we then have $v_{MAX}(q) = v_{MAX}(p)$ and $v_{MIN}(q) = v_{MIN}(p)$, and we are back in case 1.

(4) $v_{MAX}(p \vee q) = v_{MAX}(q)$ and $v_{MIN}(p \vee q) = v_{MIN}(q)$. Then $(p \vee q) \equiv_E q$, and the rest is obvious.

Parts 5b and 5c: Let $\mathcal{A} = \{A,B\}$, $p, \neg q \in A$, $\neg p, q \in B$, $v(A) = 10$, $v(B) = 0$ and $\delta = 0.5$.

Part 5d: Let $p \equiv_E q$. For symmetry reasons, we only need to treat two cases.

(1) $v_{MAX}(p) = v_{MAX}(q)$ and $v_{MIN}(p) = v_{MIN}(q)$: Trivial.

(2) $v_{MAX}(p) > v_{MAX}(q)$ and $v_{MIN}(q) > v_{MIN}(p)$: We then have:
$v_\delta(p \vee q) =$
$= \delta \cdot v_{MAX}(p \vee q) + (1-\delta) \cdot v_{MIN}(p \vee q)$
$= \delta \cdot v_{MAX}(p) + (1-\delta) \cdot v_{MIN}(p) = v_\delta(p)$

PROOF OF THE COROLLARY TO OBSERVATION 7.29. *Part 1*: Substitute $p\&q$ for p and $p\&\neg q$ for q in $(p \vee q) \equiv_i p \vee (p \vee q) \equiv_i q$, which follows from Part 1b of the observation. *Part 2* of the corollary follows in the same way from Part 2b of the observation.

PROOFS FOR CHAPTER 8

PROOF OF OBSERVATION 8.3. *Left to right*: Let H be a $\geq'$-positive predicate. Suppose that $\neg H$ does not satisfy $\geq'$-negativity. Then there are relata p and q such that $\neg Hp$, $p \geq' q$, and $\neg(\neg Hq)$. Hence, Hq, $p \geq' q$, and $\neg Hp$, contrary to the positivity of H.

Right to left: Let $\neg H$ be a $\geq'$-negative predicate. Suppose that H does not satisfy $\geq'$-positivity. Then there are relata p and q such that Hp, $q\geq'p$, and $\neg(Hq)$. Hence, $\neg Hq$, $q\geq'p$, and $\neg(\neg Hp)$, contrary to the negativity of $\neg H$.

PROOF OF OBSERVATION 8.7. It is sufficient to prove that H is $\geq'$-circumscriptive. To show that, let H^+ and H^- be monadic predicates defined as follows:

$$H^+p \leftrightarrow (\exists q)(Hq \,\&\, p \geq' q)$$
$$H^-p \leftrightarrow (\exists q)(Hq \,\&\, q \geq' p)$$

It follows directly that H^+ is $\geq'$-positive and that H^- is $\geq'$-negative. We have to show that

$$Hp \leftrightarrow H^+p \,\&\, H^-p.$$

The left-to-right direction follows directly from our definitions of H^+ and H^- and the reflexivity of $\geq'$. For the right-to-left direction, let H^+p & H^-p. It follows from H^+p that there is some r such that Hr and $p\geq'r$. It follows from H^-p that there is some s such that Hs and $s\geq'p$. We can now make use of the fact that $\geq'$ is continuous and obtain Hp, which concludes the proof.

PROOF OF OBSERVATION 8.12. *Part 1*: For the nontrivial direction, suppose to the contrary that H satisfies positivity, nonduplicity, and negation-comparability but not $Hp \rightarrow p\geq'\neg p$. Then there is some p such that Hp and $\neg(p\geq'\neg p)$. It follows from negation-comparability that $\neg p>'p$, and from positivity that $H\neg p$, contrary to ND.

The proof of Part 2 is similar.

PROOF OF OBSERVATION 8.13. Suppose to the contrary that $\langle G,B\rangle$ satisfies PN, ND, and NC, but not ME. Then, since ME does not hold, there is some p such that Gp & Bp. It follows from NC that $p\geq'\neg p \vee \neg p\geq'p$. From PN follows $\neg p\geq'p \,\&\, Gp \rightarrow G\neg p$ and $p\geq'\neg p \,\&\, Bp \rightarrow B\neg p$. It follows by sentential logic that $(Gp \,\&\, G\neg p) \vee (Bp \,\&\, B\neg p)$, contrary to ND.

PROOF OF THEOREM 8.15. *Part 1*: To see that the positivity part of PN holds, let G_Cp and $r\geq'p$. Then it holds for all q that if $q\geq'^*r$, then $q\geq'^*p$. It follows from Definition 8.14 that G_Cr. The negativity part of PN is proved analogously.

To see that ND is satisfied, suppose to the contrary that $G_C p$ and $G_C \neg p$. Due to ancestral reflexivity, $p \geq'^* p$, and since $G_C p$, Definition 8.14 yields $p >' \neg p$. In the same way, it follows from $G_C \neg p$ that $\neg p >' p$. This contradiction is sufficient to ensure that $\neg(G_C p \ \& \ G_C \neg p)$. The proof that $\neg(B_C p \ \& \ B_C \neg p)$ is similar.

To see that NC is satisfied, note that due to ancestral reflexivity, $G_C p$ implies $p >' \neg p$ and $B_C p$ implies $\neg p >' p$.

Part 2: Let Gp and $q \geq'^* p$. Then there is a series of sentences $s_0, \ldots s_n$, such that $s_0 \leftrightarrow p$, $s_n \leftrightarrow q$ and for all integers k, if $0 \leq k < n$, then $s_{k+1} \geq' s_k$. Clearly, Gs_0. From Gs_k and $s_{k+1} \geq' s_k$ it follows by PN that Gs_{k+1}. Thus, by iteration, Gs_n, i.e., Gq.

From Gq, it follows by NC that $q >' \neg q \vee \neg q \geq' q$. Suppose that $\neg q \geq' q$. Then by PN follows $G \neg q$, so that $Gq \ \& \ G \neg q$, contrary to ND. It follows that $q >' \neg q$.

Thus, if Gp, then for all q, if $q \geq'^* p$, then $q >' \neg q$. The corresponding property for Bp can be proved in the same way, thus completing the proof of the theorem.

PROOF OF OBSERVATION 8.16. First, let (a) hold. By sentential logic, (a) implies $p >' q \rightarrow Gp \vee Bp \vee Gq \vee Bq$. By PN, $p >' q$ implies $Gq \rightarrow Gp$ and $Bp \rightarrow Bq$. It follows, by sentential logic, that (b) holds.

Next, let (b) hold. It implies $\neg Gp \ \& \ \neg Bq \rightarrow \neg(p >' q)$. This formula has the substitution-instance $\neg Gq \ \& \ \neg Bp \rightarrow \neg(q >' p)$. From these two formulas, (a) follows by sentential logic.

PROOF OF THEOREM 8.17. Let $\langle G,B \rangle$ satisfy PN, ND, NC, and closeness. It follows from Part 1 of Theorem 8.15 that $\langle G_C,B_C \rangle$ satisfies PN, ND, and NC. To see that it satisfies closeness, suppose that $p >' q$. Since $\langle G,B \rangle$ satisfies closeness, we have $Gp \vee Bp$, and by Part (2) of Theorem 8.15 we have $G_C p \vee B_C p$.

PROOF OF THEOREM 8.19. For one direction, note that if $\langle G_N,B_N \rangle$ does not satisfy PN, then it cannot be identical with $\langle G_C,B_C \rangle$, which satisfies this condition.

For the other direction, let $\langle G_N,B_N \rangle$ satisfy PN. Since it also satisfies ND and NC, it follows from Theorem 8.15 that $G_N p \rightarrow G_C p$ and $B_N p \rightarrow B_C p$. It follows directly from ancestral reflexivity and Definitions 8.14 and 8.18 that $G_C p \rightarrow G_N p$ and $B_C p \rightarrow B_N p$.

PROOF OF OBSERVATION 8.20. Let $p, q, \neg p, \neg q \in \cup\mathcal{A}$.

Part 1: Let $\mathcal{A} = \{A,B,C,D\}$, with $p,\neg q \in A$, $p,q \in B$, $p,q \in C$, and $\neg p,q \in D$. Let $\geq$ be transitive and complete and such that $A{\equiv}B> C{\equiv}D$. Let $\geq'$ be an extremal preference relation that satisfies sensitivity. It follows from $\text{MAX}(q){\geq}\text{MAX}(p)$ and $\text{MIN}(q){\geq}\text{MIN}(p)$ that $q{\geq'}p$. Since $\geq'$ satisfies sensitivity, $p{>'}\neg p$ follows from $\text{MAX}(p){>}\text{MAX}(\neg p)$ and $\text{MIN}(p){\geq}\text{MIN}(\neg p)$, and similarly $\neg q{>'}q$ follows from $\text{MAX}(\neg q){\geq}\text{MAX}(q)$ and $\text{MIN}(\neg q){>}\text{MIN}(q)$. We therefore have G_Np and $q{\geq'}p$ but $\neg G_Nq$, so that G_N is not $\geq'$-positive. Similarly, B_Nq and $q{\geq'}p$ but $\neg B_Np$.

Part 2: To show the $\geq_i$-positivity of G_N, let G_Np and $q{\geq_i}p$. Then $p{>_i}\neg p$ so that $\text{MIN}(p){>}\text{MIN}(\neg p)$, and $\text{MIN}(q){\geq}\text{MIN}(p)$. By transitivity, $\text{MIN}(q){>}\text{MIN}(\neg p)$, hence $\neg q \in \text{MIN}(\neg p)$, hence $\text{MIN}(q){>}\text{MIN}(\neg q)$, hence $q{>_i}\neg q$, i.e., G_Nq.

To see that B_N is not $\geq_i$-negative, let $\mathcal{A} = \{A,B,C\}$, with $A{>}B{\equiv}C$, $\neg p, q \in A$, $p, \neg q \in B$, and $p, q \in C$. Then B_Np, $p{\geq_i}q$, and $\neg(B_Nq)$.

Part 3: To show the $\geq_x$-negativity of B_N, let B_Np and $p{\geq_x}q$. Then $\text{MAX}(\neg p){>}\text{MAX}(p){\geq}\text{MAX}(q)$, hence $\neg q \in \text{MAX}(\neg p)$ so that $\text{MAX}(\neg q){>}\text{MAX}(q)$, i.e., B_Nq.

To see that G_N is not $\geq_x$-positive, let $\mathcal{A} = \{A,B,C\}$, with $A{\equiv}B{>}C$, $p, q \in A$, $p, \neg q \in B$, and $\neg p, q \in C$. Then G_Np, $q{\geq_x}p$, and $\neg(G_Nq)$.

Part 4: Let $\mathcal{A} = \{A,B,C,D\}$, with $\neg p,\neg q \in A$, $\neg p,q \in B$, $p,\neg q \in C$, and $\neg p,\neg q \in D$. Furthermore, let $A{>}B{>}C{>}D$, and let similarity be inversely proportionate to distance in the following representation:

$A \quad\quad B \quad\quad\quad\quad\quad\quad C \quad\quad D$

Then G_N is not $\geq_f$-positive, and B_N is not $\geq_f$-negative.

PROOF OF OBSERVATION 8.21. *Part 1*: In order to prove that $Gp \rightarrow G_Np$, let Gp hold. It follows from NC that either $\neg p{\geq'}p$ or $p{>'}\neg p$. If $\neg p{\geq'}p$, then PN yields $G\neg p$, contrary to ND. We may conclude that $p{>'}\neg p$, i.e., G_Np.

To prove $B\neg p \rightarrow G_Np$, let $B\neg p$. Then by NC, either $\neg p{\geq'}p$ or $p{>'}\neg p$. Suppose that $\neg p{\geq'}p$. Then, by PN, Bp, contrary to ND. Thus not $\neg p{\geq'}p$, thus $p{>'}\neg p$, i.e., G_Np.

Part 2 can be proved in the same way as Part 1.

Part 3: One direction of the equivalence was proved in Part 1. To prove $G_Np \rightarrow Gp \vee B\neg p$, let G_Np hold. Then $p{>'}\neg p$ and, by closeness for $\langle G,B\rangle$, $Gp \vee B\neg p$.

Part 4 can be proved in the same way.

PROOF OF OBSERVATION 10.3. Left to the reader.

PROOF OF OBSERVATION 10.4. Let $p>'q$ and $\neg p\geq'\neg q$. For one direction, let Op hold. It then follows from the $\geq'$-contranegativity of O that Oq. For the other direction, let Oq hold. It then follows from the $\geq'$-positivity of O that Op.

PROOF OF OBSERVATION 10.6. Suppose to the contrary that $\geq'$ does not satisfy contraposition of strict preference. Then, since $\geq'$ is complete, there are p and q such that $p>'q$ and $\neg p\geq'\neg q$. Since $\mathcal{H}$ is a fine-grained positive set, there must be some element O of $\mathcal{H}$ such that Op and $\neg Oq$. It follows from Op and $\neg p\geq'\neg q$, since O is $\geq'$-contranegative, that Oq.

PROOF OF OBSERVATION 10.7. *Part 1:* Let Op and $(\neg p)\geq_x(\neg q)$. It follows from Op that $p \in \cap I$, hence $p>_x\neg p$, hence, $p>_x\neg p\geq_x\neg q$, so that $p>_x\neg q$, i.e., $\mathfrak{max}(p)>\mathfrak{max}(\neg q)$. Since $\mathcal{A}$ is contextually complete, it follows that $\mathfrak{max}(q)>\mathfrak{max}(\neg q)$, hence Oq.

Part 2: Let $\mathcal{A} = \{A, B, C\}$, with $\{p, \neg q\} \subseteq A$, $\{p, q\} \subseteq B$, and $\{\neg p, q\} \subseteq C$. Let $A\equiv B>C$. Then Op and $q\geq_x p$ but $\neg Oq$.

PROOF OF THEOREM 10.8. *From (1) to (2)*: Let O be a predicate that is contranegative with respect to $\geq'$, and let Op & Oq. Equivalently, $O\neg(\neg p)$ & $O\neg(\neg q)$. We can apply the substitution-instance $((\neg p)\geq'^*(\neg p\vee\neg q)) \vee ((\neg q)\geq'^*(\neg p\vee\neg q))$ of (1).

First case, $((\neg p)\geq'^*(\neg p\vee\neg q))$: Since $O\neg$ is a negative predicate, we can use $O\neg(\neg p)$ and $((\neg p)\geq'^*(\neg p\vee\neg q))$ to obtain $O\neg(\neg p\vee\neg q)$, or equivalently $O(p\&q)$.

Second case, $((\neg q)\geq'^*(\neg p\vee\neg q))$: We can use $O\neg(\neg q)$ and $((\neg q)\geq'^*(\neg p\vee\neg q))$ to obtain in the same way $O(p\&q)$.

From completeness and (2) to (1): We are going to assume that completeness holds but (1) does not hold, and prove that (2) is violated. Suppose that $\geq'$ does not satisfy $(p\geq'^*(p\vee q)) \vee (q\geq'^*(p\vee q))$. Since $\geq'$ satisfies completeness, we have either $p\geq'q$ or $q\geq'p$. Without loss of generality, we may assume that $p\geq'q$. Let W be the predicate such that for all $r \in \mathcal{L}_{\mathcal{A}}$, Wr holds if and only if $p\geq'^*r$. Clearly, W satisfies negativity. Since $\geq'$ is complete, it is reflexive, and therefore $p\geq'^*p$ and Wp. It follows from $p\geq'q$ that Wq. We have assumed that $p\geq'^*(p\vee q)$ does not hold. It follows from this and the definition of W that $\neg W(p \vee q)$. We therefore have Wp, Wq and $\neg W(p \vee q)$. Equivalently for the corresponding contranegative predicate $O\neg$: $O(\neg p)$, $O(\neg q)$ and $\neg O(\neg p \& \neg q)$, so that (2) does not hold.

PROOF OF OBSERVATION 10.9. Let $p, q \in \mathcal{L}_{\mathcal{A}}$. It follows from Theorem 7.14 that $\geq'$ is extremal and from Lemma 7.33 that either $\text{MAX}(p) \equiv \text{MAX}(p \vee q)$ or $\text{MAX}(q) \equiv \text{MAX}(p \vee q)$. Without loss of generality, we may assume that $\text{MAX}(p) \equiv \text{MAX}(p \vee q)$.

According to Lemma 7.33, either $\text{MIN}(p) \equiv \text{MIN}(p \vee q)$ or $\text{MIN}(p) > \text{MIN}(p \vee q)$. In the former case, $p \equiv' (p \vee q)$ follows since $\geq'$ is extremal. In the latter case, we have:

$p \equiv' \{\text{MIN}(p), \text{MAX}(p)\}$ (since $\geq'$ is extremal)
$\geq' \{\text{MIN}(p \vee q), \text{MAX}(p)\}$ (nonnegative response and $\text{MIN}(p) > \text{MIN}(p \vee q)$)
$\equiv' \{\text{MIN}(p \vee q), \text{MAX}(p \vee q)\}$ (positionality)
$\equiv' p \vee q$ (since $\geq'$ is extremal),

so that $p \geq' p \vee q$ in this case as well.

Hence, in both cases, $(p \geq' p \vee q) \vee (q \geq' p \vee q)$. The rest follows from Theorem 10.8.

PROOF OF OBSERVATION 10.10. From Observation 10.9.

PROOF OF THEOREM 10.11. *From (1) to (2)*: Let O be a $\geq'$-positive predicate, and let Op and Oq. According to (1), either $p \& q \geq'^* p$ or $p \& q \geq'^* q$. In both cases, $O(p \& q)$ follows from the $\geq'$-positivity of O.

From completeness and (2) to (1): We are going to assume that completeness holds but (1) does not hold, and show that then (2) is violated. Since (1) does not hold, there are p and q such that $\neg((p \& q) \geq'^* p)$ and $\neg((p \& q) \geq'^* q)$. Since $\geq'$ is complete, we can assume without loss of generality that $p \geq' q$. Let O be the predicate such that for all $r \in \mathcal{L}_{\mathcal{A}}$, Or holds if and only if $r \geq'^* q$. Then O is a $\geq'$-positive predicate. It follows from completeness that $q \geq' q$, hence Oq. It also follows from $p \geq' q$ that Op and from $\neg((p \& q) \geq'^* q)$ that $\neg O(p \& q)$. Hence (2) does not hold.

PROOF OF OBSERVATION 10.12. *Part 1*: Condition (1) of Theorem 10.11 is satisfied by $\geq_i$ since $\min(p \& q) \geq \min(p)$).

Parts 2–4: It is sufficient to show that condition (1) of Theorem 10.11 is not satisfied. Let $\mathcal{A} = \{A, B, C\}$, with $p,\neg q \in A$, $\neg p,q \in B$, and p, $q \in C$. Let $\geq$ be transitive and such that $A \equiv B > C$.

Part 5: The same example can be used as in Parts 2–4. Let $v(A) = v(B) = 10$, $v(C) = 0$, and $\delta = 0.5$.

PROOF OF THEOREM 10.13. *From (1) to (2)*: Let (1) hold. Let O be a predicate that is contranegative with respect to $\geq'$ and such that $\vDash_{\mathcal{A}} p \rightarrow q$ and Op. Then, equivalently: $\vDash_{\mathcal{A}} \neg q \rightarrow \neg p$ and $O\neg(\neg p)$. It follows from (1) that $\neg p \geq'^{*} \neg q$ and from the negativity of $O\neg$ that $O\neg(\neg q)$, hence Oq.

From (1) to (3): Let Op and $\vDash_{\mathcal{A}} p \rightarrow q$. It follows from (1) that $q \geq'^{*} p$, hence from the $\geq'$-positivity of O that Oq.

From ancestral reflexivity and (2) to (1): We are going to assume that ancestral reflexivity holds, but (1) does not hold, and prove that then (2) is also violated. Since (1) is not satisfied, there are p and q such that $\vDash_{\mathcal{A}} q \rightarrow p$ and $\neg(p \geq'^{*} q)$.

Let W be the predicate such that for all $r \in \mathcal{L}_{\mathcal{A}}$, Wr holds if and only if $p \geq'^{*} r$. Then W is $\geq'$-negative. Since $\geq'$ satisfies ancestral reflexivity, we have $p \geq'^{*} p$ and thus Wp. It follows from $\neg(p \geq'^{*} q)$ that $\neg Wq$. We therefore have $\vDash_{\mathcal{A}} q \rightarrow p$, Wp and $\neg Wq$, or equivalently for the corresponding $\geq'$-contranegative predicate O: $\vDash_{\mathcal{A}} \neg p \rightarrow \neg q$, $O(\neg p)$ and $\neg O(\neg q)$. This is sufficient to show that (2) is violated.

From ancestral reflexivity and (3) to (1): We are going to assume that ancestral reflexivity holds but (1) does not hold, and then prove that (3) does not hold. Since (1) is violated, there are p and q such that $\vDash_{\mathcal{A}} q \rightarrow p$ and $\neg(p \geq'^{*} q)$. Let O be the predicate such that for all $r \in \mathcal{L}_{\mathcal{A}}$, Or holds if and only if $r \geq'^{*} q$. Then O satisfies positivity. It follows from ancestral reflexivity that Oq and from $\neg(p \geq'^{*} q)$ that $\neg Op$.

PROOF OF OBSERVATION 10.14. *Parts 1 and 3–4*: It is sufficient to show that condition (1) of Theorem 10.13 does not hold. To show this, let $\mathcal{A} = \{A, B, C\}$ with $repr(p) = \{A, B\}$ and $repr(q) = \{A\}$, and let $\geq$ be a transitive and complete relation on $\mathcal{A}$ such that $A > B > C$. Then $\vDash_{\mathcal{A}} q \rightarrow p$, $\mathfrak{max}(q) \equiv \mathfrak{max}(p)$, and $\mathfrak{min}(q) > \mathfrak{min}(p)$. It follows that $\neg(p \geq_{i} q)$, $\neg(p \geq_{ix} q)$, and $\neg(p \geq_{xi} q)$.

Parts 2a and 2b: It follows from Observation 7.28 that $\geq_{x}$ satisfies condition (1) of Theorem 10.13.

Part 5: The same example can be used as in Parts 1 and 3–4. Let $v(A) = 10$, $v(B) = 5$, $v(C) = 0$, and $\delta = 0.5$.

PROOF OF THEOREM 10.15. *From (1) to (2)*: Let (1) hold. Let O be a predicate that is contranegative with respect to $\geq'$ and such that $\vDash_{\mathcal{A}} p \rightarrow q$ and Oq. Then, equivalently: $\vDash_{\mathcal{A}} \neg q \rightarrow \neg p$ and $O\neg(\neg q)$. It follows from (1) that $\neg q \geq'^{*} \neg p$ and from the negativity of $O\neg$ that $O\neg(\neg p)$, hence Op.

From (1) to (3): Let Oq and $\vDash_{\mathcal{A}} p \rightarrow q$. It follows from (1) that $p \geq'^{*} q$, hence from the $\geq'$-positivity of O that Op.

From ancestral reflexivity and (2) to (1): We are going to assume that ancestral reflexivity holds, but (1) does not hold, and prove that then (2) is violated. Since (1) is not satisfied, there are p and q such that $\vDash_{\mathcal{A}} p \rightarrow q$ and $\neg(p \geq'^{*} q)$.

Let W be the predicate such that for all $r \in \mathcal{L}_{\mathcal{A}}$, Wr holds if and only if $p \geq'^{*} r$. Then W is $\geq'$-negative. Since $\geq'$ satisfies ancestral reflexivity, we have $p \geq'^{*} p$ and thus Wp. It follows from $\neg(p \geq'^{*} q)$ that $\neg Wq$. We therefore have $\vDash_{\mathcal{A}} p \rightarrow q$, Wp and $\neg Wq$, or equivalently for the corresponding $\geq'$-contranegative predicate O: $\vDash_{\mathcal{A}} \neg q \rightarrow \neg p$, $O(\neg p)$ and $\neg O(\neg q)$. This is sufficient to show that (2) is violated.

From ancestral reflexivity and (3) to (1): We are going to assume that ancestral reflexivity holds but (1) does not hold, and then prove that (3) does not hold. Since (1) is violated, there are p and q such that $\vDash_{\mathcal{A}} p \rightarrow q$ and $\neg(p \geq'^{*} q)$. Let O be the predicate such that for all $r \in \mathcal{L}_{\mathcal{A}}$, Or holds if and only if $r \geq'^{*} q$. Then O satisfies positivity. It follows from ancestral reflexivity that Oq and from $\neg(p \geq'^{*} q)$ that $\neg Op$.

PROOF OF OBSERVATION 10.16. *Parts 1a and 1b:* It follows from Part 1 of Observation 7.28 that $\geq_{i}$ satisfies Part 1 of Theorem 10.15.

Parts 2–4: Let $\mathcal{A} = \{A,B,C\}$, with $\neg p, q \in A$, $p, q \in B$, $\neg p, \neg q \in C$, and $A>B>C$. Then $\vDash_{\mathcal{A}} p \rightarrow q$, $\mathfrak{max}(q) > \mathfrak{max}(p)$, and $\mathfrak{min}(q) \equiv \mathfrak{min}(p)$. It follows that $\neg(p \geq_{x} q)$, $\neg(p \geq_{ix} q)$, and $\neg(p \geq_{xi} q)$.

Part 5: The same example as for Parts 2–4 can be used, just let $v(A) = 10$, $v(B) = 5$, $v(C) = 0$, and $\delta = 0.5$.

PROOF OF THEOREM 10.17. *From (1) to (2)*: Let O be $\geq'$-contranegative and let $O(p\&q)$, i.e., $O\neg(\neg p \vee \neg q)$. According to (1), either $\neg p \vee \neg q \geq'^{*} \neg p$ or $\neg p \vee \neg q \geq'^{*} \neg q$. In the first case, $O\neg(\neg p)$ follows from the negativity of $O\neg$, and in the second case $O\neg(\neg q)$ follows in the same way.

From (2) and completeness to (1): Assuming completeness, we are going to show that if (1) is violated, then so is (2). If (1) is violated, then there are p and q such that $p >' (p \vee q)$ and $q >' (p \vee q)$. Let W be such that for all r, Wr holds if and only if $p \vee q \geq'^{*} r$. Then W is $\geq'$-negative, $W(p \vee q)$, $\neg Wp$, and $\neg Wq$. Equivalently, for the corresponding contranegative predicate O: $O(\neg p \& \neg q)$, $\neg O \neg p$, and $\neg O \neg q$.

PROOF OF OBSERVATION 10.18. Let $p, q \in \mathcal{L}_{\mathcal{A}}$. It follows from Theorem 7.14 that $\geq'$ is extremal. According to Lemma 7.33, either $\text{MIN}(p) \equiv \text{MIN}(p \vee q)$ or $\text{MIN}(q) \equiv \text{MIN}(p \vee q)$. Without loss of generality, we may assume that $\text{MIN}(p) \equiv \text{MIN}(p \vee q)$.

Furthermore, according to the same lemma, either $\text{MAX}(p \vee q) \equiv \text{MAX}(p)$ or $\text{MAX}(p \vee q) > \text{MAX}(p)$. In the former case, since $\geq'$ is extremal, we have $(p \vee q) \equiv' p$. In the later case, we have:

$p \vee q \equiv' \{\text{MIN}(p \vee q), \text{MAX}(p \vee q)\}$ (since $\geq'$ is extremal)
$\equiv' \{\text{MIN}(p), \text{MAX}(p \vee q)\}$ (positionality)
$\geq' \{\text{MIN}(p), \text{MAX}(p)\}$ (nonnegative response)
$\equiv' p$ (since $\geq'$ is extremal)

so that $p \vee q \geq' p$ in this case as well.

Hence, in both cases $(p \vee q \geq' p) \vee (p \vee q \geq' q)$. The rest follows from Theorem 10.17.

PROOF OF OBSERVATION 10.19. From Observation 10.18.

PROOF OF THEOREM 10.20. *From (1) to (2)*: Let O be $\geq'$-positive, and let $O(p \& q)$. It follows by $\geq'$-positivity from (1) that either Op or Oq.

From (2) to (1): Assuming completeness, we need to show that if (1) is violated, then so is (2). If (1) is violated, then there are p and q such that $p \& q >'^{*} p$ and $p \& q >'^{*} q$. Let O be such that for all r, Or if and only if $r \geq'^{*} (p \& q)$. Then O is $\geq'$-positive, $O(p \& q)$, $\neg Op$, and $\neg Oq$.

PROOF OF OBSERVATION 10.21. *Parts 1, 3, and 4*: Let $\mathcal{A} = \{A, B, C\}$, with $p \& q \in A$, $p \& \neg q \in B$, and $\neg p \& q \in C$, and let $A > B \equiv C$.

Part 2: From Theorem 10.20 since $\max(p) \geq \max(p \& q)$ (cf. Lemma 7.33).

Part 5: The example from Parts 1, 3, and 4 can be used. Let $v(A) = 10$, $v(B) = v(C) = 0$, and $\delta = 0.5$.

PROOF OF THEOREM 10.22. *From (1) to (2)*: Let O be $\geq'$-contranegative, and let $Op \& Oq$. By substitution in (1), we can conclude that either $\neg p \geq'^{*} \neg p \& \neg q$ or $\neg q \geq'^{*} \neg p \& \neg q$, equivalently: $\neg p \geq'^{*} \neg(p \vee q)$ or $\neg q \geq'^{*} \neg(p \vee q)$. The rest follows by $\geq'$-contranegativity.

From (2) and completeness to (1): Assuming completeness, we need to show that if (1) is violated, then so is (2). If (1) is violated, then there are p and q such that $\neg(p\geq'^*p\&q)$ and $\neg(q\geq'^*p\&q)$. Since completeness holds, either $p\geq'q$ or $q\geq'p$. If $p\geq'q$, let O be the predicate such that for all r, Or if and only if $p\geq'^*\neg r$. It is easy to verify that O is contranegative and that $O\neg p$, $O\neg q$, and $\neg O\neg(p\&q)$. The other case is proved analogously.

PROOF OF OBSERVATION 10.23. *Parts 1, 3, and 4*: Let $\mathcal{A} = \{A, B, C\}$, with $p\&q \in A$, $p\&\neg q \in B$, and $\neg p\&q \in C$, and let $A>B\equiv C$.

Part 2: From Theorem 10.22 since $\mathbf{max}(p)\geq\mathbf{max}(p\&q)$ (cf. Lemma 7.33).

Part 5: The example from Parts 1, etc., can be used. Let $v(A) = 10$, $v(B) = v(C) = 0$, and $\delta = 0{,}5$,

PROOF OF THEOREM 10.24. *From (1) to (2)*: Let O be $\geq'$-positive and let Op & Oq. According to (1), either $p\vee q\geq'^*p$, in which case $O(p\vee q)$ follows by $\geq'$-positivity from Op, or $p\vee q\geq'^*q$, in which case it follows from Oq.

From (2) and completeness to (1): Assuming completeness, we need to show that if (1) is violated, then so is (2). If (1) is violated, then there are p and q such that $\neg(p\vee q\geq'^*p)$ and $\neg(p\vee q\geq'^*q)$. Due to completeness, either $p\geq'q$ or $q\geq'p$. If $p\geq'q$, let O be such that Or holds if and only if $r\geq'^*q$. It is easy to verify that O is $\geq'$-positive and that Op and Oq but $\neg O(p\vee q)$. In the other case, let O be such that Or holds if and only if $r\geq'^*p$.

PROOF OF OBSERVATION 10.25. It follows in the same way as Observation 10.18.

PROOF OF OBSERVATION 10.26. From Observation 10.25.

PROOF OF THEOREM 10.27. *From (1) to (2)*: Let (1) be satisfied, and let O be a predicate that is contranegative with respect to $\geq'$. Let Op & $O(p\rightarrow q)$. Equivalently, $O\neg(\neg p)$ & $O\neg(\neg(p\rightarrow q))$. It follows by substitution from (1) that $(\neg p)\geq'^*(\neg q) \vee ((p\&\neg q)\geq'^*(\neg q))$.

First case, $(\neg p)\geq'^*(\neg q)$: Since $O\neg$ is a negative predicate, we can use $O\neg(\neg p)$ to obtain $O\neg(\neg q)$, or equivalently Oq.

Second case, $(p\&\neg q)\geq'^*(\neg q)$: Equivalently, $(\neg(p\rightarrow q))\geq'^*(\neg q)$. We can use $O\neg(\neg(p\rightarrow q))$ and the negativity of $O\neg$ to obtain $O\neg(\neg q)$, or equivalently Oq.

From (2) and completeness to (1): Let $\geq'$ satisfy completeness. We are going to assume that (1) does not hold and prove that then neither does (2). Since (1) does not hold, there are p and q such that $\neg(p\geq'^*q)$ and $\neg((\neg p\&q)\geq'^*q)$. Due to completeness, either $p\geq'(\neg p\&q)$ or $(\neg p\&q)\geq'p$.

Case 1, $p\geq'(\neg p\&q)$: Let W be the predicate such that for all $r \in \mathcal{L}_{\mathcal{A}}$, Wr holds if and only if $p\geq'^*r$. Then W satisfies $\geq'$-negativity. Since $\geq'$ is complete, it is reflexive, and therefore $p\geq'^*p$ and Wp. It follows from $p\geq'(\neg p\&q)$ that $W(\neg p\&q)$. Since $p\geq'^*q$ does not hold, we also have $\neg Wq$. We therefore have Wp, $W(\neg p\&q)$, and $\neg Wq$, i.e., $W\neg(\neg p)$, $W\neg(\neg p\rightarrow\neg q)$, and $\neg W\neg(\neg q)$, so that (2) does not hold.

Case 2, $(\neg p\&q)\geq'p$: Let W be the predicate such that for all $r \in \mathcal{L}_{\mathcal{A}}$, Wr holds if and only if $(\neg p\&q)\geq'^*r$. Then W satisfies negativity. Since $\geq'$ is complete, it is reflexive, and therefore $(\neg p\&q)\geq'^*(\neg p\&q)$ and $W(\neg p\&q)$. It follows from $(\neg p\&q)\geq'^*p$ that Wp. Since $(\neg p\&q)\geq'^*q$ does not hold, we also have $\neg Wq$. Just as in the previous case, we therefore have Wp, $W(\neg p\&q)$ and $\neg Wq$, so that (2) does not hold.

PROOF OF OBSERVATION 10.28. *Part 1*: From Lemma 7.33 follows $\text{MIN}(\neg p\&q) \geq \text{MIN}(q)$, hence $(\neg p\&q)\geq_i q$. The rest follows from Theorem 10.27.

Part 2: Clearly, either $p \in \text{MAX}(q)$ or $\neg p \in \text{MAX}(q)$. If $p \in \text{MAX}(q)$, then $\text{MAX}(p) \geq \text{MAX}(q)$, hence $p\geq_x q$. If $\neg p \in \text{MAX}(q)$, then $\neg p\&q \in \text{MAX}(q)$, so that $\text{MAX}(\neg p\&q) \geq \text{MAX}(q)$, and hence $(\neg p\&q)\geq_x q$. The rest follows from Theorem 10.27.

Part 3–4: Let $\mathcal{A} = \{A, B, C, D\}$ with $A > B > C > D$, $repr(p) = \{A, C\}$ and $repr(q) = \{A, B\}$. Then condition (1) of Theorem 10.27 is not satisfied for $\geq_{ix}$ or for $\geq_{xi}$.

Part 5: The same example can be used as in Parts 3–4. Let $v(A) = 2$, $v(B) = 1$, $v(C) = 0$, and $\delta = 0.5$.

PROOF OF THEOREM 10.29. *From (1) to (2)*: Let (1) hold, and let Op and $O(p\rightarrow q)$. It follows from (1) that either $q\geq'^*p$, in which case Oq follows from Op, or $q\geq'^*(\neg p\vee q)$, equivalently $q\geq'^*(p\rightarrow q)$, in which case Oq follows from $O(p\rightarrow q)$.

From completeness and (2) to (1): Let $\geq'$ satisfy completeness. We are going to assume that (1) does not hold, and then show that (2) does not hold either. Since (1) does not hold, completeness yields $p>'q$ and $(\neg p\vee q)>'q$. There are two cases:

Case *i*, $p\geq'(\neg p\vee q)$: Let O be the predicate such that for all r, Or iff $r\geq'^*(\neg p\vee q)$. It follows directly that O is positive and that Op and $O(p\rightarrow q)$. Since $\neg(q\geq'^*(\neg p\vee q))$, we also have $\neg Oq$.

Case *ii*, $\neg(p\geq'(\neg p\vee q))$: Then $(\neg p\vee q)\geq' p$. Let O be the predicate such that Or iff $r\geq'^* p$, and proceed as in Case i.

PROOF OF OBSERVATION 10.30. *Part 1*: It follows from Observation 7.28 that $q\geq_i^*(\neg p\vee q)$. The rest follows from Theorem 10.29.

Parts 2–4: Let $\mathcal{A}=\{A,B,C\}$, with $p,\neg q\in A$, $\neg p,\neg q\in B$, and $p, q\in C$. Furthermore, let $\geq$ be a transitive and complete relation over $\mathcal{A}$, such that $A\equiv B>C$. Then condition (1) of Theorem 10.29 is not satisfied.

Part 5: The same example can be used as in Parts 2–4. Let $v(A) = v(B) = 10$, $v(C) = 0$, and $\delta = 0.5$.

PROOF OF THEOREM 10.31. *From (1) to (2)*: According to (1), either $p\geq'^*(p\&q)$ or $p\geq'^*(p\&\neg q)$. If $p\geq'^*(p\&q)$, then we can use $P(p\&q)$ and the positivity of P to obtain Pp. If $p\geq'^*(p\&\neg q)$, then we can use $P(p\&\neg q)$ and the positivity of P to obtain Pp.

From (2) and completeness to (1): Let $\geq'$ satisfy completeness. We are going to assume that (1) does not hold, and prove that then neither does (2). Since (1) does not hold, there are p and q such that $\neg(p\geq'^*(p\&q))$ and $\neg(p\geq'^*(p\&\neg q))$. Due to completeness, there are two cases.

Case *i*, $(p\&q)\geq'(p\&\neg q)$: Let P be the predicate such that for all r, Pr iff $r\geq'(p\&\neg q)$. Then P is $\geq'$-positive, and it follows directly that $P(p\&q)$ and $P(p\&\neg q)$. It follows from $\neg(p\geq'^*(p\&\neg q))$ that $\neg Pp$.

Case *ii*, $(p\&\neg q)\geq'(p\&q)$: The proof proceeds in the same way as in Case i.

PROOF OF OBSERVATION 10.32. By substituting $p\&q$ for p and $p\&\neg q$ for q in $((p\vee q)\geq'^*p)\vee((p\vee q)\geq'^*q)$, we obtain $(p\geq'^*(p\&q))\vee(p\geq'^*(p\&\neg q))$. The rest follows from Theorem 10.31.

PROOF OF OBSERVATION 10.33. From Observation 10.32, which is applicable since these preference relations satisfy completeness and disjunctive interpolation (cf. Observation 7.29).

PROOF OF THEOREM 10.34. *From (1) to (2)*: Let (1) hold, let O be $\geq'$-positive, and let $Pp \leftrightarrow \neg O \neg p$ for all p. We are going to show (2) in its converse form, i.e., we will assume that $\neg Pp$ and then prove that either $\neg P(p \& q)$ or $\neg P(p \& \neg q)$.

Let $\neg Pp$, or equivalently $O \neg p$. It follows from (1), by substitution of $\neg p$ for p, that either $(\neg p \vee q) \geq'^{*} \neg p$ or $(\neg p \vee \neg q) \geq'^{*} \neg p$, hence by the positivity of O either $O(\neg p \vee q)$ or $O(\neg p \vee \neg q)$, i.e., equivalently $\neg P(p \& \neg q)$ or $\neg P(p \& q)$.

From ancestral reflexivity and (2) to (1): Let $\geq'$ satisfy ancestral reflexivity. We are going to assume that (1) does not hold, and show that then neither does (2).

Since (1) does not hold, there must be some p and q such that $\neg((p \vee q) \geq'^{*} p)$ and $\neg((p \vee \neg q) \geq'^{*} p)$. Let O be the predicate such that for all r, Or iff $r \geq'^{*} p$. Then O is a $\geq'$-positive predicate, and it follows from ancestral reflexivity that Op. Furthermore, since (1) does not hold, we also have $\neg O(p \vee q)$ and $\neg O(p \vee \neg q)$. Then, equivalently, $\neg P(\neg p)$, $P(\neg p \& \neg q)$, and $P(\neg p \& q)$, contrary to (2).

PROOF OF OBSERVATION 10.35. *Parts 1, 3, and 4*: Let $\mathcal{A} = \{A, B, C\}$, with $p, q \in A, \neg p, q \in B$, and $\neg p, \neg q \in C$. Furthermore, let $\geq$ be a transitive and complete relation over $\mathcal{A}$, such that $A > B \equiv C$. We then have $\neg((p \vee q) \geq_{i}^{*} p)$ and $\neg((p \vee \neg q) \geq_{i}^{*} p)$, and similarly for $\geq_{ix}$ and $\geq_{xi}$. The rest follows from Theorem 10.34.

Part 2: From Theorem 10.34, since $(p \vee q) \geq_{x}^{*} p$. (See Observation 7.28.)

Part 5: The same example can be used as in Part 1, etc. Let $v(A) = 2$, $v(B) = v(C) = 0$, and $\delta = 0.5$.

PROOF OF THEOREM 10.36. *From (1) to (2)*: Let Op. Then it follows directly from $O \neg(\neg p)$ and the substitution-instance $\neg p \geq' \neg(p \vee \neg p)$ of (1) that $O \neg(\neg(p \vee \neg p))$, i.e., $O(p \vee \neg p)$.

From (2) and ancestral reflexivity to (1): We are going to assume that ancestral reflexivity holds and that (1) is violated, and show that then (2) is also violated. Since (1) is violated, there is some p such that $\neg(\neg p \geq'^{*} (p \& \neg p))$, equivalently $\neg(\neg p \geq'^{*} \neg(p \vee \neg p))$. Let O be a predicate such that for all r, Or iff $\neg p \geq'^{*} \neg r$. Then O is contranegative. It follows from ancestral reflexivity that Op and from $\neg(\neg p \geq'^{*} \neg(p \vee \neg p))$ that $\neg O(p \vee \neg p)$.

PROOF OF OBSERVATION 10.37. *Parts 1 and 3*: According to Definition 7.12, if $p \in \cup\mathcal{A}$, then $\text{MIN}(p\&\neg p) > \text{MIN}(p)$. It follows from this that $\neg(p \geq_{i} (p\&\neg p))$ and $\neg(p \geq_{ix} (p\&\neg p))$.

Parts 2 and 4: Case *i*, $p \in \cup\mathcal{A}$: It follows from Definition 7.12 that $\mathfrak{max}(p) > \mathfrak{max}(p\&\neg p)$. Hence, $p \geq_{x} (p\&\neg p)$ and $p \geq_{xi} (p\&\neg p)$. The rest follows from Theorem 10.36.

Case *ii*, $p \notin \cup\mathcal{A}$: Then $\text{MAX}(p) \equiv \text{MAX}(p\&\neg p)$ and $\text{MIN}(p) \equiv \text{MIN}(p\&\neg p)$, and since $\geq'$ is extremal we again have $p \geq_{x} (p\&\neg p)$ and $p \geq_{xi} (p\&\neg p)$.

The rest follows from Theorem 10.36.

PROOF OF THEOREM 10.38. *From (1) to (2)*: Directly from the stated conditions.

From ancestral reflexivity and (2) to (1): We will assume that ancestral reflexivity is satisfied and that (1) is violated, and show that then (2) does not hold either. Since (1) does not hold, there is some p such that $\neg((p \vee \neg p) \geq'^{*} p)$. Let O be the predicate such that for all r, Or iff $r \geq'^{*} p$. It follows from ancestral reflexivity that Op and from $\neg((p \vee \neg p) \geq'^{*} p)$ that $\neg O(p \vee \neg p)$, which contradicts (2).

PROOF OF OBSERVATION 10.39. *Parts 1, 3, and 4*: Let $\mathcal{A} = \{A,B\}$ with $p \in A$ and $\neg p \in B$, and let $A > B$. Then $\text{MAX}(p) \equiv \text{MAX}(p \vee \neg p)$ and $\text{MIN}(p) > \text{MIN}(p \vee \neg p)$. It follows that $p >_{i} (p \vee \neg p)$, $p >_{ix} (p \vee \neg p)$, and $p >_{xi} (p \vee \neg p)$. The rest follows from Theorem 10.38.

Part 2: It holds for all p that $\mathfrak{max}(p \vee \neg p) \geq \mathfrak{max}(p)$, hence $(p \vee \neg p) \geq_{x} p$. The rest follows from Theorem 10.38.

Part 5: Let $\mathcal{A}$ be as in Parts 1, 3, and 4. Let $\delta = 0.5$, $v(A) = 10$, and $v(B) = 0$.

PROOF OF OBSERVATION 10.40. *Part 1*: If $p \geq' (\neg p)$, then let O be such that for all q, Oq if and only if $p \geq'^{*} \neg q$. If $(\neg p) \geq' p$, then let O be such that for all q, Oq if and only if $\neg p \geq'^{*} \neg q$.

Part 2: If $p \geq' (\neg p)$, then let O be such that for all q, Oq if and only if $q \geq'^{*} \neg p$. If $(\neg p) \geq' p$, then let O be such that for all q, Oq if and only if $q \geq'^{*} p$.

PROOF OF OBSERVATION 10.42. We are going to prove the observation in its equivalent form that if W is nonempty and $\geq'$-negative, then it is a sentence-limited $\geq'$-negative predicate based on $\geq'$ and some sentence f.

Let W be $\geq'$-negative and nonempty. Since it is nonempty, there is some sentence r_0 such that Wr_0. The series $r_0, r_1 \ldots$ is constructed as follows:

(*i*) If there is some sentence s such that Ws and $s>'r_n$, then let r_{n+1} be a sentence such that Wr_{n+1} and $r_{n+1}>'r_n$.
(*ii*) Otherwise, let $r_{n+1} = r_n$.

Due to the finiteness property given in the observation and to the fact that $\geq'$ is an ordering, there is some r_m such that $r_{m+1} = r_m$. We have to show that for all p, Wp holds iff $r_m\geq'p$. For one direction, let $\neg(r_m\geq'p)$. Then $p>'r_m$. It follows from the construction of r_m that $\neg Wp$. For the other direction, let $r_m\geq'p$. It follows from Wr_m and the negativity of W that Wp.

PROOF OF THEOREM 10.43. *From 1 to 2*: For (i), let $O(p\&q)$. Then $\mathfrak{min}(f)\geq\mathfrak{min}(\neg(p\&q))$, so that either $\mathfrak{min}(f)\geq\mathfrak{min}(\neg p)$ or $\mathfrak{min}(f)\geq \mathfrak{min}(\neg q)$, hence either Op or Oq.

For (ii), let Oq and $\vdash p\rightarrow q$. Equivalently, $\mathfrak{min}(f)\geq\mathfrak{min}(\neg q)$ and $\vdash \neg q\rightarrow\neg p$. It follows from the latter that $\mathfrak{min}(\neg q)\geq\mathfrak{min}(\neg p)$, hence by transitivity $\mathfrak{min}(f)\geq\mathfrak{min}(\neg p)$, hence Op.

For (iii): It follows from $f \in \cup\mathcal{A}$ that $\mathfrak{min}(f)\geq\mathfrak{min}(\neg\perp)$. From this, it follows that $O\perp$.

From 2 to 1: The following three properties can be derived from (i), (ii), and (iii):

(i′) If $p \in \{s \mid \neg Os\}$ and $q \in \{s \mid \neg Os\}$, then $p\&q \in \{s \mid \neg Os\}$.
(ii′) If $p \in \{s \mid \neg Os\}$ and $\vdash p\rightarrow q$, then $q \in \{s \mid \neg Os\}$.
(iii′) If $p \in \{s \mid \neg Os\}$, then $p \nvdash \perp$.

It follows from (i′) and (ii′), given that the logic is compact, that $\{s \mid \neg Os\}$ is logically closed, and then from (iii′) that it is consistent.

Let $\mathcal{A}_2$ be the set of alternatives (worlds) containing $\{p \mid \neg Op\}$ and let $\mathcal{A}_1 = \mathcal{A}\backslash\mathcal{A}_2$. Let $\geq$ be such that $\mathcal{A}_1$ and $\mathcal{A}_2$ are equivalence classes with respect to $\geq$ and that $\mathcal{A}_1>\mathcal{A}_2$. Let f be a sentence such that $\mathfrak{min}(f) \in \mathcal{A}_2$. Then:

$Op \leftrightarrow p \notin \{s \mid \neg Os\}$
$\leftrightarrow p \notin \cap\mathcal{A}_2$ (due to the logical closure and consistency of $\{s \mid \neg Os\}$)
$\leftrightarrow \neg p \in \cup\mathcal{A}_2$
$\leftrightarrow \mathfrak{min}(f)\geq\mathfrak{min}(\neg p)$
$\leftrightarrow f\geq_i\neg p$.

PROOF OF THEOREM 10.44. *From 1 to 2:* (i) Let Op & Oq, i.e., $\mathfrak{max}(f) \geq \mathfrak{max}(\neg p)$ and $\mathfrak{max}(f) \geq \mathfrak{max}(\neg q)$. Then $\mathfrak{max}(f) \geq \mathfrak{max}(\neg p \vee \neg q)$, hence $O(p\&q)$.

For (ii), let Op and $\vdash p \rightarrow q$. Then $\mathfrak{max}(f) \geq \mathfrak{max}(\neg p)$, and it follows from $\vdash \neg q \rightarrow \neg p$ that $\mathfrak{max}(\neg p) \geq \mathfrak{max}(\neg q)$. Transitivity yields $\mathfrak{max}(f) \geq \mathfrak{max}(\neg q)$, hence Oq.

(iii) Since $\mathfrak{max}(f)$ is not maximal, there is some $X \in \mathcal{A}$ such that $X > \mathfrak{max}(f)$, and clearly $\neg\bot \in X$, hence $\neg O\bot$.

(iv) It follows from the construction that $O\neg f$.

From 2 to 1: It follows directly from (i), (ii), and (iii), due to compactness, that $\{p \mid Op\}$ is consistent and logically closed. Let $\mathcal{A}_1$ consist of the alternatives (worlds) containing $\{p \mid Op\}$. It follows from (iv) that $\mathcal{A}_1$ is nonempty. Let $\mathcal{A}_2 = \mathcal{A} \backslash \mathcal{A}_1$. Let $\mathcal{A}_1$ and $\mathcal{A}_2$ be equivalence classes with respect to $\geq$, let $\mathcal{A}_1 > \mathcal{A}_2$, and let f be a sentence such that $\mathfrak{max}(f) \in \mathcal{A}_2$. Then:

$$\begin{aligned} &Op \leftrightarrow p \in \{p \mid Op\} \\ &\leftrightarrow p \in \cap\mathcal{A}_1 \\ &\leftrightarrow \neg p \notin \cup\mathcal{A}_1 \\ &\leftrightarrow \mathfrak{max}(\neg p) \in \cup\mathcal{A}_2 \\ &\leftrightarrow \mathfrak{max}(f) \geq \mathfrak{max}(\neg p) \\ &\leftrightarrow f \geq_x \neg p. \end{aligned}$$

Lemma 10.59. *Let a language $\mathcal{L}$ have a finite number of equivalence classes with respect to logical equivalence. If $C \subset B \subset A \subseteq \mathcal{L}$, A is consistent and A, B, and C are logically closed, then there are consistent and logically closed sets B' and C' such that each of $A \cup B'$, $B' \cup C'$, and $A \cup C'$ is inconsistent and such that $A \backslash (B \backslash C) = A \backslash (B' \backslash C')$.*

PROOF OF LEMMA 10.59. Let $a = \&A$, $b = \&B$, and $c = \&C$. (Note that a logically implies b.) Let $B' = \mathrm{Cn}(\{b\&\neg a\})$, and let $C' = \mathrm{Cn}(\{c\&\neg b\})$. We have $A \cap B' = \mathrm{Cn}(\{a\}) \cap \mathrm{Cn}(\{b\&\neg a\}) = \mathrm{Cn}(\{a \vee (b\&\neg a)\}) = \mathrm{Cn}(\{a \vee b\}) = \mathrm{Cn}(\{a\}) \cap \mathrm{Cn}(\{b\}) = A \cap B$. Furthermore, $A \cap B' \cap C' = \mathrm{Cn}(\{a \vee b\}) \cap \mathrm{Cn}(\{c\&\neg b\}) = \mathrm{Cn}(\{a \vee b \vee (c\&\neg b)\}) = \mathrm{Cn}(\{a \vee b \vee c\}) = \mathrm{Cn}(\{a\}) \cap \mathrm{Cn}(\{b\}) \cap \mathrm{Cn}(\{c\}) = A \cap B \cap C$. It follows that $A \backslash (B' \backslash C') = A \backslash ((A \cap B') \backslash (A \cap B' \cap C')) = A \backslash ((A \cap B) \backslash (A \cap B \cap C)) = A \backslash (B \backslash C)$. The rest of the desired properties follow directly.

PROOF OF THEOREM 10.45. *From 1 to 2*: For (i), let Op and Oq, Then $f \geq_{ix} \neg p$ and $f \geq_{ix} \neg q$. There are four cases:

(α) $\min(f)>\min(\neg p)$ and $\min(f)>\min(\neg q)$: Then $\min(f)>\min(\neg p\vee\neg q)$, hence $f\geq_{ix}(\neg p\vee\neg q)$.

(β) $\min(f)>\min(\neg p)$, $\min(f)\equiv\min(\neg q)$, and $\max(f)\geq\max(\neg q)$: Then $\min(f)>\min(\neg p\vee\neg q)$, hence $f\geq_{ix}(\neg p\vee\neg q)$.

(γ) $\min(f)\equiv\min(\neg p)$, $\max(f)\geq\max(\neg p)$, $\min(f)>\min(\neg q)$: Symmetric with case β.

(δ) $\min(f)\equiv\min(\neg p)$, $\max(f)\geq\max(\neg p)$, $\min(f)\equiv\min(\neg q)$, and $\max(f)\geq\max(\neg q)$: Then $\min(f)\equiv\min(\neg p\vee\neg q)$. Since either $\max(\neg p\vee\neg q)\equiv\max(\neg p)$ or $\max(\neg p\vee\neg q)\equiv\max(\neg q)$, it follows from transitivity that $\max(f)\geq\max(\neg p\vee\neg q)$.

Hence, in all four cases $f\geq_{ix}(\neg p\vee\neg q)$, hence $O(p\&q)$.

For (ii), let $O(p\&q)$. Then $f\geq_{ix}(\neg p\vee\neg q)$. There are two cases:

(α) $\min(f)>\min(\neg p\vee\neg q)$: Then either $\min(f)>\min(\neg p)$ or $\min(f)>\min(\neg q)$, hence either Op or Oq.

(β) $\min(f)\equiv\min(\neg p\vee\neg q)$ and $\max(f)\geq\max(\neg p\vee\neg q)$: It follows from $\max(f)\geq\max(\neg p\vee\neg q)$ that $\max(f)\geq\max(\neg p)$ and $\max(f)\geq\max(\neg q)$. Since either $\min(\neg p\vee\neg q)\equiv\min(\neg p)$ or $\min(\neg p\vee\neg q)\equiv\min(\neg q)$, either $\min(f)\geq\min(\neg p)$ or $\min(f)\geq\min(\neg q)$. Hence, either Op or Oq.

For (iii), suppose to the contrary that Oq. From $\vdash p\rightarrow q$ follows $\vdash \neg q\rightarrow\neg p$, hence $\min(\neg q)\geq\min(\neg p)$. $\neg Op$ and Os yield $\neg p>_{ix}f$ and $f\geq_{ix}\neg s$, hence $\neg p>_{ix}\neg s$. Due to $\vdash s\rightarrow p$, i.e., $\vdash \neg p\rightarrow\neg s$, we have $\max(\neg s)\geq\max(\neg p)$. It follows from $\neg p>_{ix}\neg s$ and $\max(\neg s)\geq\max(\neg p)$ that $\min(\neg p)>\min(\neg s)$. Due to $\vdash r\rightarrow s$, i.e., $\vdash \neg s\rightarrow\neg r$, we have $\min(\neg s)\geq\min(\neg r)$. It follows from $\min(\neg q)\geq\min(\neg p)$, $\min(\neg p)>\min(\neg s)$, and $\min(\neg s)\geq\min(\neg r)$ that $\min(\neg q)>\min(\neg r)$. However, it follows from $\neg Or$ and Oq that $\neg r>_{ix}f$ and $f\geq_{ix}\neg q$, hence $\neg r>_{ix}\neg q$, hence $\min(\neg r)\geq\min(\neg q)$. Contradiction.

For (iv): If $\min(f)$ is nonminimal or $\min(f)>\min(\neg\bot)$, then $f\geq_{ix}\neg\bot$, hence $O\bot$. If $\max(f)$ is maximal, i.e., $\max(f)\geq\max(\neg\bot)$, then this can be combined with $\min(f)\geq\min(\neg\bot)$ to obtain $f\geq_{ix}\neg\bot$, hence $O\bot$.

From 2 to 1: The following construction will be used in the proof:

Let $Z=\{p\mid\neg Op\}$. Let $A=\mathrm{Cn}(Z)$, $B=\mathrm{Cn}(A\backslash Z)$, and $C=\{p\in B\mid\neg Op\}$. We need to show (A) that A is consistent, (B) that $Z=A\backslash(B\backslash C)$, and (C) that C is logically closed.

For (A), suppose to the contrary that A is inconsistent. Then, due to compactness, there is a finite and inconsistent subset of Z, but this is impossible due to (ii) and (iv).

For (B), first let $p\in Z$. Then clearly $p\in A$, and if $p\in B$ then $p\in C$.

Hence, $p \in A\backslash(B\backslash C)$. For the other direction, suppose that $p \in A\backslash(B\backslash C)$ and $p \notin Z$, i.e., Op. Then $p \in A\backslash Z$ so that $p \in B$. Since $Op, p \notin C$, hence $p \in B\backslash C$, contrary to the conditions.

For (C), we need to show that (Cα) if $\vdash p \rightarrow q$ and $p \in C$ then $q \in C$, and (Cβ) if $p, q \in C$ then $p \& q \in C$.

For (Cα), let $\vdash p \rightarrow q$ and $p \in C$. Then $p \in B$ and $\neg Op$. It follows from the logical closure of B that $q \in B$. In order to prove that $q \in C$, it remains to show that $\neg Oq$. It follows from $p \in B$ that there are $s_1, \ldots s_n \in A$ such that $\vdash s_1 \& \ldots s_n \rightarrow p$, and $Os_1, \ldots$, and Os_n. Due to (i), we then have $O(s_1 \& \ldots s_n)$. Letting $s = s_1 \& \ldots . \& s_n$, we then have $\vdash s \rightarrow p$, Os and (since A is logically closed) $s \in A$. Due to the construction of A, it follows that there are $r_1, \ldots r_m$ such that $\neg Or_1, \ldots \neg Or_m$, and $\vdash r_1 \& \ldots \& r_m \rightarrow s$. Let $r = r_1 \& \ldots \& r_m$. Due to (ii), $\neg Or$. From $\vdash r \rightarrow s$, $\vdash s \rightarrow p$, $\neg Or$, Os, $\vdash p \rightarrow q$ and $\neg Op$, it follows, due to (iii), that $\neg Oq$.

For (Cβ), let $p, q \in C$. Then $p, q \in B$ so that $p \& q \in B$, and (ii) yields $\neg O(p \& q)$, hence $p \& q \in C$.

For the construction, there are two cases, depending on whether or not $C \subset B \subset A$.

Case 1, $C \subset B \subset A$: Due to Lemma 10.59, there are consistent and logically closed sets B' and C' such that each of $A \cup B'$, $B' \cup C'$, and $A \cup C'$ is inconsistent and such that $A\backslash(B\backslash C) = A\backslash(B'\backslash C')$. We can now proceed to make the construction for the proof. Let $\mathcal{A}_1 = \{X \in \mathcal{A} \mid B' \subseteq X\}$, $\mathcal{A}_3 = \{X \in \mathcal{A} \mid C' \subseteq X\}$, $\mathcal{A}_4 = \{X \in \mathcal{A} \mid A \subseteq X\}$, and $\mathcal{A}_2 = \mathcal{A}\backslash(\mathcal{A}_1 \cup \mathcal{A}_3 \cup \mathcal{A}_4)$. Let $\mathcal{A}_1$, $\mathcal{A}_2$, $\mathcal{A}_3$, and $\mathcal{A}_4$ be equivalence classes with respect to $\geq$, and let $\mathcal{A}_1 > \mathcal{A}_2 > \mathcal{A}_3 > \mathcal{A}_4$. Let $\mathfrak{max}(f) \in \mathcal{A}_2$ and $\mathfrak{min}(f) \in \mathcal{A}_3$. Then $\mathfrak{min}(f)$ is nonminimal and:

$$
\begin{aligned}
& Op \leftrightarrow p \notin A\backslash(B\backslash C) \\
& \leftrightarrow p \notin \cap\mathcal{A}_4\backslash((\cap\mathcal{A}_1)\backslash(\cap\mathcal{A}_3)) \\
& \leftrightarrow p \notin \cap\mathcal{A}_4 \vee (p \notin \cap\mathcal{A}_3 \;\&\; p \in \cap\mathcal{A}_4 \;\&\; p \in \cap\mathcal{A}_1) \\
& \leftrightarrow \neg p \in \cup\mathcal{A}_4 \vee (\neg p \in \cup\mathcal{A}_3 \;\&\; \neg p \notin \cup\mathcal{A}_4 \;\&\; \neg p \notin \cup\mathcal{A}_1) \\
& \leftrightarrow \mathfrak{min}(f) > \mathfrak{min}(\neg p) \vee (\mathfrak{min}(f) \equiv \mathfrak{min}(\neg p) \;\&\; \mathfrak{max}(f) \geq \mathfrak{max}(\neg p) \\
& \leftrightarrow f \geq_{\text{ix}} \neg p
\end{aligned}
$$

Case 2: In this case, $C \subseteq B \subseteq A$ but not $C \subset B \subset A$. Hence, either $A = B$ or $B = C$. If $A = B$, then $Z = A\backslash(A\backslash C) = C$. If $B = C$, then $A = A\backslash(B\backslash B) = A$. Hence, in both cases $Z = \{p \mid \neg Op\}$ is logically closed. It follows in the same way as in Theorem 10.43 that O is sententially contranegative on a maximin relation $\geq_{\text{i}}$. In other words, there is some f

such that for all p, Op iff $\mathfrak{min}(f) \geq \mathfrak{min}(\neg p)$. Let f' be a sentence such that $\mathfrak{max}(f') \equiv \mathfrak{max}(\top)$ and $\mathfrak{min}(f') \equiv \mathfrak{min}(f)$. Then:

$Op \leftrightarrow \mathfrak{min}(f) \geq \mathfrak{min}(\neg p)$
$\leftrightarrow \mathfrak{min}(f') > \mathfrak{min}(\neg p) \vee \mathfrak{min}(f') \equiv \mathfrak{min}(\neg p)$
$\leftrightarrow \mathfrak{min}(f') > \mathfrak{min}(\neg p) \vee (\mathfrak{min}(f') \equiv \mathfrak{min}(\neg p) \;\&\; \mathfrak{max}(f') \geq \mathfrak{max}(\neg p))$
$\leftrightarrow f' \geq_{\text{ix}} \neg p$.

PROOF OF THEOREM 10.46. *From 1 to 2*: For (i), let Op and Oq. Then $f \geq_{\text{xi}} \neg p$ and $f \geq_{\text{xi}} \neg q$. There are four cases:

(α) $\mathfrak{max}(f) > \mathfrak{max}(\neg p)$ and $\mathfrak{max}(f) > \mathfrak{max}(\neg q)$: Since either $\mathfrak{max}(\neg p) \equiv \mathfrak{max}(\neg p \vee \neg q)$ or $\mathfrak{max}(\neg q) \equiv \mathfrak{max}(\neg p \vee \neg q)$, it follows by transitivity that $\mathfrak{max}(f) > \mathfrak{max}(\neg p \vee \neg q)$, hence $f \geq_{\text{xi}} (\neg p \vee \neg q)$.

(β) $\mathfrak{max}(f) > \mathfrak{max}(\neg p)$, $\mathfrak{max}(f) \equiv \mathfrak{max}(\neg q)$, and $\mathfrak{min}(f) \geq \mathfrak{min}(\neg q)$: Then $\mathfrak{max}(f) \geq \mathfrak{max}(\neg p \vee \neg q)$ and $\mathfrak{min}(f) \geq \mathfrak{min}(\neg p \vee \neg q)$. Hence, $f \geq_{\text{xi}} (\neg p \vee \neg q)$.

(γ) $\mathfrak{max}(f) \equiv \mathfrak{max}(\neg p)$, $\mathfrak{min}(f) \geq \mathfrak{min}(\neg p)$, $\mathfrak{max}(f) > \mathfrak{max}(\neg q)$: Symmetric with case β.

(δ) $\mathfrak{max}(f) \equiv \mathfrak{max}(\neg p)$, $\mathfrak{min}(f) \geq \mathfrak{min}(\neg p)$, $\mathfrak{max}(f) \equiv \mathfrak{max}(\neg q)$, and $\mathfrak{min}(f) \geq \mathfrak{min}(\neg q)$: Since either $\mathfrak{max}(\neg p) \equiv \mathfrak{max}(\neg p \vee \neg q)$ or $\mathfrak{max}(\neg q) \equiv \mathfrak{max}(\neg p \vee \neg q)$, it follows by transitivity that $\mathfrak{max}(f) \geq \mathfrak{max}(\neg p \vee \neg q)$. Since either $\mathfrak{min}(\neg p) \equiv \mathfrak{min}(\neg p \vee \neg q)$ or $\mathfrak{min}(\neg q) \equiv \mathfrak{min}(\neg p \vee \neg q)$, we also have $\mathfrak{min}(f) \geq \mathfrak{min}(\neg p \vee \neg q)$. Hence, $f \geq_{\text{xi}} (\neg p \vee \neg q)$.

Thus, in all four cases $f \geq_{\text{xi}} (\neg p \vee \neg q)$, so that $O(p \& q)$.

For (ii), let $O(p \& q)$. Then $f \geq_{\text{xi}} (\neg p \vee \neg q)$. There are two cases:

(α) $\mathfrak{max}(f) > \mathfrak{max}(\neg p \vee \neg q)$: Then $\mathfrak{max}(f) > \mathfrak{max}(\neg p)$ and $\mathfrak{max}(f) > \mathfrak{max}(\neg q)$, hence Op and Oq.

(β) $\mathfrak{max}(f) \equiv \mathfrak{max}(\neg p \vee \neg q)$ and $\mathfrak{min}(f) \geq \mathfrak{min}(\neg p \vee \neg q)$: Then $\mathfrak{max}(f) \geq \mathfrak{max}(\neg p)$ and $\mathfrak{max}(f) \geq \mathfrak{max}(\neg q)$. Since either $\mathfrak{min}(\neg p \vee \neg q) \equiv \mathfrak{min}(\neg p)$ or $\mathfrak{min}(\neg p \vee \neg q) \equiv \mathfrak{min}(\neg q)$, either $\mathfrak{min}(f) \geq \mathfrak{min}(\neg p)$ or $\mathfrak{min}(f) \geq \mathfrak{min}(\neg q)$. Hence, either Op or Oq.

For (iii), suppose to the contrary that $\neg Oq$. Due to $\vdash r \rightarrow s$, i.e., $\vdash \neg s \rightarrow \neg r$, we have $\mathfrak{max}(\neg r) \geq \mathfrak{max}(\neg s)$. $\neg Os$ and Op yield $\neg s >_{\text{xi}} f$ and $f \geq_{\text{xi}} \neg p$, hence $\neg s >_{\text{xi}} \neg p$. Due to $\vdash s \rightarrow p$, i.e., $\vdash \neg p \rightarrow \neg s$, we have $\mathfrak{min}(\neg p) \geq \mathfrak{min}(\neg s)$. Hence, $\mathfrak{max}(\neg s) > \mathfrak{max}(\neg p)$. From $\vdash p \rightarrow q$ follows $\vdash \neg q \rightarrow \neg p$, hence $\mathfrak{max}(\neg p) \geq \mathfrak{max}(\neg q)$. $\neg Oq$, i.e., $\neg q >_{\text{xi}} f$, implies $\mathfrak{max}(\neg q) \geq \mathfrak{max}(f)$. In summary, $\mathfrak{max}(\neg r) \geq \mathfrak{max}(\neg s) > \mathfrak{max}(\neg p) \geq \mathfrak{max}(\neg q) \geq \mathfrak{max}(f)$, hence $\mathfrak{max}(\neg r) > \mathfrak{max}(f)$. However, it follows from Or that $\mathfrak{max}(f) \geq \mathfrak{max}(\neg r)$. This contradiction concludes the proof of (iii).

For (iv), suppose to the contrary that $O\bot$, i.e., that $f \geq_{xi} \neg\bot$. Then $\max(f)$ is maximal, contrary to our conditions.

For (v), note that $O\neg f$.

From 2 to 1: The following construction will be used in the proof:

Let $Z = \{p \mid Op\}$. Let $A = \text{Cn}(Z)$, $B = \text{Cn}(A \backslash Z)$, and $C = \{p \in B \mid Op\}$. We need to show (A) that A is consistent, (B) that $Z = A\backslash(B\backslash C)$, and (C) that C is logically closed.

For (A), suppose to the contrary that A is inconsistent. Then, due to compactness, there is a finite and inconsistent subset of Z, but this is impossible due to (i) and (iv).

For (B), first let $p \in Z$. Then clearly $p \in A$, and if $p \in B$ then $p \in C$. Hence, $p \in A\backslash(B\backslash C)$. For the other direction, suppose that $p \in A\backslash(B\backslash C)$ and $p \notin Z$, i.e., $\neg Op$. Then $p \in A\backslash Z$ so that $p \in B$. Since $\neg Op$, $p \notin C$, hence $p \in B\backslash C$, contrary to the conditions.

For (C), we need to show that (Cα) if $\vdash p \rightarrow q$ and $p \in C$ then $q \in C$, and (Cβ) if $p, q \in C$ then $p\&q \in C$.

For (Cα), let $\vdash p \rightarrow q$ and $p \in C$. Then $p \in B$ and Op. It follows from the logical closure of B that $q \in B$. In order to prove that $q \in C$, it remains to show that Oq. It follows from $p \in B$ that there are $s_1, \ldots s_n \in A$ such that $\vdash s_1\& \ldots s_n \rightarrow p$, and $\neg Os_1, \ldots,$ and $\neg Os_n$. Due to (ii), $\neg O(s_1\& \ldots s_n)$. Letting $s = s_1\&. \ldots \&s_n$, we then have $\vdash s \rightarrow p$, $\neg Os$ and (since A is logically closed) $s \in A$. Due to the construction of A, there are $r_1, \ldots r_m$ such that $Or_1, \ldots Or_m$, and $\vdash r_1\& \ldots \&r_m \rightarrow s$. Let $r = r_1\& \ldots \&r_m$. Due to (i), Or, hence we have $\vdash r \rightarrow s$, $\vdash s \rightarrow p$, Or, and $\neg Os$. From this, $\vdash p \rightarrow q$ and Op it follows, due to (iii), that Oq.

For (Cβ), let $p, q \in C$. Then $p, q \in B$ so that $p\&q \in B$. Due to (i), $O(p\&q)$, thus $p\&q \in C$.

We can now proceed with the construction. There are two cases, depending on whether or not $C \subset B \subset A$.

Case 1, $C \subset B \subset A$: It follows from Lemma 10.59 that there are consistent and logically closed sets B' and C' such that each of $A \cup B'$, $B' \cup C'$, and $A \cup C'$ is inconsistent and that $A\backslash(B\backslash C) = A\backslash(B'\backslash C')$. Let $\mathcal{A}_1 = \{X \in \mathcal{A} \mid A \subseteq X\}$, $\mathcal{A}_2 = \{X \in \mathcal{A} \mid C' \subseteq X\}$, $\mathcal{A}_4 = \{X \in \mathcal{A} \mid B' \subseteq X\}$, and $\mathcal{A}_3 = \mathcal{A}\backslash(\mathcal{A}_1 \cup \mathcal{A}_2 \cup \mathcal{A}_4)$. Let $\mathcal{A}_1$, $\mathcal{A}_2$, $\mathcal{A}_3$, and $\mathcal{A}_4$ be equivalence classes with respect to $\geq$, and let $\mathcal{A}_1 > \mathcal{A}_2 > \mathcal{A}_3 > \mathcal{A}_4$. Let $\max(f) \in \mathcal{A}_2$ and $\min(f) \in \mathcal{A}_4$. We then have:

$Op \leftrightarrow p \in A\backslash(B\backslash C)$
$\leftrightarrow p \in \cap\mathcal{A}_1\backslash((\cap\mathcal{A}_4)\backslash(\cap\mathcal{A}_2))$

$\leftrightarrow (p \in \cap\mathcal{A}_1\ \&\ p \in \cap\mathcal{A}_2) \vee (p \in \cap\mathcal{A}_1 \notin \cap\mathcal{A}_2\ \&\ p \notin \cap\mathcal{A}_4)$
$\leftrightarrow (\neg p \notin \cup\mathcal{A}_1\ \&\ \neg p \notin \cup\mathcal{A}_2) \vee (\neg p \notin \cup\mathcal{A}_1\ \&\ \neg p \in \cup\mathcal{A}_2\ \&\ \neg p \in \cup\mathcal{A}_4)$
$\leftrightarrow (\mathfrak{max}(f) > \mathfrak{max}(\neg p)) \vee ((\mathfrak{max}(f) \equiv \mathfrak{max}(\neg p)\ \&\ \mathfrak{min}(f) \geq \mathfrak{min}(\neg p))$
$\leftrightarrow f >_{xi} \neg p$

Case 2: Not $C \subset B \subset A$: Since $C \subseteq B \subseteq A$, either $A = B$ or $B = C$. If $A = B$, then $Z = A \backslash (A \backslash C) = C$. If $B = C$, then $Z = A \backslash (B \backslash B) = A$. Hence, in both cases $Z = \{p \mid Op\}$ is logically closed. It follows in the same way as in the proof of Theorem 10.44, and using (iv), that there is a sentence f such that $\mathfrak{max}(f)$ is nonmaximal and that for all p, Op iff $\mathfrak{max}(f) \geq \mathfrak{max}(\neg p)$. Let f' be a sentence such that $\mathfrak{max}(f') \equiv \mathfrak{min}(f') \equiv \mathfrak{max}(f)$. Then:

$Op \leftrightarrow \mathfrak{max}(f) \geq \mathfrak{max}(\neg p)$
$\leftrightarrow \mathfrak{max}(f') > \mathfrak{max}(\neg p) \vee \mathfrak{max}(f') \equiv \mathfrak{max}(\neg p)$
$\leftrightarrow \mathfrak{max}(f') > \mathfrak{max}(\neg p) \vee ((\mathfrak{max}(f') \equiv \mathfrak{max}(\neg p)\ \&\ \mathfrak{min}(f') \geq \mathfrak{min}(\neg p))$
$\leftrightarrow f' \geq_{xi} \neg p$.

PROOF OF THEOREM 10.48. *From 1 to 2:*

(i): Due to the symmetry between p and q, there are two cases:

Case 1, $v_{MAX}(\neg p) \geq v_{MAX}(\neg q)$ and $v_{MIN}(\neg p) \geq v_{MIN}(\neg q)$: Then $v_{MAX}(\neg p \vee \neg q) = v_{MAX}(\neg p)$ and $v_{MIN}(\neg p \vee \neg q) = v_{MIN}(\neg q)$, hence $v_{MIN}(\neg p) \geq v_{MIN}(\neg p \vee \neg q)$. It follows that $\neg p \geq_E (\neg p \vee \neg q)$, and since $f \geq_E \neg p$ that $f \geq_E (\neg p \vee \neg q)$, hence $O(p\&q)$.

Case 2, $v_{MAX}(\neg p) \geq v_{MAX}(\neg q)$ and $v_{MIN}(\neg q) > v_{MIN}(\neg p)$: Then $v_{MAX}(\neg p \vee \neg q) = v_{MAX}(\neg p)$ and $v_{MIN}(\neg p \vee \neg q) = v_{MIN}(\neg p)$, hence $f \geq_E \neg p \geq_E (\neg p \vee \neg q)$, hence $O(p\&q)$.

(ii): Let $O(p\&q)$. Then $f \geq_E (\neg p \vee \neg q)$. We have either $v_{MAX}(\neg p \vee \neg q) = v_{MAX}(\neg p)$ or $v_{MAX}(\neg p \vee \neg q) = v_{MAX}(\neg q)$. For symmetry reasons, we may, without loss of generality, assume that $v_{MAX}(\neg p \vee \neg q) = v_{MAX}(\neg p)$. There are two cases.

Case 1, $v_{MIN}(\neg p \vee \neg q) = v_{MIN}(\neg p)$: Then clearly $\neg p \vee \neg q \equiv_E \neg p$, and it follows from $f \geq_E (\neg p \vee \neg q)$ that $f \geq_E \neg p$, hence Op.

Case 2, $v_{MIN}(\neg p \vee \neg q) = v_{MIN}(\neg q)$: Since $v_{MAX}(\neg p \vee \neg q) \geq v_{MAX}(\neg q)$, we have $f \geq_E (\neg p \vee \neg q) \geq_E \neg q$, hence Oq.

(iii): Suppose to the contrary that W^+p, W^+q, $W(p \vee r)$, $\neg W(p \vee s)$, $\neg W(q \vee r)$, and $W(q \vee s)$. It follows from W^+p and $\neg W(p \vee s)$ that $v_\delta(p \vee s) = \delta \cdot v_{MAX}(s) + (1-\delta) \cdot v_{MIN}(p \vee s) > v_\delta(f)$ and also that $v_{MAX}(s) > v_\delta(f)$. It then follows from W^+q and $W(q \vee s)$ that $v_\delta(q \vee s) = \delta \cdot v_{MAX}(s) + (1-\delta) \cdot$

$v_{\mathrm{MIN}}(q \vee s) \leq v_{\delta}(f)$. Combining these, we find that $v_{\mathrm{MIN}}(p \vee s) > v_{\mathrm{MIN}}(q \vee s)$, from which it follows that $\neg(v_{\mathrm{MIN}}(q) > v_{\mathrm{MIN}}(p))$. In the same way, it follows from $W^{+}q$ and $\neg W(q \vee r)$ that $v_{\delta}(q \vee r) = \delta \cdot v_{\mathrm{MAX}}(r) + (1-\delta) \cdot v_{\mathrm{MIN}}(q \vee r) > v_{\delta}(f)$ and $v_{\mathrm{MAX}}(r) > v_{\delta}(f)$. It then follows from $W^{+}p$ and $W(p \vee r)$ that $v_{\delta}(p \vee r) = \delta \cdot v_{\mathrm{MAX}}(r) + (1-\delta) \cdot v_{\mathrm{MIN}}(p \vee r) \leq v_{\delta}(f)$. Combining these, we obtain $v_{\mathrm{MIN}}(q \vee r) > v_{\mathrm{MIN}}(p \vee r)$, hence $\neg(v_{\mathrm{MIN}}(p) > v_{\mathrm{MIN}}(q))$. We can conclude from $\neg(v_{\mathrm{MIN}}(q) > v_{\mathrm{MIN}}(p))$ and $\neg(v_{\mathrm{MIN}}(p) > v_{\mathrm{MIN}}(q))$ that $v_{\mathrm{MIN}}(p) = v_{\mathrm{MIN}}(q)$.

It follows from $v_{\mathrm{MIN}}(p) = v_{\mathrm{MIN}}(q)$ that $v_{\mathrm{MIN}}(p \vee r) = v_{\mathrm{MIN}}(q \vee r)$. From this, $W(p \vee r)$, and $\neg W(q \vee r)$, it follows that $v_{\mathrm{MAX}}(q \vee r) > v_{\mathrm{MAX}}(p \vee r)$. However, this is impossible since $v_{\mathrm{MAX}}(p) \leq v_{\delta}(f)$, $v_{\mathrm{MAX}}(q) \leq v_{\delta}(f)$, and $v_{\mathrm{MAX}}(r) > v_{\delta}(f)$.

(iv): Suppose to the contrary that $P^{+}p$, $P^{+}q$, $P(p \vee r)$, $\neg P(p \vee s)$, $\neg P(q \vee r)$, and $P(q \vee s)$. It follows from $P^{+}p$ and $\neg P(p \vee s)$ that $v_{\delta}(p \vee s) = \delta \cdot v_{\mathrm{MAX}}(p \vee s) + (1-\delta) \cdot v_{\mathrm{MIN}}(s) \leq v_{\delta}(f)$ and also that $v_{\mathrm{MIN}}(s) < v_{\delta}(f)$. It then follows from $P(q \vee s)$ that $v_{\delta}(q \vee s) = \delta \cdot v_{\mathrm{MAX}}(q \vee s) + (1-\delta) \cdot v_{\mathrm{MIN}}(s) > v_{\delta}(f)$. Combining these, we find that $v_{\mathrm{MAX}}(q \vee s) > v_{\mathrm{MAX}}(p \vee s)$, from which it follows that $\neg(v_{\mathrm{MAX}}(p) > v_{\mathrm{MAX}}(q))$. In the same way, it follows from $P^{+}q$ and $\neg P(q \vee r)$ that $v_{\delta}(q \vee r) = \delta \cdot v_{\mathrm{MAX}}(q \vee r) + (1-\delta) \cdot v_{\mathrm{MIN}}(r) \leq v_{\delta}(f)$ and $v_{\mathrm{MIN}}(r) < v_{\delta}(f)$. It then follows from $P^{+}p$ and $P(p \vee r)$ that $v_{\delta}(p \vee r) = \delta \cdot v_{\mathrm{MAX}}(p \vee r) + (1-\delta) \cdot v_{\mathrm{MIN}}(r) > v_{\delta}(f)$. Combining these, we obtain $v_{\mathrm{MAX}}(p \vee r) > v_{\mathrm{MAX}}(q \vee r)$, hence $\neg(v_{\mathrm{MAX}}(q) > v_{\mathrm{MAX}}(p))$. We can conclude from $\neg(v_{\mathrm{MAX}}(p) > v_{\mathrm{MAX}}(q))$ and $\neg(v_{\mathrm{MAX}}(q) > v_{\mathrm{MAX}}(p))$ that $v_{\mathrm{MAX}}(p) = v_{\mathrm{MAX}}(q)$.

It follows from $v_{\mathrm{MAX}}(p) = v_{\mathrm{MAX}}(q)$ that $v_{\mathrm{MAX}}(p \vee r) = v_{\mathrm{MAX}}(q \vee r)$. From this, $P(p \vee r)$ and $\neg P(q \vee r)$, it follows that $v_{\mathrm{MIN}}(p \vee r) > v_{\mathrm{MIN}}(q \vee r)$. However, this is impossible since $v_{\mathrm{MIN}}(p) > v_{\delta}(f)$, $v_{\mathrm{MIN}}(q) > v_{\delta}(f)$, and $v_{\mathrm{MIN}}(r) < v_{\delta}(f)$.

From 2 to 1: As a *first step in the construction*, we will introduce the following relation on the set of maximal consistent sentences: (A sentence p is maximal consistent iff there is no sentence q such that both $p \& q$ and $p \& \neg q$ are consistent.)

$p \underline{\gg} q$ iff it holds for all r that if $W(p \vee r)$ then $W(q \vee r)$.

That $\underline{\gg}$ is transitive follows from the definition. We also need to show that it is complete, i.e., that if p and q are maximal consistent sentences, then either $p \underline{\gg} q$ or $q \underline{\gg} p$. Suppose to the contrary that there are p and q such that this is not the case. Then $\neg(p \underline{\gg} q)$ and $\neg(q \underline{\gg} p)$. Hence, there are r and s such that $W(p \vee r)$, $\neg W(q \vee r)$, $W(q \vee s)$ and $\neg W(p \vee s)$. It follows from (i) and (ii) that $Wp \vee Wr$, $\neg Wq \vee \neg Wr$, $Wq \vee Ws$, and $\neg Wp \vee \neg Ws$. It follows truth-functionally that either $Wp \& Wq$ or $\neg Wp \& \neg Wq$. We can treat these cases separately.

First case, Wp & Wq. Since p and q are maximal consistent sentences, $W^{+}p$ and $W^{+}q$. We get a contradiction directly from (iii).

Second case, $\neg Wp$ & $\neg Wq$. Since p and q are maximal consistent sentences, we then have: $P^{+}p$ and $P^{+}q$, $\neg P(p\vee r)$, $P(q\vee r)$, $\neg P(q\vee s)$, and $P(p\vee s)$. We get a contradiction directly from (iv).

The *second step of the construction* consists in assigning numerical values to each maximal consistent sentence (and thus to each alternative, i.e., world), as follows: (1) To each maximal consistent sentence p such that Wp, let $v(p)$ be equal to –1 times the number of equivalence classes of sentences above p in the $\underline{\gg}$ ordering that contain a sentence α such that $W\alpha$. (Hence, if the maximal consistent sentences α with the properties $W\alpha$ and $\alpha\gg p$ are divided between 5 equivalence classes, then $v(p) = -5$.)

(2) For each maximal consistent sentence p such that $\neg Wp$, list all maximal consistent sentences α such that $W(p\vee\alpha)$, identify among them a sentence that is highest in the $\underline{\gg}$ ordering, and call it $c(p)$. Then let $v(p) = ((\delta-1)/\delta)\cdot v(c(p))$. If $\neg W(p\vee\alpha)$ for all α, then let $v(p) = ((1-\delta)/\delta)\cdot(m+1)$, where m is the number of equivalence classes with respcct to $\underline{\gg}$ of sentences α such that $W\alpha$.

(3) Let f be any maximal consistent sentence such that Wf and that $\neg Wt$ for all t such that $t\gg f$. Then $v_{\delta}(f) = 0$.

Verification of the construction: We are going to show that for all s, Os iff $v_{\delta}(f)\geq v_{\delta}(\neg s)$, or equivalently, that for all s, Ws iff $0\geq v_{\delta}(s)$. The verification consists of two steps. The *first* step is a demonstration that for all s, Ws iff $W(p\vee r)$, where p and r are maximal consistent sentences such that $v(p) = v_{MAX}(s)$ and $v(r) = v_{MIN}(s)$.

Since $\underline{\gg}$ is complete, s is equivalent with $p\vee q_1\vee \ldots q_n\vee r$, where $p, q_1, \ldots q_n$, and r are maximal consistent sentences such that $p\underline{\gg}q_1\underline{\gg}\ldots\underline{\gg}q_n\underline{\gg}r$. We need to show that (A) $v(p) = v_{MAX}(s)$, (B) $v(r) = v_{MIN}(s)$, and (C) $W(p\vee r)$ iff $W(p\vee q_1\vee \ldots \vee q_n\vee r)$.

For (A), suppose that this is false. Then there is some $t\in\{q_1, \ldots q_n, r\}$ such that $v(t)>v(p)$ and $p\underline{\gg}t$.

Case 1, Wp: It follows from $W(p\vee p)$ and $p\underline{\gg}t$ that $W(p\vee t)$. Applying $p\underline{\gg}t$ once more we obtain $W(t\vee t)$, i.e., Wt. It follows from Wp, Wt, $v(t)>v(p)$, and the construction of $\underline{\gg}$ that $t\gg p$. Contradiction.

Case 2, $\neg Wp$: It follows from $v(t)>v(p)$ that $\neg Wt$. It also follows from $v(t)>v(p)$ that there is some x such that $\neg W(t\vee x)$ and $W(p\vee x)$. Hence, $\neg(p\underline{\gg}t)$. Contradiction.

For (B), suppose to the contrary that this is false. Then there is some $t\in\{p, q_1 \ldots q_n\}$ such that $v(r)>v(t)$ and $t\underline{\gg}r$.

Case 1, Wr: It follows from $v(r)>v(t)$ that $r\gg t$. Contradiction.

Case 2, $\neg Wr$: It follows from $t\underline{\gg}r$ that $\neg Wt$. Due to $v(r)>v(t)$, there is some x such that $\neg W(r\vee x)$ and $W(t\vee x)$, hence $\neg(t\underline{\gg}r)$. Contradiction.

For (C): For one direction, let $W(p\vee r)$. It follows from $p\underline{\gg}q_1$ that $W(q_1\vee r)$, then from $q_1\underline{\gg}q_2$ that $W(q_2\vee r)$, etc., all the way to $W(q_n\vee r)$. We can use (i) repeatedly to obtain $W((p\vee r)\vee(q_1\vee r)\vee \ldots \vee(q_n\vee r))$, or equivalently $W(p\vee q_1\vee \ldots \vee q_n\vee r)$.

For the other direction, let $\neg W(p\vee r)$. Suppose that for some q_k, $W(p\vee q_k\vee r)$. It follows from (ii) that Wq_k, and then from $q_k\underline{\gg}r$ that $W(q_k\vee p) \rightarrow W(p\vee r)$, hence $\neg W(q_k\vee p)$. It follows from $\neg W(p\vee r)$ and $\neg W(q_k\vee p)$, via (ii), that $\neg W(p\vee q_k\vee r)$. We can conclude from this contradiction that for all q_k, $\neg W(p\vee q_k\vee r)$. Hence, with repeated use of (ii), $\neg W(p\vee q_1\vee \ldots \vee q_n\vee r)$.

This concludes the first verification step. In the *second* step, now to follow, we have to show that $W(p\vee q)$ iff $0\geq v_\delta(p\vee q)$, with p and q defined as in the first step of the verification. There are five cases.

Case (a), Wp. It follows from $W(p\vee p)$ and $p\underline{\gg}q$ that $W(p\vee q)$. It follows from $W(p\vee q)$ and $p\underline{\gg}q$ that $W(q\vee q)$, i.e., Wq. Hence, $0\geq v(p)$ and $0\geq v(q)$, hence $v_\delta(p\vee q) = \delta\cdot v(p)+(1-\delta)\cdot v(q) \leq 0$.

Case (b), $\neg Wp$ and $\neg Wq$: It follows from (ii) that $\neg W(p\vee q)$. Furthermore, it follows from $\neg Wp$ and $\neg Wq$ that $v(p)>0$ and $v(q)>0$. Hence, $v_\delta(p\vee q) = \delta\cdot v(p)+(1-\delta)\cdot v(q) > 0$.

Case (c), $\neg Wp$, Wq, and $c(p)\underline{\gg}q$. Then it follows from $W(p\vee c(p))$ that $W(p\vee q)$. It also follows from $c(p)\underline{\gg}q$ that $v(c(p))\geq v(q)$, so that $v_\delta(p\vee q) = \delta\cdot v(p)+(1-\delta)\cdot v(q) = (\delta\cdot(\delta-1)/\delta)\cdot v(c(p))+(1-\delta)\cdot v(q) = (1-\delta)(v(q)-v(c(p))) \leq 0$.

Case (d), $\neg Wp$, Wq, and $q\gg c(p)$. Then it follows from the definition of v that $\neg W(p\vee q)$. Furthermore, it follows from $q\gg c(p)$ (cf. the previous case) that $v_\delta(p\vee q) = (1-\delta)(v(q)-v(c(p))) > 0$.

Case (e), $\neg Wp$, Wq, and $c(p)$ is undefined. Then $\neg W(p\vee\alpha)$ holds for all α such that $W\alpha$, hence $\neg W(p\vee q)$. Furthermore, according to the construction, $v(p)>((\delta-1)/\delta)\cdot v(\alpha)$ for all sentences α. It follows that $v_\delta(p\vee q) = \delta\cdot v(p) + (1-\delta)\cdot v(q) > \delta\cdot((\delta-1)/\delta)\cdot v(q) + (1-\delta)\cdot v(q)) = 0$.

PROOF OF OBSERVATION 10.49. (i) and (ii) of Theorem 10.48 follow directly from the SDL postulates. For (iii), it is sufficient to show that W^+q, $W(p\vee r)$, and $\neg W(q\vee r)$ are incompatible. In SDL, it follows from $W(p\vee r)$, i.e., $O(\neg p\& \neg r)$, that $O\neg r$. It follows from this and Wq that $O(\neg q\&\neg r)$, contrary to $\neg W(q\vee r)$. For (iv), it is sufficient to show that $P(p\vee r)$, $\neg P(p\vee s)$, and $\neg P(q\vee r)$ are incompatible. It follows in SDL from $\neg P(p\vee s)$ that $O\neg p$ and similarly from $\neg P(q\vee r)$ that $O\neg r$. Then $O(\neg p\&\neg r)$, contrary to $P(p\vee r)$.

PROOF OF OBSERVATION 10.51. *Part 1*: Suppose to the contrary that $O_{C}p$ and $O_{C}\neg p$. By substitution of $\neg p$ for s in the definition of $O_{C}p$, we obtain, since $\geq'$ satisfies ancestral reflexivity, $p>'\neg p$. In the same way, $O_{C}\neg p$ yields $\neg p>'p$. This contradiction is sufficient to show that $O_{C}p$ and $O_{C}\neg p$ cannot both hold.

Part 2: For one direction, suppose that $Op\rightarrow O_{C}p$ for all p. Then it follows directly from Part 1 that O satisfies consistency.

For the other direction, suppose that it does *not* hold that $Op\rightarrow O_{C}p$ for all p. Then there is some p such that Op and $\neg O_{C}p$. It follows from $\neg O_{C}p$ that there is some s such that $\neg p\geq'^{*}s$ and $\neg(\neg s>'s)$. Due completeness, $s\geq'\neg s$. It follows from $\neg p\geq'^{*}s$ and $s\geq'\neg s$ that $\neg p\geq'^{*}\neg s$. Since O is contranegative, it follows from Op and $\neg p\geq'^{*}s$ that $O\neg s$. Similarly, it follows from Op and $\neg p\geq'^{*}\neg s$ that Os.

PROOF OF OBSERVATION 10.53. It follows from Theorem 7.14 and Observation 10.9 that $\geq'$ is extremal and that O satisfies agglomeration.

One direction is obvious since O is nonobeyable if $O\perp$. For the other direction, let O be nonobeyable. Note that since $\mathcal{L}_{\mathcal{A}}$ has a finite number of equivalence classes with respect to logical equivalence, we can form a conjunctive sentence $\&\{p \mid Op\}$ that is equivalent with $\{p \mid Op\}$. Since O is nonobeyable, $\&\{p \mid Op\}$ is not represented in $\mathcal{A}$. By agglomeration, $O(\&\{p \mid Op\})$. Since $\&\{p \mid Op\} \notin \cup\mathcal{A}$, the minima and maxima of $\neg\&\{p \mid Op\}$ and $\neg\perp$ coincide. Since $\geq'$ is extremal, we have $\neg\&\{p \mid Op\} \geq' \neg\perp$, and it follows from the contranegativity of O that $O\perp$.

PROOF OF OBSERVATION 10.55. *Part 1*: Let $\geq'$ satisfy ancestral reflexivity and suppose that $O_{T}\perp$. By substituting $\perp$ for p in the definition of O_{T}, we obtain $(\forall s)(\mathrm{T}\geq'^{*}s \rightarrow \mathrm{T}>'s)$ that has the instance $\mathrm{T}\geq'^{*}\mathrm{T} \rightarrow \mathrm{T}>'\mathrm{T}$. This is impossible since $\geq'$ satisfies ancestral reflexivity and $>'$ is irreflexive.

Part 2: For one direction, let $Op\rightarrow O_{T}p$ for all p. It follows from Part 1 that $\neg O\perp$. For the other direction, suppose that there is some $\geq'$-contranegative O that satisfies $\neg O\perp$, and such that for some p, Op and $\neg O_{T}p$. It follows from $\neg O_{T}p$ that there is some s such that $\neg p\geq'^{*}s$ and $\neg(\mathrm{T}>'s)$. Since $\geq'$ is complete, $s\geq'\mathrm{T}$. From this and $\neg p\geq'^{*}s$ follows $\neg p\geq'^{*}\mathrm{T}$, i.e., $\neg p\geq'^{*}\neg\perp$. Since O is contranegative, it follows from this and Op that $O\perp$, contrary to the conditions.

PROOF OF OBSERVATION 10.56. *Part 1*:
$O_{\mathrm{T}}p \leftrightarrow (\forall s)(\neg p \geq'^* s \rightarrow \mathrm{T} >' s)$
$\leftrightarrow (\forall s)\neg(\neg p \geq'^* s \geq' \mathrm{T})$ (completeness)
$\leftrightarrow \neg(\neg p \geq'^* \mathrm{T})$
$\rightarrow \neg(\neg p \geq' \mathrm{T})$
$\leftrightarrow \mathrm{T} >' \neg p$ (completeness)

Part 2: If $\geq'$ is transitive, then $\rightarrow$ can be replaced by $\leftrightarrow$ on the fifth line of the proof of Part 1.

PROOF OF OBSERVATION 10.57. Suppose to the contrary that, under the conditions given, $O_{\mathrm{T}}p$ and $\neg O_{\mathrm{C}}p$. Since $\geq'$ satisfies completeness, we can use Observation 10.56 and $O_{\mathrm{T}}p$ to obtain $\mathrm{T} >' \neg p$. It follows from $\neg O_{\mathrm{C}}p$ that there is some s such that $\neg p \geq'^* s$ and $\neg(\neg s > s)$, hence by completeness $s \geq' \neg s$. Transitivity yields $\neg p \geq' s$ and $\neg p \geq' \neg s$. From this and $\mathrm{T} >' \neg p$ follows $\mathrm{T} >' s$ and $\mathrm{T} >' \neg s$, contrary to our assumption that $s \geq' \mathrm{T} \vee \neg s \geq' \mathrm{T}$.

PROOF OF OBSERVATION 10.58. That (1) and (2) are equivalent follows from Observation 10.55.

(2)-to-(3): Let $\neg O_{\mathrm{C}}\perp$. Then:
$(\exists s)\neg(\neg\perp \geq'^* s \rightarrow \neg s >' s)$ (definition of O_{C})
$(\exists s)(\mathrm{T} \geq'^* s \;\&\; s >' \neg s)$ (completeness)
$(\exists s)(\mathrm{T} \geq' s \;\&\; \mathrm{T} \geq' \neg s)$ (completeness and transitivity)

(3)-to-(2): Let $\mathrm{T} \geq' s$ and $\mathrm{T} \geq' \neg s$. It follows from completeness that either $s \geq' \neg s$ or $\neg s \geq' s$. Without loss of generality, we may assume that $s \geq' \neg s$. We then have:
$\mathrm{T} \geq' s$ and $s \geq' \neg s$
$\neg\perp \geq'^* s$ and $\neg(\neg s >' s)$
$\neg(\forall s)(\neg\perp \geq'^* s \rightarrow \neg s >' s)$
$\neg O_{\mathrm{C}}\perp$.

PROOF FOR CHAPTER 11

PROOF OF OBSERVATION 11.3. *Part 1*: Since $\geq'$ is transitive and complete, so is $\geq_p'$. We can therefore rewrite the definition of O_{T}:

$O_{\mathrm{T}}(q, p) \leftrightarrow (\forall s)(\neg q \geq_p'^* s \rightarrow \mathrm{T} >_p' s)$
$\leftrightarrow (\forall s)\neg(\neg q \geq_p'^* s \geq_p' \mathrm{T})$ (completeness)
$\leftrightarrow \neg(\neg q \geq_p' \mathrm{T})$ (transitivity)

$\leftrightarrow \top >_p' \neg q$ (completeness)
$\leftrightarrow p >' p \& \neg q$

Part 2: Suppose to the contrary that $O_\top(\bot, p)$. Then, according to Part 1, $p >' p \& \neg \bot$, i.e., $p >' p$, which is impossible.

Part 3: Suppose to the contrary that $O_\top(q, \bot)$. Then $\bot >' \bot \& \neg q$, i.e., $\bot >' \bot$, which is impossible.

Part 4: $O_\top(q, \top) \leftrightarrow \top >' \top \& \neg q$
$\leftrightarrow \top >' \neg q$
$\leftrightarrow O_\top(q)$ (Observation 10.56)

Part 5: Let $O_\top(q_1, p)$ & $O_\top(q_2, p)$. Then $p >' p \& \neg q_1$ and $p >' p \& \neg q_2$. It follows from completeness that either $p \& \neg q_1 \geq' p \& \neg q_2$ or $p \& \neg q_2 \geq' p \& \neg q_1$. In the first case, disjunctive interpolation yields $p \& \neg q_1 \geq' (p \& \neg q_1) \vee (p \& \neg q_2)$, and in the second case it yields $p \& \neg q_2 \geq' (p \& \neg q_1) \vee (p \& \neg q_2)$. In both cases, transitivity yields $p >' ((p \& \neg q_1) \vee (p \& \neg q_2))$, equivalently $p >' (p \& \neg (q_1 \& q_2))$, from which it follows that $O_\top(q_1 \& q_2, p)$.

Part 6: (Various parts of Lemma 7.33 are used throughout this part of the proof without explicit reference.) Let $O_\top(q, p_1)$ & $O_\top(q, p_2)$. Then $p_1 >' p_1 \& \neg q$ and $p_2 >' p_2 \& \neg q$.

For $\geq_{\mathbf{x}}$ and $\geq_{\mathbf{xi}}$: It can be concluded from $p_1 >' p_1 \& \neg q$ that $\mathfrak{max}(p_1) > \mathfrak{max}(p_1 \& \neg q)$. (Note, in the case of $\geq_{\mathbf{xi}}$, that $\mathfrak{min}(p_1 \& \neg q) \geq \mathfrak{min}(p_1)$.) In the same way, we obtain $\mathfrak{max}(p_2) > \mathfrak{max}(p_2 \& \neg q)$. Furthermore, since $repr(p_1 \& \neg q) \neq \emptyset \neq repr(p_2 \& \neg q)$ and $\mathcal{A}$ is contextually complete, either $\mathfrak{max}((p_1 \vee p_2) \& \neg q) \equiv \mathfrak{max}(p_1 \& \neg q)$ or $\mathfrak{max}((p_1 \vee p_2) \& \neg q) \equiv \mathfrak{max}(p_2 \& \neg q)$. Without loss of generality, we may assume that $\mathfrak{max}((p_1 \vee p_2) \& \neg q) \equiv \mathfrak{max}(p_1 \& \neg q)$. Since $\mathfrak{max}(p_1 \vee p_2) \geq \mathfrak{max}(p_1)$, we have in summary $\mathfrak{max}(p_1 \vee p_2) \geq \mathfrak{max}(p_1) > \mathfrak{max}(p_1 \& \neg q) \equiv \mathfrak{max}((p_1 \vee p_2) \& \neg q)$, hence by transitivity $\mathfrak{max}(p_1 \vee p_2) > \mathfrak{max}((p_1 \vee p_2) \& \neg q)$. Hence, $p_1 \vee p_2 >_{\mathbf{x}} (p_1 \vee p_2) \& \neg q$, respectively $p_1 \vee p_2 >_{\mathbf{xi}} (p_1 \vee p_2) \& \neg q$, and (in both cases) $O_\top(q, p_1 \vee p_2)$.

For $\geq_{\mathbf{ix}}$: Since $\mathfrak{min}(p_1 \& \neg q) \geq \mathfrak{min}(p_1)$, it follows from $p_1 >_{\mathbf{ix}} p_1 \& \neg q$ that $\mathfrak{min}(p_1) \equiv \mathfrak{min}(p_1 \& \neg q)$ and $\mathfrak{max}(p_1) > \mathfrak{max}(p_1 \& \neg q)$. Similarly, $\mathfrak{min}(p_2) \equiv \mathfrak{min}(p_2 \& \neg q)$ and $\mathfrak{max}(p_2) > \mathfrak{max}(p_2 \& \neg q)$. It follows from $\mathfrak{min}(p_1) \equiv \mathfrak{min}(p_1 \& \neg q)$ and $\mathfrak{min}(p_2) \equiv \mathfrak{min}(p_2 \& \neg q)$, since $repr(p_1 \& \neg q) \neq \emptyset \neq repr(p_2 \& \neg q)$ and $\mathcal{A}$ is contextually complete, that $\mathfrak{min}(p_1 \vee p_2) \equiv \mathfrak{min}((p_1 \& \neg q) \vee (p_2 \& \neg q))$. In the same way, it is possible to conclude from $\mathfrak{max}(p_1) > \mathfrak{max}(p_1 \& \neg q)$ and $\mathfrak{max}(p_2) > \mathfrak{max}(p_2 \& \neg q)$ that $\mathfrak{max}(p_1 \vee p_2) > \mathfrak{max}((p_1 \& \neg q) \vee (p_2 \& \neg q))$. Hence, $(p_1 \vee p_2) >_{\mathbf{ix}} ((p_1 \& \neg q) \vee$

$(p_2\&\neg q))$, that is $p_1 \vee p_2 >_{ix} (p_1 \vee p_2)\&\neg q$, from which it follows that $O_T(q, p_1 \vee p_2)$.

PROOFS FOR CHAPTER 12

PROOF OF THEOREM 12.7. Since $a(N,S) = a(I(N),S)$, we can without loss of generality assume that N contains no individual variables, so that $N = I(N)$ and $\mathcal{N} = I(\mathcal{N})$.

It follows from Definition 12.5 that $a(N,S) \subseteq \mathrm{Cn}(I(\mathcal{N}) \cup S) \cap \mathcal{L}_{NC}$. For the other direction, we need the following definitions:

A normative rule $p \Rightarrow \delta$ is *S-detachable* iff $p \in S$.

For any set N of normative rules, $dis(N)$, the *disjunctive closure* of N, is the smallest set of normative rules such that $N \subseteq dis(N)$ and that if $p_1 \Rightarrow \delta_1, p_2 \Rightarrow \delta_2 \in dis(N)$, then $p_1 \vee p_2 \Rightarrow \delta_1 \vee \delta_2 \in dis(N)$. The disjunctive closure of a "translated" set $\mathcal{N}$ with $\rightarrow$ replacing $\Rightarrow$ is defined analogously.

We are going to show that, whenever $\mathcal{N}$ is nonempty:

(*i*) If $\beta \in \mathrm{Cn}(\mathcal{N} \cup S) \cap \mathcal{L}_{NC}$, then there is some $\mathcal{N}'' \subseteq dis(\mathcal{N})$ such that (a) all elements of $\mathcal{N}''$ are S-detachable, and (b) β follows logically (by Cn) from the set of consequents of $\mathcal{N}''$.

It is easy to verify that if (*i*) holds and $\beta \in \mathrm{Cn}(\mathcal{N} \cup S) \cap \mathcal{L}_{NC}$, then $\beta \in a(N,S)$. The proof of (*i*) proceeds by induction on the number of elements of $\mathcal{N}$. Since the desired result follows trivially if β is a tautology, it will be assumed that β is nontautologous.

Initial step: Let $\mathcal{N}$ consist of only one element, $p \rightarrow \alpha$, with $p \in \mathcal{L}_D$ and $\alpha \in \mathcal{L}_{NC}$. In order to prove (*i*), we will assume that $\beta \in \mathrm{Cn}(\mathcal{N} \cup S)$ and $\beta \in \mathcal{L}_{NC}$. Then $\beta \in \mathrm{Cn}(S \cup \{p \rightarrow \alpha\})$. Due to the compactness of Cn (cf. Definition 12.1) and the logical closure of S, there is some $s \in S$ such that $\beta \in \mathrm{Cn}(\{s\&(p \rightarrow \alpha)\})$. It follows from this that $\beta \in \mathrm{Cn}(\{s\&\neg p\})$. Since $s\&\neg p \in \mathcal{L}_D$, $\beta \in \mathcal{L}_N$, β is nontautologous and $\mathcal{L}_N$ is separable from $\mathcal{L}_D$, we then have $\perp \in \mathrm{Cn}(\{s\&\neg p\})$, hence $p \in \mathrm{Cn}(\{s\})$, hence $p \in \mathrm{Cn}(S)$.

It also follows from $\beta \in \mathrm{Cn}(\{s\&(p \rightarrow \alpha)\})$ that $\beta \in \mathrm{Cn}(\{s\&\alpha\})$, hence $\alpha \rightarrow \beta \in \mathrm{Cn}(\{s\})$. Since S is consistent, so is s. Furthermore, $s \in \mathcal{L}_D$ and $\alpha \rightarrow \beta \in \mathcal{L}_N$. Since $\mathcal{L}_N$ is separable from $\mathcal{L}_D$, we may conclude that $\alpha \rightarrow \beta \in \mathrm{Cn}(\emptyset)$.

Now let $\mathcal{N}'' = \{p \rightarrow \alpha\}$. Then $\mathcal{N}'' = \mathcal{N} \subseteq dis(\mathcal{N})$. That condition (a) of (*i*) is satisfied follows from $p \in \mathrm{Cn}(S)$. That condition (b) is satisfied follows from $\alpha \rightarrow \beta \in \mathrm{Cn}(\emptyset)$.

Induction step: Let (*i*) hold for all sets of normative rules with at most n elements, $n \geq 1$. Let $\mathcal{N}$ be such a set, and let $p_x \rightarrow \alpha_x$ be a normative rule. In order to prove that (*i*) holds for $\mathcal{N} \cup \{p_x \rightarrow \alpha_x\}$, we will assume that $\beta \in \text{Cn}(\mathcal{N} \cup \{p_x \rightarrow \alpha_x\} \cup S)$. It follows directly that

(1) $(p_x \rightarrow \alpha_x) \rightarrow \beta \in \text{Cn}(S \cup \mathcal{N})$

From this follows $\alpha_x \rightarrow \beta \in \text{Cn}(S \cup \mathcal{N})$. According to the induction hypothesis, there is some $\mathcal{N}_1 \subseteq dis(\mathcal{N})$, all elements of which are S-detachable, and such that $\alpha_x \rightarrow \beta$ follows from the set of consequents of $\mathcal{N}_1$.

It also follows from (1) that $\beta \in \text{Cn}(S \cup \{\neg p_x\} \cup \mathcal{N})$. According to the induction hypothesis, there is some $\mathcal{N}_2 \subseteq dis(\mathcal{N})$, each element of which is $S \cup \{\neg p_x\}$-detachable and such that β follows from the consequents of $\mathcal{N}_2$. Let $\mathcal{N}_2 = \{p_1 \rightarrow \alpha_1, \ldots p_s \rightarrow \alpha_s\}$. That each element $p_k \rightarrow \alpha_k$ of $\mathcal{N}_2$ is $S \cup \{\neg p_x\}$-detachable means that $p_k \in \text{Cn}(S \cup \{\neg p_x\})$, or equivalently (since S is logically closed), $p_k \vee p_x \in S$.

Let $\mathcal{N}_3 = \{p_1 \vee p_x \rightarrow \alpha_1 \vee \alpha_x, \ldots p_s \vee p_x \rightarrow \alpha_s \vee \alpha_x\}$. We have just shown that all antecedents of $\mathcal{N}_3$ are elements of S. It also follows directly that $\mathcal{N}_3 \subseteq dis(\mathcal{N} \cup \{p_x \rightarrow \alpha_x\})$.

Finally, let $\mathcal{N}'' = \mathcal{N}_1 \cup \mathcal{N}_3$. Since both $\mathcal{N}_1$ and $\mathcal{N}_3$ are subsets of $dis(\mathcal{N} \cup \{p_x \rightarrow \alpha_x\})$, so is $\mathcal{N}''$. We have already seen that all antecedents of $\mathcal{N}_1$ and $\mathcal{N}_3$ are elements of S, hence condition (a) of (i) is satisfied. It remains to show that condition (b) is also satisfied.

We already know that the consequents of $\mathcal{N}_2$ imply β, i.e., $\beta \in \text{Cn}(\{\alpha_1 \& \ldots \& \alpha_s\})$. It follows that:

$$\beta \in \text{Cn}(\{\alpha_1 \vee \alpha_x, \ldots \alpha_s \vee \alpha_x\} \cup \{\alpha_x \rightarrow \beta\}).$$

Since $\{\alpha_1 \vee \alpha_x, \ldots \alpha_s \vee \alpha_x\}$ are the consequents of $\mathcal{N}_3$, and $\alpha_x \rightarrow \beta$ follows from the consequents of $\mathcal{N}_1$, we may conclude that β follows from the consequents of $\mathcal{N}'' = \mathcal{N}_1 \cup \mathcal{N}_3$, so that condition (b) holds. This completes the proof.

PROOF OF OBSERVATION 12.8. *Parts 1* and *2*: Directly from the definitions.

Part 3: Due to Theorem 12.7, and simplifying notation by assuming that material implication represents the rule conditional, what we need to show is that $a(N,S) = \text{Cn}(I(a(N,S)) \cup S) \cap \mathcal{L}_{\text{NC}}$. For one direction, let $\alpha \in a(N,S)$. Since $a(N,S) = I(a(N,S))$, we then have $\alpha \in \text{Cn}(I(a(N,S)) \cup S)$. It also follows from $\alpha \in a(N,S)$ that $\alpha \in \mathcal{L}_{NC}$.

For the other direction, let $\alpha \in \text{Cn}(I(a(N,S)) \cup S) \cap \mathcal{L}_{NC}$. Then:

$\alpha \in \text{Cn}(I(a(N,S)) \cup S)$
$= \text{Cn}(a(N,S) \cup S)$
$= \text{Cn}(((\text{Cn}(I(N) \cup S)) \cap \mathcal{L}_{NC}) \cup S)$
$\subseteq \text{Cn}(I(N) \cup S)$.

We also have $\alpha \in \mathcal{L}_{NC}$, hence $\alpha \in \text{Cn}(I(N) \cup S) \cap \mathcal{L}_{NC} = a(N,S)$.

Part 4: Again we simplify notation by assuming that material implication represents the rule conditional. Let $N \subseteq N' \subseteq N \cup a(N,S)$. The inclusion $a(N,S) \subseteq a(N',S)$ follows from left monotony. For the other direction, let $\alpha \in a(N',S)$. Then:

$\alpha \in a(N',S)$
$\subseteq a(N \cup a(N,S),S)$ (left monotony and $N' \subseteq N \cup a(N, S)$)
$\subseteq \text{Cn}(I(N \cup a(N,S)) \cup S)$
$= \text{Cn}(I(N) \cup (\text{Cn}(I(N) \cup S) \cap \mathcal{L}_{NC}) \cup S)$
$\subseteq \text{Cn}(I(N) \cup (\text{Cn}(I(N) \cup S) \cup S)$
$= \text{Cn}(\text{Cn}(I(N) \cup S))$
$= \text{Cn}(I(N) \cup S)$

We also have $\alpha \in \mathcal{L}_{NC}$, and we may conclude that $\alpha \in a(N,S)$.

Part 5: Let $a(N,S) \subseteq N' \subseteq N \cup a(N,S)$. For one direction:

$a(N,S) = a(a(N,S),S)$ (left idempotence)
$\subseteq a(N',S)$ (left monotony and $a(N,S) \subseteq N'$)

For the other direction:

$a(N',S)$
$\subseteq a(N \cup a(N,S), S)$ (left monotony and $N' \subseteq N \cup a(N,S)$)
$= a(N,S)$ (left stability)

PROOF OF OBSERVATION 12.12. This follows directly from Definitions 12.10 and 12.11, given the result $c(N,S) = \text{Cn}(I(N)^*\&S) \cap \mathcal{L}_{NC}$ that was proved above in the text.

References

Adams, Robert Merrihew (1974) "Theories of Actuality." *Noûs* 8:211–231.

Alchourrón, Carlos (1993) "Philosophical Foundations of Deontic Logic and the Logic of Defeasible Conditionals," pp. 43–84 in John-Jules Ch. Meyer and Roel J. Wieringa (eds.). *Deontic Logic in Computer Science*. John Wiley & Sons, Chichester.

Alchourrón, Carlos, Peter Gärdenfors, and David Makinson (1985) "On the Logic of Theory Change: Partial Meet Functions for Contraction and Revision." *Journal of Symbolic Logic* 50:510–530.

Alchourrón, Carlos and David Makinson (1981) "Hierarchies of Regulation and Their Logic," pp. 125–148 in Risto Hilpinen (ed.). *New Studies in Deontic Logic*. D. Reidel Publishing, Dordrecht.

Al-Hibri, Azizah (1980) "Conditionality and Ross's Deontic Distinction." *Southwestern Journal of Philosophy* 11:79–87.

Anand, Paul (1993) "The Philosophy of Intransitive Preference." *Economic Journal* 103:337–346.

Åqvist, Lennart (1967) "Good Samaritans, Contrary-to-Duty Imperatives, and Epistemic Obligations." *Noûs* 1:361–379.

——— (1968) "Chisholm-Sosa Logics of Intrinsic Betterness and Value." *Noûs* 2:253–270.

Armstrong, W. E. (1950) "A Note on the Theory of Consumer's Behaviour." *Oxford Economic Papers* 2:119–122.

Barbera, S., C. R. Barrett, and Prasanta K. Pattanaik (1984) "On Some Axioms for Ranking Sets of Alternatives." *Journal of Economic Theory* 33:301–308.

Beck, L. W. (1941) "The Formal Properties of Ethical Wholes." *Journal of Philosophy* 38:5–14.

Becker, Gary S. (1992) "Habits, Addictions, and Traditions." *Kyklos* 45:327–346.

Bergström, Lars (1966) *The Alternatives and Consequences of Actions*. Stockholm Studies in Philosophy 4, Stockholm University, Stockholm.

Bowles, Samuel (1998) "Endogenous Preferences: The Cultural Consequences of Markets and Other Economic Institutions." *Journal of Economic Literature* 36:75–111.

Bradbury, H. et al. (1984) "Intransitivity of Preferences of Four-Year-Old Children." *Journal of Genetic Psychology* 145:145–146.

Brandt, Richard B. (1959) *Ethical Theory*. Prentice-Hall, Englewood Cliffs, N.J.

———- (1965) "The Concepts of Obligation and Duty." *Mind* 73:374–393.

Brogan, Albert P. (1919) "The Fundamental Value Universal." *Journal of Philosophy, Psychology, and Scientific Methods* 16:96–104.
Broome, John (1978) "Choice and Value in Economics." *Oxford Economic Papers* 30:313–333.
Brown, Mark A. (1996) "A Logic of Comparative Obligation." *Studia Logica* 57:117–137.
Bulygin, Eugenio and Carlos Alchourrón (1977) "Unvollständigkeit, Widersprüchlichkeit und Unbestimmtheit der Normenordnungen." pp. 20–32 in Amedeo Conte, Risto Hilpinen, and Georg Henrik von Wright (eds.). *Deontische Logik und Semantik*. Athenaion, Wiesbaden.
Burros, R. H. (1976) "Complementary Relations in the Theory of Preference." *Theory and Decision* 7:181–190.
Carlson, Eric (1997) "The Intrinsic Value of Non-Basic States of Affairs." *Philosophical Studies* 85:95–107.
Carnap, Rudolf (1956) *Meaning and Necessity*, Second edition. Phoenix Books, Chicago.
Castañeda, Hector-Neri (1958) Review of Sören Halldén, "On the Logic of 'Better.'" *Philosophy and Phenomenological Research* 19:266.
——— (1969) "Ought, Value, and Utilitarianism." *American Philosophical Quarterly* 6:257–275.
——— (1981) "The Paradoxes of Deontic Logic: The Simplest Solution to All of Them in One Fell Swoop," pp. 37–95 in Risto Hilpinen (ed.). *New Studies in Deontic Logic*. Reidel, Dordrecht.
——— (1984) "Philosophical Refutations," pp. 227–258 in James H. Fetzer (ed.). *Principles of Philosophical Reasoning*. Rowman & Allanfield, Totowa, N.J.
Cave, Eric M. (1998) "Habituation and Rational Preference Revision." *Dialogue* 37:219–234.
Chisholm, Roderick M. (1963) "Supererogation and Offence: A Conceptual Schema for Ethics." *Ratio* 5:1–14.
——— (1964) "The Ethics of Requirement." *American Philosophical Quarterly* 1:147–153.
Chisholm, Roderick M. and Ernest Sosa (1966a) "On the Logic of 'Intrinsically Better.'" *American Philosophical Quarterly* 3:244–249.
——— (1966b) "Intrinsic Preferability and the Problem of Supererogation." *Synthese* 16:321–331.
Coval, S. C. and J. C. Smith (1982) "Rights, Goals, and Hard Cases." *Law and Philosophy* 1:451–480.
Cresswell, M. J. (1971) "A Semantics for a Logic of 'Better.'" *Logique et Analyse* 14:775–782.
Danielsson, Sven (1968) *Preference and Obligation*, Filosofiska Föreningen, Uppsala.
——— (1997) "Harman's Equation and the Additivity of Intrinsic Value," pp. 23–34 in Lars Lindahl, Paul Needham, and Rysiek Sliwinski (eds.). *For Good Measure: Philosophical Essays Dedicated to Jan Odelstad on the Occasion of His Fiftieth Birthday*. Uppsala Philosophical Studies 46, Uppsala.

——— (1998) "Numerical Representations of Value-Orderings: Some Basic Problems," pp. 114–122 in Christoph Fehige and Ulla Wessels (eds.). *Preferences*. Walter de Gruyter, Berlin.

Davidson, Donald, J. C. C. McKinsey, and Patrick Suppes (1955) "Outline of a Formal Theory of Value I." *Philosophy of Science* 22:140–160.

Davis, John W. (1966) "Is There a Logic for Ethics?" *Southern Journal of Philosophy* 4:1–8.

Day, Richard H. (1986) "On Endogenous Preferences and Adaptive Economizing," pp. 153–170 in Richard H. Day and Gunnar Eliasson (eds.). *The Dynamics of Market Economics*, North-Holland, Amsterdam.

Fehige, Christoph and Ulla Wessels (1998) "Preferences – an Introduction," pp. xx–xliii in Christoph Fehige and Ulla Wessels (eds.). *Preferences*. Walter de Gruyter, Berlin.

Feinberg, Joel (1970) "The Nature and Value of Rights." *Journal of Value Inquiry* 4:243–257.

Fishburn, Peter C. (1972) "Even-Chance Lotteries in Social Choice Theory." *Theory and Decision* 3:18–40.

Fitch, Frederick B. (1967) "A Revision of Hohfeld's Theory of Legal Concepts." *Logique et Analyse* 10:269–276.

Føllesdal, Dagfinn and Risto Hilpinen (1970) "Deontic Logic: An Introduction," pp. 1–35 in Risto Hilpinen (ed.). *Deontic Logic: Introductory and Systematic Readings*. Reidel, Dordrecht.

Foot, Philippa (1983) "Moral Realism and Moral Dilemma." *Journal of Philosophy* 80:379–398.

Forrester, J. W. (1984) "Gentle Murder, or the Adverbial Samaritan." *Journal of Philosophy* 81:193–197.

Forrester, Mary (1975) "Some Remarks on Obligation, Permission, and Supererogation." *Ethics* 85:219–226.

Frankena, W. K. (1952) "The Concept of Universal Human Rights," pp. 189–207 in *Science, Language and Human Rights*, American Philosophical Association, Eastern Division, vol. 1, Philadelphia.

Freeman, James B. (1973) "Fairness and the Value of Disjunctive Actions." *Philosophical Studies* 24:105–111.

Fuhrmann, André (1991a) "Theory Contraction through Base Contraction." *Journal of Philosophical Logic* 20:175–203.

——— (1991b) "Reflective Modalities and Theory Change." *Synthese* 81:115–134.

Fuhrmann, André and Sven Ove Hansson (1994) "A Survey of Multiple Contractions." *Journal of Logic, Language and Information* 3:39–76.

Gabbay, Dov (1985) "Theoretical Foundations for Nonmonotonic Reasoning in Expert Systems," pp. 439–457 in Krzysztof R. Apt (ed.). *Logics and Models of Concurrent Systems*. Springer-Verlag, Berlin.

Garcia, J. L. A. (1986) "Evaluator Relativity and the Theory of Value." *Mind* 95:242–245.

——— (1989) "The Problem of Comparative Value." *Mind* 98:277–283.

Gärdenfors, Peter (1973) "Positionalist Voting Functions." *Theory and Decision* 4:1–24.

Gibbard, Allan and William L. Harper (1978) "Counterfactuals and Two Kinds of Expected Utility." *Foundations and Applications of Decision Theory* 1:125–162.
Glaister, S. M. (2000) "Recovery Recovered." *Journal of Philosophical Logic* 29:171–206.
Goble, Lou (1989) "A Logic of *Better.*" *Logique et Analyse* 32:297–318.
——— (1990) "A Logic of *Good, Should* and *Would.*" *Journal of Philosophical Logic* 19:169–199.
Goldstein, William M. and Hillel J. Einhorn (1987) "Expression Theory and the Preference Reversal Phenomena." *Psychological Review* 94:236–254.
Greenspan, P. S. (1975) "Conditional Oughts and Hypothetical Imperatives." *Journal of Philosophy* 72:259–276.
Guendling, John E. (1974) "Modal Verbs and the Grading of Obligations." *Modern Schoolman* 51:117–138.
Gupta, Rajender Kumar (1959) "Good, Duty and Imperatives." *Methodos* 11:161–167.
Halldén, Sören (1957) *On the Logic of 'Better.'* Library of Theoria, Lund.
Halldin, C. (1986) "Preference and the Cost of Preferential Choice." *Theory and Decision* 21:35–63.
Hansson, Bengt (1968) "Fundamental Axioms for Preference Relations." *Synthese* 18:423–442.
——— (1969) "An Analysis of Some Deontic Logics." *Noûs* 3:373–398.
——— (1970a) "An Analysis of Some Deontic Logics," pp. 121–147 in Risto Hilpinen (ed.). *Deontic Logic: Introductory and Systematic Readings*. Reidel, Dordrecht. [Reprint of Bengt Hansson 1969.]
——— (1970b) "Deontic Logic and Different Levels of Generality." *Theoria* 36:241–248.
Hansson, Sven Ove (1986a) "Individuals and Collective Actions." *Theoria* 52:87–97.
——— (1986b) "A Note on the Typology of Rights." pp. 47–57 in Paul Needham and Jan Odelstad (eds.). *Changing Positions: Essays Dedicated to Lars Lindahl on the Occasion of His Fiftieth Birthday*, Philosophical Studies no. 38. Department of Philosophy, Uppsala University, Uppsala.
——— (1988a) "Rights and the Liberal Paradoxes." *Social Choice and Welfare* 5:287–302.
——— (1988b) "Deontic Logic Without Misleading Alethic Analogies – Part I." *Logique et Analyse* 31:337–353.
——— (1988c) "Deontic Logic Without Misleading Alethic Analogies – Part II." *Logique et Analyse* 31:355–370.
——— (1989a) "A New Semantical Approach to the Logic of Preferences." *Erkenntnis* 31:1–42.
——— (1989b) "New Operators for Theory Change." *Theoria* 55:114–132.
——— (1990a) "Defining 'Good' and 'Bad' in Terms of 'Better.'" *Notre Dame Journal of Formal Logic* 31:136–149.
——— (1990b) "A Formal Representation of Declaration-Related Legal Relations." *Law and Philosophy* 9:399–416.

——— (1990c) "Preference-Based Deontic Logic (PDL)." *Journal of Philosophical Logic* 19:75–93.
——— (1991a) "The Revenger's Paradox." *Philosophical Studies* 61:301–305.
——— (1991b) "Norms and Values." *Crítica* 23:3–13.
——— (1992a) "A Procedural Model of Voting." *Theory and Decision* 32:269–301.
——— (1992b) "Similarity Semantics and Minimal Changes of Belief." *Erkenntnis* 37:401–429.
——— (1992c) "A Dyadic Representation of Belief," pp. 89–121 in Peter Gärdenfors (ed.). *Belief Revision*. Cambridge University Press, Cambridge.
——— (1992d) "In Defense of the Ramsey Test." *Journal of Philosophy* 89:522–540.
——— (1993a) "Money-Pumps, Self-Torturers and the Demons of Real Life." *Australasian Journal of Philosophy* 71:476–485.
——— (1993b) "A Note on Anti-Cyclic Properties of Complete Binary Relations." *Reports on Mathematical Logic* 27:41–44.
——— (1993c) "The False Promises of Risk Analysis." *Ratio* 6:16–26.
——— (1993d) "Reversing the Levi Identity." *Journal of Philosophical Logic* 22:637–669.
——— (1993e) "A Resolution of Wollheim's Paradox." *Dialogue* 32:681–687.
——— (1994) "Philosophical Craftsmanship." *Metaphilosophy* 25:316–325.
——— (1995) "Changes in Preference." *Theory and Decision* 38:1–28.
——— (1996a) "What Is *Ceteris Paribus* Preference?" *Journal of Philosophical Logic* 25:307–332.
——— (1996b) "Extending Rabinowicz' Account of Stability," pp. 150–157 in Sten Lindström, Rysiek Sliwinski, and Jan Österberg (eds.). *Odds and Ends*. Uppsala Philosophical Studies 45. Department of Philosophy, Uppsala University, Uppsala.
——— (1996c) "Legal Relations and Potestative Rules." *Archiv für Rechts- und Sozialphilosophie* 82:266–274.
——— (1996d) "Decision-Making under Great Uncertainty." *Philosophy of the Social Sciences* 26:369–386.
——— (1996e) "Social Choice with Procedural Preferences." *Social Choice and Welfare* 13:215–230.
——— (1997a) "Decision-Theoretic Foundations for Axioms of Rational Preference." *Synthese* 109:401–412.
——— (1997b) "Semi-Revision." *Journal of Non-Classical Logic* 7:151–175.
——— (1997c) "The Limits of Precaution." *Foundations of Science* 2:293–306.
——— (1997d) "Situationist Deontic Logic." *Journal of Philosophical Logic* 26:423–448.
——— (1998a) "Should We Avoid Moral Dilemmas?" *Journal of Value Inquiry* 32:407–416.
——— (1999a) "Adjusting Scientific Practices to the Precautionary Principle." *Human and Ecological Risk Assessment* 5:909–921.
——— (1999b) "But What Should I Do?" *Philosophia* 27:433–440.
——— (1999c) *A Textbook of Belief Dynamics*. Kluwer, Dordrecht.

——— (1999d) "Recovery and Epistemic Residue." *Journal of Logic, Language and Information* 8:421–428.
——— (2000a) "Formalization in Philosophy." *Bulletin of Symbolic Logic* 6:162–175.
——— (2000b) "Preference Logic," to be published in Dov Gabbay (ed.) *Handbook of Philosophical Logic*, second ed., vol. 8.
Hansson, Sven Ove and David Makinson (1997) "Applying Normative Rules with Restraint," pp. 313–332 in M. L. Dalla Chiara et al. (eds.). *Logic and Scientific Method*. Kluwer, Dordrecht.
Hansson, Sven Ove and Hans Rott (1998) "A Plea for Accuracy." *Journal of Applied Non-Classical Logics* 8:221–224.
Harman, Gilbert (1967) "Toward a Theory of Intrinsic Value." *Journal of Philosophy* 64:792–804.
——— (1977) *The Nature of Morality*. Oxford University Press, New York.
Hart, H. L. A. (1954) "Definition and Theory in Jurisprudence." *Law Quarterly Review* 70:37–60.
Herrestad, Henning (1996) *Formal Theories of Rights*. Juristforbundets Forlag, Oslo.
Herzberger, H. G. (1973) "Ordinal Preference and Rational Choice." *Econometrica* 412:187–237.
Hilpinen, Risto (1985) "Normative Conflicts and Legal Reasoning," pp. 191–208 in E. Bulygin et al. (eds.). *Man, Law and Modern Forms of Life*. Reidel, Dordrecht.
Hohfeld, Wesley Newcomb ([1919], 1964) *Fundamental Legal Conceptions* (ed. W. W. Cook). Yale University Press, New Haven, Conn.
Hourani, George F. (1985) *Reason and Tradition in Islamic Ethics*. Cambridge University Press, Cambridge.
Houston, David A., et al. (1989) "The Influence of Unique Features and Direction of Comparison on Preferences." *Journal of Experimental Social Psychology* 25:121–141.
Humphreys, Patrick (1983) "Decision Aids: Aiding Decisions," pp. 14–44 in Lennart Sjöberg, Tadeusz Tyszka, and James A. Wise (eds.). *Human Decision Making*. Doxa, Bodafors, Sweden.
Irwin, Francis W. (1961) "On Desire, Aversion, and the Affective Zero." *Psychological Review* 68:293–300.
Jackson, Frank (1985) "On the Semantics and Logic of Obligation." *Mind* 94:177–195.
Jackson, Frank and Robert Pargetter (1987) "The Two Puzzles about Conditional Obligation." *Philosophical Papers* 16:75–83.
Jamison, D. T. and J. L. Lau (1973) "Semiorders and the Theory of Choice." *Econometrica* 41:901–912.
Jeffrey, Richard C. J. (1977) "A Note on the Kinematics of Preference." *Erkenntnis* 11:135–141.
Jennings, R. E. (1967) "Preference and Choice as Logical Correlates." *Mind* 76:556–567.
——— (1985) "Can There Be a Natural Deontic Logic?" *Synthese* 65:257–273.

Johnson, Eric J. et al. (1988) "Information Displays and Preference Reversals." *Organizational Behavior and Human Decision Processes* 42:1–21.
Jones, Andrew J. I. and Ingmar Pörn (1985) "Ideality, Sub-Ideality and Deontic Logic." *Synthese* 65:275–290.
Kamba, Walter J. (1974) "Legal Theory and Hohfeld's Analysis of a Legal Right." *Juridical Review* 19:249–262.
Kamm, Frances Myrna (1985) "Supererogation and Obligation." *Journal of Philosophy* 82:118–138.
Kamp, Hans (1973) "Free Choice Permission." *Proceedings of the Aristotelian Society* 74:57–74.
Kanger, Stig (1957) "New Foundations for Ethical Theory." Reprinted in Risto Hilpinen (ed.). *Deontic Logic: Introductory and Systematic Readings*. Synthese Library, Dordrecht, 1971, pp. 36–58.
——— (1972) "Law and Logic." *Theoria* 38:105–132.
Kanger, Stig and Helle Kanger (1966) "Rights and Parliamentarism." *Theoria* 32:85–115.
Kannai, Yakar and Bezalel Peleg (1984) "A Note on the Extension of an Order on a Set to the Power Set." *Journal of Economic Literature* 32:172–175.
Kapitan, Tomis (1986) "Deliberation and the Presumption of Open Alternatives." *Philosophical Quarterly* 36:230–251.
Kirchsteiger, Georg and Clemens Puppe (1996) "Intransitive Choices Based on Transitive Preferences: The Case of Menu-Dependent Information." *Theory and Decision* 41:37–58.
Klein, Ewan (1980) "A Semantics for Positive and Comparative Adjectives." *Linguistics and Philosophy* 4:1–45.
Kron, Aleksandar and Veselin Milovanović (1975) "Preference and Choice." *Theory and Decision* 6:185–196.
Ladd, John (1957) *The Structure of a Moral Code*. Harvard University Press, Cambridge, Mass.
Langford, C. H. (1942) "The Notion of Analysis in Moore's Philosophy," pp. 319–342 in P. A. Schilpp (ed.). *The Philosophy of G.E. Moore*. The Library of Living Philosophers, vol. 4. Northwestern University, Evanston,Ill.
Lee, R. (1984) "Preference and Transitivity." *Analysis* 44:129–134.
Lemmon, E. J. (1962) "Moral Dilemmas." *Philosophical Review* 71:139–158.
Lenk, Hans (1978) "Varieties of Commitment." *Theory and Decision* 9:17–37.
Levi, Isaac (1974) "On Indeterminate Probabilities." *Journal of Philosophy* 71:391–418.
——— (1977) "Subjunctives, Dispositions and Chances." *Synthese* 34:423–455.
——— (1980) *The Enterprise of Knowledge*. MIT Press, Cambridge, Mass.
——— (1986) *Hard Choices: Decision Making under Unresolved Conflict*. Cambridge University Press, Cambridge.
——— (1988) "Iterations of Conditionals and the Ramsey Test." *Synthese* 76:49–81.
——— (1991) *The Fixation of Belief and Its Undoing*. Cambridge University Press, Cambridge.
Lewis, David (1973a) *Counterfactuals*. Harvard University Press, Cambridge, Mass.

——— (1973b) "Causation." *Journal of Philosophy* 70:556–567.

——— (1974) "Semantic Analyses for Dyadic Deontic Logic," pp. 1–14 in Sören Stenlund (ed.). *Logical Theory and Semantic Analysis*. Reidel, Dordrecht.

——— (1981) "Ordering Semantics and Premise Semantics for Counterfactuals." *Journal of Philosophical Logic* 10:217–234.

Lindahl, Lars (1977) *Position and Change. A Study in Law and Logic*. Reidel, Dordrecht.

——— (1994) "Stig Kanger's Theory of Rights," pp. 889–912 in Dag Prawitz, Brian Skyrms, and Dag Westerståhl (eds.). *Logic, Methodology and Philosophy of Science*, vol. 9. Kluwer, Dordrecht.

Lindström, Sten and Wlodek Rabinowicz (1989) "On Probabilistic Representation of Non-Probabilistic Belief Revision." *Journal of Philosophical Logic* 18: 69–101.

——— (1991) "Epistemic Entrenchment with Incomparabilities and Relational Belief Revision," pp. 93–126 in André Fuhrmann (ed.). *The Logic of Theory Change*. Springer-Verlag, Berlin.

Livesey, Steven J. (1986) "The Oxford Calculatores, *Quantification of Qualities, and Aristotle's Prohibition of Metabasis*." *Vivarium* 24:50–69.

Lyons, David (1969) "Rights, Claimants and Beneficiaries." *American Philosophical Quarterly* 6:173–185.

MacCormick, Neil (1982) "Rights, Claims and Remedies." *Law and Philosophy* 1:337–357.

MacIntosh, Duncan (1992) "Preference-Revision and the Paradoxes of Instrumental Rationality." *Canadian Journal of Philosophy* 22:503–530.

McArthur, Robert P. (1981) "Anderson's Deontic Logic and Relevant Implication." *Notre Dame Journal of Formal Logic* 22:145–154.

McClennen, Edward F. (1990) *Rationality and Dynamic Choice: Foundational Explorations*. Cambridge University Press, Cambridge.

McConnell, Terrance C. (1978) "Moral Dilemmas and Consistency in Ethics." *Canadian Journal of Philosophy* 8:269–287.

McLaughlin, R. N. (1955) "Further Problems of Derived Obligation." *Mind* 64:400–402.

McMullin, Ernan (1985) "Galilean Idealization." *Studies in History and Philosophy of Science* 16:247–273.

McPherson, Michael S. (1982) "Mill's Moral Theory and the Problem of Preference Change." *Ethics* 92:252–273.

Makinson, David (1984) "Stenius' Approach to Disjunctive Permission." *Theoria* 50:138–147.

——— (1986) "On the Formal Representation of Rights Relations." *Journal of Philosophical Logic* 15:403–425.

——— (1993) "General Patterns in Nonmonotonic Reasoning," pp. 35–110 in Dov Gabbay et al. (eds.). *Handbook of Logic in Artificial Intelligence and Logic Programming*, vol. 3. Oxford University Press, Oxford.

——— (1999) "On a Fundamental Problem of Deontic Logic," pp. 29–53 in P. McNamara, P. Prakken, and H. Prakken (eds.). *Norms, Logics, and Information Systems: New Studies on Deontic Logic and Computer Science*. IOS Press, Amsterdam.

Marcus, Ruth Barcan (1960) "Extensionality." *Mind* 69:55–62.
Meinong, Alexius (1894) *Psychologisch-ethische Untersuchungen zur Werth-Theorie*. Leuschner & Lubensky, Graz.
Mendola, Joseph (1987) "The Indeterminacy of Options." *American Philosophical Quarterly* 24:125–136.
Merrill, G. H. (1978) "Formalization, Possible Worlds and the Foundations of Modal Logic." *Erkenntnis* 12:305–327.
Meyer, J.-J. Ch. (1987) "A Simple Solution to the 'Deepest' Paradox in Deontic Logic." *Logique et Analyse* 117–118:81–90.
Miller, David (1974) "Popper's Qualitative Theory of Verisimilitude." *British Journal for the Philosophy of Science* 25:166–177.
Mish'alani, James K. (1969) "'Duty,' 'Obligation' and 'Ought.'" *Analysis* 30:33–40.
Mistri, Maurizio (1996) "Temporary Equilibria of the Consumer and Changing Preferences in an Evolutionary Context." *Rivista Internazionale di Scienze Economiche e Commerciali* 43:331–346.
Mitchell, Edwin T. (1950) *A System of Ethics*. Scribner's, New York.
Moore, George Edward ([1903] 1951) *Principia Ethica*. Cambridge University Press, Cambridge.
——— (1912) *Ethics*. Oxford University Press, London.
Nickel, James W. (1982) "Are Human Rights Utopian?" *Philosophy and Public Affairs* 11:246–264.
Nozick, Robert (1969) "Newcomb's Problem and Two Principles of Choice." pp. 114–146 in N. Rescher (ed.). *Essays in Honor of Carl G. Hempel*. Reidel, Dordrecht.
Oldfield, Edward (1977) "An Approach to a Theory of Intrinsic Value." *Philosophical Studies* 32:233–249.
Olsson, Erik (1997) *Coherence*. Ph.D. thesis, Uppsala University, Uppsala.
Packard, Dennis J. (1979) "Preference Relations." *Journal of Mathematical Psychology* 19:295–306.
——— (1987) "Difference Logic for Preference." *Theory and Decision* 22:71–76.
Pattanaik, Prasanta and Yongsheng Xu (1990) "On Ranking Opportunity Sets in Terms of Freedom of Choice." *Recherches Economiques de Louvain* 56:383–390.
Pollock, John L. (1983) "How Do You Maximize Expectation Value?" *Noûs* 17:409–421.
Pörn, Ingmar (1970) *The Logic of Power*. Basil Blackwell, Oxford.
Price, Huw (1986) "Against Causal Decision Theory." *Synthese* 67:195–212.
Prior, A. N. (1954) "The Paradoxes of Derived Obligation." *Mind* 63:64–65.
——— (1958) "Escapism," pp. 135–146 in A. I. Melden (ed.). *Essays in Moral Philosophy*. University of Washington Press, Seattle.
Quinn, Warren S. (1974) "Theories of Intrinsic Value." *American Philosophical Quarterly* 11:123–132.
Rabinowicz, Wlodek (1989) "Stable and Retrievable Options." *Philosophy of Science* 56:624–641.
——— (1995) "Global Belief Revision Based on Similarities Between Worlds," pp. 80–105 in Sven Ove Hansson and Wlodek Rabinowicz (eds.). *Logic for a*

Change: Essays Dedicated to Sten Lindström on the Occasion of His Fiftieth Birthday. Uppsala Prints and Preprints in Philosophy 1995:9.

——— (2000) "Money Pump with Foresight," in Michael J. Almeida (ed.). *Imperceptible Harms and Benefits*. Kluwer, Dordrecht.

Ramsey, F. P. ([1931] 1950) *The Foundations of Mathematics and Other Logical Essays*. Kegan Paul, Trench, Trubner & Co., London.

Raub, Werner and Thomas Voss (1990) "Individual Interests and Moral Institutions: An Endogenous Approach to the Modification of Preferences," pp. 81–117 in Michael Hechter et al. (eds.). *Social Institutions*. De Gruyter, New York.

Rawling, Piers (1990) "The Ranking of Preference." *Philosophical Quarterly* 40:495–501.

Rawls, John (1971) *A Theory of Justice*. Oxford University Press, Oxford.

Raz, Joseph (1970) *The Concept of a Legal System*. Oxford University Press, Oxford.

——— (1975) "Permissions and Supererogation." *American Philosophical Quarterly* 12:161–168.

Regan, Donald H. (1983) "Against Evaluator Relativity: A Response to Sen." *Philosophy and Public Affairs* 12:93–112.

Rescher, Nicholas (1967) "Semantic Foundations for the Logic of Preference," pp. 37–62 in Nicholas Rescher (ed.). *The Logic of Decision and Action*. University of Pittsburgh Press, Pittsburgh.

——— (1968) *Topics in Philosophical Logic*. Reidel, Dordrecht.

Reynolds, J. F. and D. C. Paris (1979) "The Concept of 'Choice' and Arrow's Theorem." *Ethics* 89:354–371.

Robinson, Richard (1971) "Ought and Ought Not." *Philosophy* 46:193–202.

Ross, Alf (1941) "Imperatives and Logic." *Theoria* 7:53–71.

Ross, David (1930) *The Right and the Good*. Clarendon Press, Oxford.

——— (1939) *Foundations of Ethics*. Clarendon Press, Oxford.

Rott, Hans (1989) "Conditionals and Theory Change: Revisions, Expansions, and Additions." *Synthese* 81:91–113.

Routley, Richard and Val Plumwood (1984) "Moral Dilemmas and the Logic of Deontic Notions." *Discussion Papers in Environmental Philosophy* 6, Philosophy Department, Australian National University.

Royakkers, Lambèr (1996) *Representing L∃g∀l Rules in Deontic Logic*. Ph.D. Thesis, Katholieke Universiteit, Brabant.

Russell, Bertrand ([1914] 1969) *Our Knowledge of the External World*. George Allen & Unwin, London.

Saito, Setsuo (1973) "Modality and Preference Relation." *Notre Dame Journal of Formal Logic* 14:387–391.

Sartre, Jean Paul (1946) *L'existentialisme est un humanisme*. Éditions Nagel, Paris.

Savage, L. J. (1954) *The Foundations of Statistics*. John Wiley & Sons, New York.

Schkade, D. A. and E. J. Johnson (1989) "Cognitive Processes in Preference Reversals." *Organizational Behavior and Human Decision Processes* 44:203–231.

Schlechta, Karl (1997) "Non-Prioritized Belief Revision Based on Distances Between Models." *Theoria* 63:34–53.

Schoemaker, Paul J. H. (1982) "The Expected Utility Model: Its Variants, Purposes, Evidence and Limitations." *Journal of Economic Literature* 20:529–563.
Schotch, Peter K. and Raymond E. Jennings (1981) "Non-Kripkean Deontic Logic," pp. 149–162 in Risto Hilpinen (ed.). *New Studies in Deontic Logic*. Reidel, Dordrecht.
Schumm, G. F. (1987) "Transitivity, Preference, and Indifference." *Philosophical Studies* 52:435–437.
Segerberg, Krister (1995) "Belief Revision from the Viewpoint of Doxastic Logic." *Bulletin of the IGPL* 3: 535–553.
Sen, Amartya (1971) "Choice Functions and Revealed Preference." *Review of Economic Studies* 38:307–317.
——— (1973) *Behaviour and the Concept of Preference*. London School of Economics, London.
——— (1982) "Rights and Agency." *Philosophy and Public Affairs* 11:3–39.
——— (1983) "Evaluator Relativity and Consequentialist Evaluation." *Philosophy and Public Affairs* 12:113–132.
——— (1991) "Welfare, Preference and Freedom." *Journal of Econometrics* 50:15–29.
——— (1993) "Internal Consistency of Choice." *Econometrica* 61:495–521.
Shafer-Landau, Russ (1997) "Moral Rules." *Ethics* 107:584–611.
Simon, Herbert Alexander (1957) *Models of Man*. John Wiley & Sons, New York.
Sinnott-Armstrong, Walter (1988) *Moral Dilemmas*. Basil Blackwell, Oxford.
Sloman, Aaron (1970) "'Ought' and 'Better.'" *Mind* 79:385–394.
Sobel, Jordan Howard (1997) "Cyclical Preferences and World Bayesianism." *Philosophy of Science* 64:42–73.
Spohn, Wolfgang (1975) "An Analysis of Hansson's Dyadic Deontic Logic." *Journal of Philosophical Logic* 4:237–252.
——— (1977) "Where Luce and Krantz Do Really Generalize Savage's Decision Model." *Erkenntnis* 11:113–134.
——— (1978) *Grundlagen der Entscheidungstheorie*. Monographien Wissenschaftstheorie und Grundlagenforschung, vol. 8, Munich.
——— (1997) "Über die Gegenstände des Glaubens," pp. 291–321 in Georg Meggle and Julian Nida-Rümelin (eds.). *Analyomen 2*. Walter de Gruyter, Berlin.
——— (1999) "Strategic Rationality." *Forschungsberichte der DFG-Forschergruppe Logik in der Philosophie*, no. 24, Universität Konstanz, Konstanz.
——— (2000) "Vier Begründungsbegriffe." *Forschungsberichte der DFG-Forschergruppe Logik in der Philosophie*, no. 49, Universität Konstanz, Konstanz.
Stenius, Erik (1982) "Ross' Paradox and Well-Formed Codices." *Theoria* 48:49–77.
Stevenson, Charles L. (1944) *Ethics and Language*. Yale University Press, New Haven, Conn.
Stocker, Michael (1987) "Moral Conflicts: What They Are and What They Show." *Pacific Philosophical Quarterly* 68:104–123.

Sugden, Robert (1998) "The Metric of Opportunity." *Economics and Philosophy* 14:307–337.
Temkin, Larry S. (1996) "A Continuum Argument for Intransitivity." *Philosophy and Public Affairs* 25:175–210.
——— (1997) "Rethinking the Good, Moral Ideals and the Nature of Practical Reasoning," pp. 290–345 in Jonathan Dancy (ed.). *Reading Parfit*. Blackwell Publishers, Oxford.
Tersman, Folke (1993) *Reflective Equilibrium: An Essay in Moral Epistemology*. Almqvist & Wiksell International, Stockholm.
Tichy, P. (1974) "On Popper's Definitions of Verisimilitude." *British Journal for the Philosophy of Science* 25:155–160.
Toda, Masanao (1976) "The Decision Process: A Perspective." *International Journal of General Systems* 3:79–88.
Toda, Masanao and Emir H. Shuford (1965) "Utility, Induced Utilities, and Small Worlds." *Behavioral Science* 10:238–254.
Trapp, Rainer W. (1985) "Utility Theory and Preference Logic." *Erkenntnis* 22:301–339.
Tullock, G. (1964) "The Irrationality of Intransitivity." *Oxford Economic Papers* 16:401–406.
Tversky, Amos (1969) "Intransitivity of Preferences." *Psychological Review* 76:31–48.
Tversky, Amos, S. Sattath, and P. Slovic (1988) "Contingent Weighting in Judgment and Choice." *Psychological Review* 95:371–384.
Tyson, P. D. (1986) "Do Your Standards Make Any Difference? Asymmetry in Preference Judgments." *Perceptual and Motor Skills* 63:1059–1066.
Ullmann-Margalit, E. and S. Morgenbesser (1977) "Picking and Choosing." *Social Research* 44:757–785.
van Fraassen, Bas C. (1973) "Values and the Heart's Command." *Journal of Philosophy* 70:5–19.
Vermazen, Bruce (1977) "The Logic of Practical 'Ought' Sentences." *Philosophical Studies* 32:1–71.
von Dalen, Dirk (1974) "Variants of Rescher's Semantics for Preference Logic and Some Completeness Theorems." *Studia Logica* 33:163–181.
von Kutschera, Franz (1975) "Semantic Analyses of Normative Concepts." *Erkenntnis* 9:195–218.
von Weizäcker, Carl Christian (1971) "Notes on Endogenous Changes of Tastes." *Journal of Economic Theory* 3:345–372.
von Wright, Georg Henrik (1951) "Deontic Logic." *Mind* 60:1–15.
——— (1963a) *The Logic of Preference*. Edinburgh University Press, Edinburgh.
——— (1963b) *Varieties of Goodness*. Routledge & Kegan Paul, London.
——— (1968) "An Essay in Deontic Logic and the General Theory of Action." *Acta Philosophica Fennica* 21:1–110.
——— (1972) "The Logic of Preference Reconsidered." *Theory and Decision* 3:140–169.
——— (1973) "Deontic Logic Revisited." *Rechtstheorie* 4:37–46.

——— (1981) "On the Logic of Norms and Actions," pp. 3–35 in Risto Hilpinen (ed.). *New Studies in Deontic Logic*. Reidel, Dordrecht.
——— (1998) "Deontic Logic – As I See It." Paper presented at the Fourth International Workshop on Deontic Logic in Computer Science, Bologna.
Wallace, John (1972) "Positive, Comparative, Superlative." *Journal of Philosophy* 69:773–782.
Wellman, Carl (1975) "Upholding Legal Rights." *Ethics* 86:49–60.
Wheeler, S. C. (1972) "Attributives and their Modifiers." *Noûs* 6:310–334.
Williams, Bernard (1965) "Ethical Consistency." *Aristotelian Society, Supplementary Volume* 39:103–124. Reprinted in *Problems of the Self: Philosophical Papers 1956–1972,* Cambridge University Press, London, 1973, pp. 166–186.
Williams, Peter C. (1978) "Losing Claims of Rights." *Journal of Value Inquiry* 12:178–196.
Williamson, T. (1988) "First-Order Logics for Comparative Similarity." *Notre Dame Journal of Formal Logic* 29:457–481.
Woleński, Jan (1980) "A Note on Free Choice Permissions." *Archiv für Rechts-und Sozialphilosophie* 66:507–510.
Wollheim, Richard (1962) "A Paradox in the Theory of Democracy," pp. 71–87 in Peter Laslett and W. G. Runciman. *Philosophy, Politics and Society* (Second Series). Blackwell, Oxford.
Zadeh, L. A. (1977) "Linguistic Characterization of Preference Relations as a Basis for Choice in Social Systems." *Erkenntnis* 11:383–410.
Zimmerman, Michael J. (1986) "Subsidiary Obligation." *Philosophical Studies* 50:65–75.

Index of Symbols

SET THEORY

$\langle x,y \rangle$ ordered pair, n-tuple
$\{x,y\}$ set
$\{x \mid Px\}$ set of objects with specified property
$\in$ membership
$\subseteq$ subset
$\subset$ proper subset
$\cap$ intersection
$\cup$ union
$\setminus$ difference
Δ symmetrical difference [Section 4.2]
$\emptyset$ empty set
$\mathcal{P}(\ )$ power set
$\times$ Cartesian product

LANGUAGES

$\mathcal{L}$ language (in general), language for states of affairs [Definitions 5.1 and 12.1]
$\mathcal{L}_{\mathcal{U}}$ object language for exclusionary preferences [Definition 3.1]
$\mathcal{L}_{\mathcal{A}}$ language for states of affairs represented in $\mathcal{A}$ [Definition 5.7]
$\mathcal{L}_{\mathrm{D}}$ descriptive language [Definition 12.1]
$\mathcal{L}_{\mathrm{N}}$ normative language [Definition 12.1]
$\mathcal{L}_{\mathrm{DC}}$ variable-free fraction of descriptive language [Definition 12.1]
$\mathcal{L}_{\mathrm{NC}}$ variable-free fraction of normative language [Definition 12.1]
$\mathcal{L}_{\mathrm{Dc}}$ set of declarative sentences [Definition 13.1]
I_{C} set of individual constants [Definition 12.1]

I_V set of individual variables [Definition 12.1]
$s(\ )$ set of substitution instances [Definition 3.3]
$I(\ )$ set of instantiations [Definition 12.2]
$[\ ]$ placeholder [Section 13.1]

LOGIC: CONSEQUENCE

Cn consequence operator [Definitions 5.2 and 12.1]
Cn_0 truth-functional consequence operator [Definition 3.3]
Cn_T consequence operator determined by T [Definition 3.3]
$\vdash$ consequence relation [Definition 5.2]
$\nvdash$ negation of consequence relation
$\vDash_{\mathcal{A}}$ model-theoretical implication [Definition 5.8]
$\nvDash_{\mathcal{A}}$ negation of model-theoretical implication
$\dashv\vDash_{\mathcal{A}}$ model-theoretical equivalence [Definition 5.8]

LOGIC: CONNECTIVES, ETC.

$\neg$ negation [Definitions 3.1 and 5.1]
$\vee$ disjunction [Definitions 3.1 and 5.1]
& conjunction [Definitions 3.1 and 5.1]
$\rightarrow$ implication [Definitions 3.1 and 5.1]
$\leftrightarrow$ equivalence [Definitions 3.1 and 5.1]
$\perp$ contradiction
$\top$ tautology
$\Box\!\!\rightarrow$ counterfactual conditional [Section 11.3]
$\Rightarrow$ rule connective [Section 12.1]
$/$ and if possible not [Definition 6.1]
$/_{\mathcal{A}}$ and if $\mathcal{A}$-possible not [Definition 6.2]
$\forall$ universal quantifier
$\exists$ existential quantifier
$\perp$ remainder set [Section 12.4]

ALTERNATIVES

$\mathcal{A}$ alternative set [Definitions 3.1 and 5.3]
$\mathcal{I}$ set of ideal alternatives [Definition 10.1]
$repr_{\mathcal{A}}(\ )$ set of alternatives represented by a sentence [Definition 7.1]
$\mathcal{U}$ universe of alternatives [Definition 3.1]
$|\ |$ reflexive domain [Definitions 3.5 and 3.7]

S situation [Definition 12.4]
N set of normative rules [Section 12.1]

PREFERENCE RELATIONS

$\geq$ better than or equal in value to (exclusionary) [Section 2.2]
$>$ better than (exclusionary) [Section 2.2.]
$\equiv$ equal in value to (exclusionary) [Section 2.2.]
$\geq'$ combinative preference relation [Section 7.1]
$\geq_p'$ *p*-conditionalization of $\geq'$ [Section 11.4]
$\geq_f$ *f*-extension of $\geq$ [Definition 6.5]
$>_f$ strict part of $\geq_f$ [Definition 6.9]
$\gg_f$ *f*-extension of $>$ [Definition 6.9]
$\equiv_f$ symmetric part of $\geq_f$ [Definition 6.10]
$\cong_f$ *f*-extension of $\equiv$ [Definition 6.10]
$\geq_{w,v}$ weighted preference relation [Definition 7.3]
$\geq_w$ weighted preference relation (abbreviated notation) [Definition 7.3]
$\geq_v$ value-based preference relation [Definition 7.4]
$\geq_{\mathbf{i}}$ maximin preference relation [Definition 7.17]
$\geq_{\mathbf{x}}$ maximax preference relation [Definition 7.17]
$\geq_{\mathbf{ix}}$ interval maximin preference relation [Definition 7.18]
$\geq_{\mathbf{xi}}$ interval maximax preference relation [Definition 7.20]
$\geq_{\mathrm{E}}$ max-min weighted preference relation [Definition 7.22]
R preference relation [Definition 3.8]
$\mathbf{R}$ preference model [Definition 3.9]
$*$ repeated application, ancestral [Section 2.2]

MONADIC VALUE AND NORM PREDICATES

H monadic predicate [Section 8.1]
H^+ positive monadic predicate [Definition 8.5]
H^- negative monadic predicate [Definition 8.5]
G good [Definition 8.9]
B bad [Definition 8.9]
G_C canonical good [Definition 8.14]
B_C canonical bad [Definition 8.14]
G_N negation-related good [Definition 8.18]
B_N negation-related bad [Definition 8.18]
G_I indifference-related good [Definition 8.22]

B_I	indifference-related bad [Definition 8.22]
O	prescriptive predicate [Section 9.1]
O_C	canonical contranegative predicate [Definition 10.50]
$O_\top$	maxiconsistent contranegative predicate [Definitions 10.54 and 11.2]
P	permissive predicate [Section 9.1]
P^+	predicate of absolute permission [Definition 10.47]
W	prohibitive predicate [Section 9.1]
W^+	predicate of absolute prohibition [Definition 10.47]

PREDICATES OF AGENCY

E_i	action predicate [Section 13.1]
Dc	declaration predicate [Definition 13.1]

OPERATIONS OF CHANGE AND APPLICATION

$*$	revision [Section 12.4]
$*_{\mathcal{B}}$	revision (with priority-index) [Definition 4.3]
$\div_{\mathcal{B}}$	contraction (with priority-index) [Definition 4.5]
$\ominus$	subtraction [Definition 4.8]
$\oplus$	addition [Definition 4.10]
a	unrestrained application [Definition 12.6]
c	consistent application [Definition 12.11]
ç	obeyable application [Definition 12.16]
$\circledast$	promulgation [Section 12.7]
⨸	derogation [Section 12.8]

SELECTION MECHANISMS

$\uparrow$	restriction [Definitions 3.2 and 3.7]
$\downarrow$	exception [Definitions 3.2 and 3.7]
$[\ ]$	set of validated sentences [Definition 3.8]
μ	measure [Definition 4.1]
$\sqsubseteq$	ordering of pairs [Definition 4.1]
$\sqsubset$	strict part of $\sqsubseteq$ [Definition 4.1]
S	three-place similarity relation [Section 6.4]
T	four-place similarity relation [Definition 6.12]
$\hat{T}$	strict part of T [Definition 6.12]
$\mathfrak{max}$	set of maximal alternatives [Definition 7.11]

$\mathfrak{min}$	set of minimal alternatives [Definition 7.11]
MAX	arbitrary maximal alternative [Definition 7.32; in Appendix]
MIN	arbitrary minimal alternative [Definition 7.32; in Appendix]
w	weight assignment [Definition 7.2]
W	combined weight [Definition 7.30; in Appendix]
δ	max-min weight [Definition 7.22]
v	value assignment [Definition 7.2]
v_{MAX}	maximal value [Definition 7.21]
v_{MIN}	minimal value [Definition 7.21]
v_{δ}	max-min weighted value [Definition 7.22]
EU	probability-weighted value [Definition 7.30; in Appendix]
f	representation function [Definition 6.4]
γ	selection function [Definition 12.9]
$cons$	set of maximal consistent applied subsets [Definition 12.10]

General Index